INFORMATION SYSTEMS DEVELOPMENT:
Methodologies, Techniques and Tools

2nd Edition

INFORMATION SYSTEMS SERIES

Consulting Editors

D. E. AVISON
BA, MSc, PhD, FBCS
Professor of Information Systems,
Department of Accounting
and Management Science,
Southampton University, UK

G. FITZGERALD
BA, MSc, MBCS
Cable & Wireless Professor of
Business Information Systems,
Department of Computer Science,
Birkbeck College, University of London, UK

This series of student and postgraduate texts covers a wide variety of topics relating to information systems. It is designed to fulfil the needs of the growing number of courses on, and interest in, computing and information systems which do not focus on the purely technological aspects, but seek to relate these to business and organisational context.

INFORMATION SYSTEMS SERIES

INFORMATION SYSTEMS DEVELOPMENT: Methodologies, Techniques and Tools

2nd Edition

D. E. AVISON BA, MSc, PhD, FBCS
Professor of Information Systems
Department of Accounting and Management Science
Southampton University, Southampton, UK

G. FITZGERALD BA, MSc, MBCS
Cable & Wireless Professor of Business Information Systems
Department of Computer Science
Birkbeck College, University of London, UK

The McGraw-Hill Companies

London · New York · St Louis · San Francisco · Auckland
Bogotá · Caracas · Lisbon · Madrid · Mexico · Milan
Montreal · New Delhi · Panama · Paris · San Juan
São Paulo · Singapore · Sydney · Tokyo · Toronto

Published by
McGraw-Hill Book Company Europe
Shoppenhangers Road, Maidenhead, Berkshire, SL6 2QL, England
Telephone 01628 23432
Facsimile 01628 770224

British Library Cataloguing in Publication Data
Avison, D.E.
Information Systems Development:
Methodologies, Techniques and Tools. -
2Rev.ed. - (Information Systems Series)
I. Title II. Fitzgerald, G. III. Series
004.024658

ISBN 0-07-709233-3

Library of Congress Cataloging -in-publication Data
The publishers have applied for a listing
Avison, D.E.
Information systems development: methodologies, techniques and tools by
D.E. Avison and G. Fitzgerald. 2nd Edition.
p.cm.
Bibliography: p
Includes index
ISBN 0-07-709233-3
1 System design. 2 system analysis.
I Fitzgerald, G. (Guy) II. Title
1995

Second edition first published 1995

    3 4 5 CUP 2 1 0 9

Printed and bound in Great Britain at the University Press, Cambridge

To Léone Marty and the
memory of Vincent Marty

To the memory of Maude
and Edward Fitzgerald

Contents

CHAPTER 4: TECHNIQUES

CHAPTER 5: TOOLS

Series Foreword

The Information Systems Series is a series of student and postgraduate texts covering a wide variety of topics relating to information systems. The focus of the series is the use of computers and the flow of information in business and large organisations. The series is designed to fill the needs of the growing number of courses on information systems and computing which do not focus on purely technical aspects but which rather seek to relate information systems to their commercial and organisational context.

The term 'information systems' has been defined as the effective design, delivery, use and impact of information technology in organisations and society. Utilising this broad definition it is clear that the subject is interdisciplinary. Thus the series seeks to integrate technological disciplines with management and other disciplines, for example, psychology and sociology. These areas do not have a natural home and were until comparatively recently, rarely represented by single departments in universities and colleges. To put such books in a purely computer science or management series restricts potential readership and the benefits that such texts can provide. The series on information systems provides such a home.

The titles are mainly for student use, although certain topics will be covered at greater depth and be more research oriented for postgraduate study.

The series includes the following areas, although this is not an exhaustive list: information systems development methodologies, office information systems, management information systems, decision-support systems, information modelling and databases, systems theory, human aspects and the human-computer interface, application systems, technology strategy, planning and control, expert systems, knowledge acquisition and its representation.

A mention of the books so far published in the series gives a 'flavour' of the richness of the information systems world. *Information and Data Modelling* (David Benyon) concerns itself with one very important aspect, the world of data, in some depth; *Information Systems Development: A*

Database Approach, second edition (David Avison) provides a coherent methodology which has been widely used to develop adaptable computer systems using databases; *Structured Systems Analysis and Design Methodology, second edition* (Geoff Cutts) looks at one particular information systems development methodology in detail; *Software Engineering for Information Systems* (Donald McDermid) discusses software engineering in the context of information systems: *Information Systems Research: Issues, Techniques and Practical Guidelines* (Robert Galliers - Editor) provides a collection of papers on key information systems issues which will be of special interest to researchers; *Multiview: An Exploration in Information Systems Development* (David Avison and Trevor Wood-Harper) looks at an approach to information systems development which combines human and technical considerations; *Relational Database Design* (Paul Beynon-Davies) offers a practical discussion of relational database design. Recent titles include *Business Management and Systems Analysis* (Eddie Moynihan), *Systems Analysis, Systems Design* (David Mason and Leslie Willcocks), *Decision Support Systems* (Paul Rhodes), *Why Information Systems Fail* (Chris Sauer).

The second edition of *Information Systems Development: Methodologies, Techniques and Tools* reflects the information explosion in recent years, the many technological and research advances in information systems and the exponential growth of such systems in commerce, industry, research institutions and the academic community. The new edition provides an extensive and comprehensive coverage of methodologies, techniques and tools but at the same time treating each discretely. It integrates technical, social, organisational and economic views of information systems development. Also it provides frameworks to assess information systems methodologies while remaining independent of any particular methodology or supplier.

The first edition has been extensively used throughout the world for second and final year first degree courses and by masters level students in computer science, information science and business courses at universities and colleges.

Preface and Reader's Guide

The structure of the first edition of this book was designed to separate out the different elements of information systems development. We have kept this basic framework in the second edition. Following the introductory chapters, Chapter 3 looks at themes in information systems development, Chapter 4 examines the various techniques used. Similarly, Chapter 5 considers the tools that help the process, and Chapter 6 the various methodologies. The final chapter reviews the methodologies, techniques and tools and, amongst other things, provides a framework for comparison.

There are a number of advantages of this approach. It allows the description of the fundamental aspects of a methodology not to be confused by the techniques and tools. It enables techniques and tools that are used in a number of methodologies to be described once only, but in some detail. Most importantly, it gives the reader flexibility to read the book in a way which suits that individual's needs (and the lecturer to plan the course in a way that suits the needs of the particular group of students rather than the exigencies of the text book chosen).

Although we have continued with this structure, many things have changed since 1985, and the new edition reflects these changes. All the chapters have been enhanced and every section has been revised. Most chapters have new sections. For example, new themes have been added to Chapter 3. These include strategic information systems, business process re-engineering, object-oriented systems development and expert systems. Rich pictures, root definitions, conceptual models, object orientation, structure diagrams and matrices have been added to the techniques chapter. Methodologies new to Chapter 6 include Yourdon Systems Method, Merise, Object-oriented Analysis, Process Innovation, Rapid Application Development, KADS and Euromethod.

The book can, of course, be read in a linear fashion, from start to finish, and many readers will use the book in this manner. However, we anticipate that many people will adopt a more contingent approach, dipping in or reading in depth at various points, jumping around between chapters and sections, depending on their requirements or experience. The book is designed to provide this flexibility and the provision of frequent cross-references adds to this.

For various categories of reader, we suggest possible paths through the book:

Those new to information systems development methodologies
The reader is assumed to have some knowledge of information systems development as provided in book form by Shah. H. U. and Avison. D. E. (1995) *The Information Systems Development Cycle: A First Course in Information Systems*, which has been designed as a 'prequel' to this book. For those coming to the concept of methodologies for the first time, wishing to know what they are and how they have evolved, we advise reading Chapters 1 to 3, the introductory sections of each of the following chapters, followed by the mainstream methodologies of Chapter 6, for example, SSADM (section 6.5) and any associated techniques and tools (referred to by cross references) from Chapters 4 and 5.

The knowledgeable general reader
This category of reader may be interested in the detail of methodologies, and might read Chapter 3, followed by 7, and turn to specific methodologies, and associated techniques and tools in Chapters 6, 4 and 5. Section 2.2, which describes the conventional life cycle approach and the following two sections which discuss its strengths and weaknesses, might be read before reading specific methodologies in Chapter 6, as a continuing point of reference for comparative purposes.

Those interested in a specific methodology
These readers will, no doubt, first turn to the particular section in Chapter 6 that describes that methodology. This will guide them to any relevant themes that the methodology embodies in Chapter 3. It will also guide the reader through the associated techniques and tools described in Chapters 4 and 5. For example, if the reader wants to know about Information Engineering (IE), they might look first at section 6.4, and then at the themes of planning and data modelling (sections 3.5 and 3.8), perhaps adding sections 3.6, 3.7 and 3.11, which are also relevant to IE. The many associated techniques are described in sections 4.5 through to 4.12, and the tools in sections 5.5 and 5.7 in particular. If the methodology of specific interest is not covered in Chapter 6, and they cannot all be covered, the reader is advised to read Chapter 3 and then choose from Chapter 6 that

methodology or those methodologies having similar characteristics. We have chosen methodologies for Chapter 6 which represent the major types and/or are well-used.

Those interested in specific development techniques
These readers are advised to read the sections in Chapter 4 which describe those techniques they wish to learn about and then the methodologies in Chapter 6 which use those techniques. Occasionally, a particular technique may be described with its associated methodology in Chapter 6 rather than Chapter 4. This occurs if a particular technique is not easily described without its context or, more likely, it is common to only one methodology. A-graphs and I-graphs, for example, are used only in ISAC and are described along with that methodology in section 6.9.

Those interested in general information systems development concepts
These readers might begin with the first three chapters, and then dip into Chapters 4, 5 and 6, in particular, the introductory sections, followed by the first few sections of the final chapter. Those who dislike using methodologies may particularly enjoy section 7.9, which gives some arguments against their use.

Finally ...lecturers and readers who have used the first edition
The structure remains the same so readers and lecturers (who are, we hope, also readers) having used the first edition need not be told how to use the book. However, they may appreciate our thanks for their contributions which have helped to shape this second edition. We are grateful to you all. We would, in particular, like to thank Sharon Dingley, Mats Lundeberg, Gilbert Mansell, Eddie Moynihan, Enid Mumford, Bob Wood and Trevor Wood-Harper, as well as colleagues from five countries who belong to the Ibiscus group: Niels Bjørn-Andersen, Viviane Bourdin, Maurizio Cavallari, Daniel Evans, André Lemaître, João Mascarenhas, Erik de Vries and Leslie Willcocks. However, very many lecturers and students have helped us immensely. Perhaps we can return this favour. We have case studies and mini case studies, questions for discussion, examination questions and overhead projector slides. *Bona fide* lecturers may contact the publishers, McGraw-Hill Book Company Europe Limited, or David Avison at Southampton University, for information about how to obtain these.

Chapter 1
Introduction

1.1 INFORMATION SYSTEMS

This text is about information systems and ways of developing information systems which will prove useful to organisations. In this chapter, we introduce the nature of information systems and illustrate this with examples and types of information system. The following two sections stress the human and organisation aspects to counteract the stress placed on the technology which is a feature of many texts on information systems. We then introduce the idea of a methodology supporting information systems development and finally provide some tentative definitions to many of the concepts discussed in more detail in later chapters.

The 'system' part of 'information system' represents a way of seeing the set of interacting components, such as:

- People (for example, analysts and business users)
- Objects (for example, computer hardware devices)
- Procedures (for example, those suggested in an information systems development methodology).

All this must take place within a boundary that separates those components relevant to the system (for example, to do with the payment of employees in a payroll system) and those concerned with the environment around the system (other information systems, customers, suppliers, governments and so on).

An information system in an organisation provides facts useful to its members and clients which should help it operate effectively. This information could concern its customers, suppliers, products, equipment and so on. Information systems in a bank might concern the payment of its

1

employees, the operation of its customer accounts, or the efficient running of its branches.

All organisations will have information systems. The organisation might be a commercial business, church, hospital, university or library. This book concerns itself with formalised information systems. By formalised, we do not mean 'mathematical' (which is one interpretation of the word 'formal' in an information systems context (see section 3.12)). We use the term to distinguish information systems discussed here from less formalised information systems such as the 'grapevine', which consists of rumour, gossip, ideas and preferences. These informal information systems are also valid information systems and tend to be intuitive or qualitative. Business discussions at the golf club or over lunch are also valid for disseminating information. However, organisations also need to develop formalised systems which will provide information on a regular basis and in a pre-defined manner, and these are the concern of this book.

We are mainly concerned with computer-based information systems, for the computer can process data (the basic facts) speedily and accurately, and provide information when and where required, which is complete and at the correct level of detail, so that it is useful for some purpose. By comparison, manual systems may be slower (information must be timely to be of value). Further, manual systems are likely to be less accurate because checking procedures can be tedious, and not failure-proof (inaccurate 'information' can lead to poor decision making). In some circumstances, these disadvantages are not a major problem and manual systems are adequate. In other circumstances, a manual system is no longer adequate and it may be replaced by a computer system.

In today's information systems, the basic 'data' to be processed can include pictures, sounds and text as well as raw factual data, such as the numbers which might represent the orders placed for a particular product in an order processing system. The computer system might be used to store data or convert the data to useful information by producing reports or handling management enquiries. This does not mean that the computer information system is 'purely' a computer system - there will be many manual (or clerical) aspects - it means that part of the system is likely to use a computer. Nor does it mean that the computer technology is the most important aspect of the information system (in the same way that a typewriter or word processor is not the most important aspect of the system whereby authors produce fiction). Thus, although technological aspects do feature in the book, it is not the major emphasis, and nor should it be. Otherwise the book would be about information technology (IT) and not about information systems.

The information systems of an organisation will be required to help it analyse the business, along with its environment, and formulate and check that it achieves its goals. These goals might be related to profitability, long-term survival, expansion, greater market share, and employee and customer satisfaction. The information system may help the organisation to achieve improved efficiency of its operations and effectiveness through better managerial decisions. Information systems are sometimes regarded as providing competitive advantage. Another way of looking at this would be to say that without good information systems, a business would be at a considerable competitive disadvantage. They are an important organisational resource.

1.2 EXAMPLES AND TYPES OF INFORMATION SYSTEM

Although many information systems are unique to particular types of organisation, such as a system for placing and paying out bets in a betting office; a system to register voters for a local authority; a ticket reservation system for an airline; a system for recording lottery choices; or a loan system for a library, many information systems are common to a wide range of types of organisation. These include payroll, invoicing, project planning, decision-support and other systems outlined below. Even systems designed for a particular type of organisation can be of a more general type. Thus, reservation systems, for example, can be found in theatres and libraries as well as in airline company offices, and a variation could be used for any organisation where people need to keep appointments.

Some examples might be helpful:
- A payroll system is an information system. All organisations have employees and they will normally be paid wages or salary. The raw data of a payroll system for waged employees includes the number of hours worked by the employees, their rates of pay, and deductions, such as tax, union subscriptions and national insurance. The information system might produce payslips, which contain information about gross wage, net wage and details of the deductions made, reports for management about the staff payroll and the records of employees, and information for the tax office. The management information reports produced might help management make decisions relating to increasing or decreasing staff levels and appraising the performance of employees.

- A sales ledger system is an information system. It is a system about the accounts of customers. The raw data of a sales ledger system relate to sales of the business to customers and remittances from them. (There is a parallel purchase or bought ledger system relating to the business purchases from their suppliers). The information system will provide statements of any balances owing, and could, for example, produce analyses of debtors' balances according to area, sales representative and customer group.

- A project planning and control system is an information system. The raw data will include the work breakdown structure, that is, the time and resource requirements associated with the various activities which make up a project. The system schedules projects so that completion is at the earliest possible date, with the least drain on resources, and provides reports on progress during the life of the project. These reports enable management to act on projects that are behind schedule or where costs are above predicted amounts. A building firm will use a project planning and control system to plan the building of roads, bridges and houses. Of course the development of an information system itself is also a 'project' and has also to be planned and controlled. The analysts involved may well use a project planning and control system to ensure that it is implemented on schedule, without costing more than that predicted and yet providing all the benefits expected.

- A conferencing system supported by video communications technology is an information system. Through video screens, it will enable people in different sites to be linked in a way that simulates a conventional meeting in one site. It might be used to help sales teams based in different sites to co-ordinate their activities.

Information systems can be categorised into types. These include transaction processing systems, decision-support systems, expert systems and office automation systems, all of which are outlined below:

- Transaction processing systems, which are probably the most common information systems, process the individual transactions in a business, such as the employee data which is used in a payroll system, the data about stock replenishment in a stock control system, or the customer order data which is used in a sales order processing system. Very often they concern the day-to-day operations of the organisation. Some organisations categorise them into the marketing, manufacturing, financial and personnel domains of the business.

- Decision-support systems aid the decision making of management. Such systems may use the whole range of facts about the organisation, or part of the organisation, or sometimes relate to aspects external to the organisation, that is, its environment, to provide information to aid the decision-maker. The system is designed to enable managers to retrieve information which will help them make decisions about, for example, where to build a factory, whether to buy out competitors, which products to sell, the prices of products and the salaries of employees.
- Expert systems attempt to simulate the role of the human expert. Their usefulness is derived from the reasoning ability of the system to use its knowledge base of the particular domain to provide solutions or guidance to problem solvers in particular situations. An expert system might be used, for example, to diagnose the reasons for failure in a business or technical process.
- Office automation systems include the various applications found in an office, such as, word processing, electronic mail, voice mail, meeting management, facsimile transmission and the like. Emphasis in these systems is placed on how the technology can fit well with the office staff using it and the business objectives.

There are not only other distinct types, but also variations on these 'themes'. A data processing system is a rather out-of-date name for a transaction processing system and was a term that covered all standard business computer applications of the 1960s and 1970s, and most of the manual systems that preceded computer technology. The decision-support system theme has a number of variants. These include management information systems, which concentrate on summary information; executive information systems, which stress the presentation of information to senior managers, usually by the use of computer graphics; information retrieval systems, which usually provide information from one database source, but provide it quickly and efficiently; and computer supported co-operative work, which supports collaborative decision making (and also other work, such as, writing, planning and negotiating). Likewise, intelligent knowledge-based systems are similar to expert systems, but the term has a wider remit, for example, covering machine translation of natural languages.

1.3 THE HUMAN DIMENSION

Many books on information systems stress the technological aspects. This text also stresses the human and organisational aspects of information systems development. In the early days of computerised business systems, technology was important particularly because of its relative cost. The 'new technology' made a large impact on the business. Most important, technological failure was the main cause in the failure of business data processing systems.

Today, the failure of information systems is rarely due to technological failure. The technology is on the whole reliable and well tried. Failure is much more likely to be caused by human and organisational problems. For example, a lack of planning may lead to much greater costs and project abandonment. Poor training may result in people not co-operating with the information system leading to failure and project abandonment. The same result may be caused by the system not making use of the users' business knowledge. Failure may also be due to poor methods, techniques and tools. We will consider the need for an information systems development methodology in a later section and distinguish between techniques and tools.

In this section we look at the people that will be involved in the development of a computer information system. The term 'users' is often a catch-all for anyone who works with the system who is not part of the technical team and unlikely to be an expert in computing. The implication is that they are a homogeneous group of individuals. But there are many different types of user. If we consider the information systems described in the last section, typical users of the transaction processing systems might include data control clerks who check the transaction data entered into the system. Managers, such as heads of department, who check the summary information to ensure that the weekly objectives of the business are being met, might be typical users of management information systems. Top management interested in using information for long-range planning might use executive information systems. Secretaries producing documents might be users of office information systems. Problem-solvers will use the expert systems available. These people are all 'users' but very different. As we shall see, it is desirable that all users of a system are involved in the development process of that system. They all have a stake in the success of the information system, indeed, they are frequently referred to as stakeholders. There will be other stakeholders, for example, customers and even 'the public at large'.

Each of these users might be categorised. For example, regular users, secretarial staff or middle managers, might prepare data for input to the computer system or interpret results from it on a regular basis - frequently daily, weekly or monthly. They need to be trained so that they use the system efficiently and are well motivated. Casual users, such as executives, will have varied use of the system. They are unlikely to have the time and inclination to train in the use of each information system that they may wish to consult. The human-computer interface needs to be particularly supportive for these users. External users include auditors, such as value added tax (VAT) officials going through the computerised company accounts, and customers, such as borrowers searching through the author database in a library.

Professional users are the technologists. Organisations differ, but they may include operations staff, operating the equipment and managing the tapes and disks; keying-in operators, ensuring data recorded on forms are converted into computer-readable media; systems analysts and designers, who analyse the present way of working and design improved methods; information advisers, who deal directly with non-professional users to discuss their requirements; database administrators, who ensure the data are collected and used effectively; and computer programmers, who code the detailed computer instructions. Once the systems are operational, their role changes to that of maintaining the efficient and effective operation of the information system. The term 'professional users' is not entirely appropriate because other users are, presumably, professionals, although not computing or information systems experts. Even the term 'technological' is not ideal, because it might imply someone trained in electronics which is not the case here as we shall see.

In the information systems development process, some users might be part of a group, such as the information systems strategy group, the steering committee and the development team. We will examine this in more detail when looking at the organisational aspects. Although we have said that it is desirable that all users are involved in the development process, systems development has in the past been dominated by professional users, in particular computer programmers and computer systems analysts. Some of the approaches to information systems development discussed in this book attempt to redress the balance. Certainly, it helps if these professional users see their task as supporting the other users of the system.

1.4 ORGANISATIONAL ASPECTS

Although methodologies, techniques and tools are all a necessary part of the infrastructure to develop information systems, so are management aspects. Information systems development as a whole and each individual information systems project need to be managed. Organisations differ, but a common arrangement is to have an information systems strategy group, a steering committee and a systems development team.

The aim of the information systems strategy group is to develop a plan for information systems development in the organisation and ensure that the plan is carried out and tuned as circumstances change. It is a high level group, usually meeting each month, representing top management, heads of the various divisions, and head of the information systems function. Information systems strategy is discussed later.

The steering committee will oversee each project within the overall plan, ensuring that the wishes of the information systems strategy group are met, and setting its own standards for the project including performance requirements, approving the personnel working on the project and approving the final system. Project control, such a major element affecting the success or failure of the information systems project, will largely rest on the steering committee. The head of the information systems function will frequently be a member of both groups. The steering committee may also specify the information systems development approach to be used in a particular project, although this may be a standard for all information systems development in the organisation. The steering committee is very likely to include the head of the department affected by the information system being developed and may also include the finance director, personnel manager and, possibly, outside consultants, as well as the systems development project team leader.

The systems development team will concern itself with the day-to-day development of the information system and include the analysts, programmers and users working on the project. The composition of the team will differ as the system progresses through the stages of the systems development life cycle (see section 2.2), although there will normally be one project team leader ensuring continuity throughout.

1.5 THE NEED FOR A METHODOLOGY

As we have suggested above, there have always been information systems, although it is only in the recent past that they have used computers. If firms have employees, there needs to be some sort of system to pay them. If firms manufacture products, then there will be a system to order the raw materials from the suppliers and another to plan the production of the goods using the raw materials. Companies need to have a system to deal with orders from customers, another to ensure that products are transported, and yet another to send invoices to the customers and to process payments.

In the time before computers, these systems were largely manual. The word 'largely' is appropriate, because the manual workers would use adding machines, typewriters, and other mechanical or electrical aids to help the system run as efficiently as possible. The use of computers represents only an extension (although a significant extension) of this process. If a manual system proved inadequate in some way, a solution which involves the use of computers may well be contemplated, for example, in situations where:

- Increasing workloads have overloaded the manual system
- Suitable staff are expensive and difficult to recruit
- There is a change in the type of work
- There are frequent errors.

The early applications of computers - say, until the 1960s - were largely implemented without the aid of an explicit information systems development methodology. In these early days, the emphasis of computer applications was towards programming, and the skills of programmers were particularly appreciated. The systems developers were therefore technically trained but were not necessarily good communicators. This often meant that the needs of the users in the application area were not well established, with the consequence that the information system design was sometimes inappropriate for the application.

Few programmers would follow any formal methodology. Frequently they would use rule-of-thumb and rely on experience. Estimating the date on which the system would be operational was difficult, and applications were frequently behind schedule. Programmers were usually overworked, and frequently spent a very large proportion of their time correcting and enhancing the applications which were operational.

Typically, a member of the user department would come to the programmers asking for a new report or a modification of one that was already supplied. This might occur because the present system did not work as specified or because of changes in the organisation and its environment. Often these changes had undesirable and unexpected effects on other parts of the system, which also had to be corrected. This vicious circle would continue, causing frustration to both programmers and users. This was not a methodology, it was only an attempt to survive the day.

As computers were used more and more and management was demanding more appropriate systems for their expensive outlay, this state of affairs could not go on. There were three main changes:

1.   There was a growing appreciation of that part of the development of the system that concerns analysis and design and therefore of the role of the systems analyst as well as that of the programmer.

2.   There was a realisation that as organisations were growing in size and complexity, it was desirable to move away from one-off solutions to a particular problem and towards a more integrated information system.

3.   There was an appreciation of the desirability of an accepted methodology for the development of information systems.

As Utterback and Abernathy (1975) argue, innovation has three elements: improved process, in this context better information systems development methodologies; improved product, in this context better information systems; and improved organisation, in this context better support for decision-making. We look next at the process of developing an information system.

1.6 REQUIREMENTS OF AN INFORMATION SYSTEMS METHODOLOGY

It was to answer the problems discussed in the previous section that methodologies were devised and adopted by many computer data processing installations. An information systems development methodology can be defined as:

> a collection of procedures, techniques, tools, and documentation aids which will help the systems developers in their efforts to implement a new information system. A methodology will consist of phases, themselves consisting of sub-phases, which will guide the systems developers in their choice of the techniques that might be appropriate at each stage of the project and also help them plan, manage, control and evaluate information systems projects.

But a methodology is more than merely a collection of these things. It is usually based on some 'philosophical' view, otherwise it is merely a method, like a recipe. Methodologies may differ in the techniques recommended or the contents of each phase, but sometimes their differences are more fundamental. Some methodologies emphasise the human aspects of developing an information system, others aim to be scientific in approach, others pragmatic, and others attempt to automate as much of the work of developing a project as possible. These differences may be best illustrated by their different assumptions, stemming from their 'philosophy' which, when greatly simplified, might be that, for example:

•   A system which makes most use of computers is a good solution

- A system which produces the most appropriate documentation is a good solution
- A system which is the cheapest to run is a good solution
- A system which is implemented earliest is a good solution
- A system which is the most adaptable is a good solution
- A system which makes the best use of the techniques and tools available is a good solution
- A system which is liked by the people who are going to use it is a good solution.

This book is about methodologies, the differences between them, why these differences exist, and which methodology might be appropriate in given circumstances. As we shall see, methodologies differ greatly, often addressing different objectives. These objectives could be:

1. *To record accurately the requirements for an information system.* The methodology should help users specify their requirements or systems developers investigate and analyse user requirements, otherwise the resultant information system will not meet the needs of the users.

2. *To provide a systematic method of development so that progress can be effectively monitored.* Controlling large-scale projects is not easy, and a project which does not meet its deadlines can have serious cost and other implications for the organisation. The provision of checkpoints and well-defined stages in a methodology should ensure that project-planning techniques can be effectively applied.

3. *To provide an information system within an appropriate time limit and at an acceptable cost.* Unless the time spent using some of the techniques included in methodologies is limited, it is possible to devote an enormous amount of time attempting to achieve perfection.

4. *To produce a system which is well documented and easy to maintain.* The need for future modifications to the information system is inevitable as a result of changes taking place in the organisation and its environment. These modifications should be made with the least effect on the rest of the system. This requires good documentation.

5. *To provide an indication of any changes which need to be made as early as possible in the development process.* As an information system progresses from analysis through design to implementation, the costs associated with making changes increases. Therefore the earlier changes are effected, the better.

6. *To provide a system which is liked by those people affected by that system.* The people affected by the information system, that is, the

stakeholders, may include clients, managers, auditors and users. If a system is liked by the stakeholders, it is more likely that the system will be used and be successful.

Having stated that this book is about information systems development methodologies, not all organisations use a standard methodology. They might have developed their own or adapted one to be more appropriate for their own circumstances. Many organisations may only use some aspects of a standard methodology. Other organisations use no methodology at all. The ways that organisations use (or do not use) information systems development methodologies will be another theme of the book.

1.7 SOME DEFINITIONS

We have already introduced many of the topics in this text and we wish at this point to provide some definitions. One problem in this area is that many of the terms used in this section are used differently and inconsistently elsewhere. We will attempt to be consistent in our usage or explain where usage differs. This is a fairly new discipline, and differences of opinion are to be expected, but it does not make our task easy. Many concepts are complex and will be developed later in the text.

The first two terms have already been mentioned: information and data. Data represent unstructured facts. When three 'strings' of data '250785', '78700199' and '19873' are associated, they could be used to give the information that a person whose identity number is 19873 possesses a driving licence (number 78700199), even though that person is under the minimum legal age for driving motor vehicles. The string of data 250785 is interpreted as the date of birth, 25 July 1985, showing that the holder is only 10 years old in January 1996. The essential difference between data and information, therefore, is that data are not interpreted, whereas information has a meaning and use to a particular recipient in a particular context. The information comes from selecting data, summarising it and presenting it in such a way that it is useful to the recipient. Too often, this process is not well refined and vast amounts of data are output. This is often referred to as 'information overload', although strictly speaking, it is not information but data, because it is not useful.

Some writers equate knowledge with information. Buckingham *et al.* (1987b) define information as 'explicit knowledge'. In other words, information expresses what is meant clearly, with nothing left implied. Knowledge may also be seen as accumulated information. People with knowledge know how to use the information presented and should become more competent in completing their tasks.

The distinction between data and information is context-dependent. Let us look at another example where a line manager analyses the departmental figures and presents the results to the planning department. For the line manager, the results are an interpretation of events and are therefore information rather than data. They have meaning for the line manager. For the central planners, these figures are the raw input for their own analyses, not yet interpreted, and are therefore data rather than information. Thus information is such because it is relevant and understandable to some person or group.

Having given a preliminary definition of information, we need to define what is meant by a system. This is more difficult because it is a term which is used widely in many different fields of activity. Thus the ecological system is a view of the world that includes the relationship between flora and fauna which we call the balance of nature, and the educational system could be viewed as our understanding of the relationship between teachers, students, books and colleges whose purpose is to pass on knowledge to all members of the community. Systems are related to each other. Telephone bills are produced by a billing system, forwarded by a postal system, and paid using a banking system. The banking system will have a customer service system, a cheque processing system, and a payroll system, amongst others. Smaller systems within larger systems are called subsystems. An information system will also have subsystems within it. An airline information system may have subsystems to report on the status of passengers, report on flights, and to analyse costs and profits.

Systems also have a purpose. For example, many information systems are designed to provide relevant information to users for decision-making. Information needs to be presented at the right time, at the appropriate level of detail and of sufficient accuracy to be of use to its recipient. This will help to ensure that the corporate information resource is utilised fully.

Buckingham *et al.* (1987b) define an information system as:

> a system which assembles, stores, processes and delivers information relevant to an organisation (or to society), in such a way that the information is accessible and useful to those who wish to use it, including managers, staff, clients and citizens. An information system is a human activity (social) system which may or may not involve the use of computer systems.

We have already introduced the term methodology, but readers may wish to distinguish this from a technique or tool. We defined a methodology as a collection of procedures, techniques, tools, and documentation aids which will help the systems developers or business users in their efforts to implement a new information system. It will consist of phases, themselves consisting of sub-phases, which will guide the systems developers in their choice of the techniques that might be appropriate at each stage of the project and also help them plan, manage, control and evaluate information systems projects. A methodology represents a way to develop information systems systematically. A methodology should have a sound theoretical basis, although it will be based on the 'philosophy', 'interests', 'viewpoint' - 'bias' if you like - of the people who developed it, for no-one is completely objective.

Techniques and tools feature in each methodology. Particular techniques and tools may feature in a number of methodologies. A technique is a way of doing a particular activity in the information systems development process and any particular methodology may recommend techniques to carry out many of these activities.

Each technique may involve the use of one or more tools which represent some of the artefacts used in information systems development. A non computer-orientated example may help. Two techniques used in the making of meringues are (1) separating the whites of eggs from the yolks and (2) beating the whites. The methodology may recommend the use of tools in these processes, for example, an egg separator and a whisk. In this text, tools are usually automated, that is, computer tools, normally software to help the development of an information system. Indeed, some of these have been designed specifically to support activities in a particular methodology. Others are more general purpose and are used in different methodologies.

An information systems development methodology, in attempting to make effective use of information technology, may also attempt to make effective use of the techniques and tools available. Information systems development methodologies are also about balancing these technical specialisms with behavioural (people-orientated) specialisms. As we shall see in the text, there are many views as to where this balance lies and how the balance is achieved in any methodology. At one extreme are the methodologies aiming at full automation of information systems development as well as the information system itself. However, even in these systems people need to interact with the system. At the other extreme, perhaps, are attempts at full user participation in the information systems development project and client-led design. Even here, user solutions may make full use of the technology and there are a growing number of tools designed to aid users develop their own information systems. The balance between technological aspects and people aspects is one which we will return to as it is a continual theme in information systems development.

FURTHER READING

Alter, S. (1992) *Information Systems: A Management Perspective.* Addison-Wesley, Reading, Mass.

McLeod, R. (1993) *Management Information Systems: A Study of Computer-Based Information Systems.* 5th ed., Macmillan, New York.

Kendall, K. and Kendall, J. (1993) *Systems Analysis and Design.* 2nd ed., Prentice Hall, Englewood Cliffs, New Jersey.

These are three of a large number of American texts which cover the basic field of systems analysis attractively and thoroughly. Most texts of this type tend to emphasise technological aspects.

Moynihan, E. (1993) *Business Management and Systems Analysis.* McGraw-Hill, Maidenhead.

This text provides a hybrid approach to the subject, treating the business and computer sides with equal importance and emphasising the necessary integration of the two in the information systems development process.

Angell, I. O. and Smithson, S. (1991) *Information Systems Management: Opportunities and Risks.* Macmillan, Basingstoke.

This text emphasises the management of information systems within business organisations, suggesting that the real problems associated with information systems are human, managerial, social and organisational, rather than technological.

Benyon, D. (1990) *Information and Data Modelling.* McGraw-Hill, Maidenhead.

Liebenau, J. and Backhouse, J. (1990) *Understanding Information: an Introduction.* Macmillan, Basingstoke.

The natures of data and information are explored in these texts.

Chapter 2
The Traditional Systems Development Life Cycle

2.1 AN HISTORICAL PERSPECTIVE

This chapter describes the basic information systems development approach which was prevalent up to the 1980s. It is still used and also provides the basis of many of today's well-known methodologies. Following an historical perspective which looks at information systems development since the 1950s, we discuss features of this conventional approach, commonly referred to as the systems development life cycle (SDLC) and look at its potential strengths and weaknesses. The chapter therefore provides a springboard for Chapter 3, which describes themes, be they humanistic, scientific, pragmatic or holistic, which address some of the potential weaknesses of the SDLC. The specific methodologies discussed in Chapter 6 have their roots in the SDLC itself and/or one or more of these themes.

We look first at that period from the 1950s and covering most of the 1960s when there was no well-accepted formalised methodology to develop data processing systems. The very early days of computing were associated with scientific applications. Later, when computers were installed in business environments, there were few practical guidelines which gave help on their use for commercial applications. These business applications were orientated towards the basic operational activities of the firm. They might include keeping files and producing reports and documents. Typical examples of each would be keeping customer records, reporting on the sales of the company and producing invoices. Another application might be producing the company payroll. These applications involve the basic data processing processes of copying, retrieving, filing, sorting, checking, analysing, calculating and communicating.

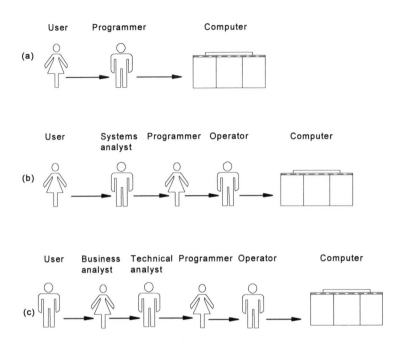

Fig. 2.1: The changing roles of people in data processing

In these early days, the people who implemented computer systems were computer programmers who were not necessarily good communicators nor understood the users' requirements. Their main expertise lay in the technological aspects. It was often difficult for users to communicate their needs to the programmers (see figure 2.1 (a)). Further, because of inexperience or misplaced optimism, applications that were developed were frequently more costly and arrived later than expected. Projects were seen more as short-term exercises or one-off solutions to sort out problems than long-term, well-planned implementation strategies for new applications.

The users were frequently dissatisfied with the operational systems because their needs had not been clearly identified and resolved by the new system. Although many systems with limited objectives may have worked well in this mode, particularly where the programmer knew the application and computer system inside-out, changes made to larger and more complex programs usually required many programmers working on the programs and making changes. Frequently these changes introduced new problems elsewhere in the system. Therefore it might take a long time to make seemingly trivial changes.

Documentation, if it ever existed, was usually out of date, and the programmer did not have the time to update it. Documentation standards were resisted by computer programmers because they were said to restrict creativity, increase workloads and thus increase the overall development time. It is possibly true that documentation standards were resented, but a lengthening of the development schedule caused by documenting systems could well have reduced the time devoted to maintaining these systems following their implementation.

Most companies became over reliant on the few programmers who knew the systems inside-out. They were well paid by the company because they were very difficult to replace. New entrants found it difficult to take over because there was little documentation and no uniform practices and techniques. Sometimes it was necessary to rewrite programs from scratch to effect small changes in user requirements, because it was impossible to understand the programs without documentation, particularly if the original programmers had left the organisation.

There were few courses giving the education and training necessary for the analysis and design work associated with developing data processing systems. This was a particular problem as applications became more complex. Most courses that were available were technical and designed to enable people to use and program the computer. Trainees entering the company would hope to learn from those presently doing the job. Even the best programmers were not necessarily good teachers and in any case they were likely to be much stronger in the programming, as against the analysis, part of their job.

As a reaction to this, there was a growing appreciation of the importance of that part of the development of the system that concerns analysis and design. There was a change of emphasis in many data processing departments. Some job titles changed to programmer-analysts, analyst-programmers and systems analysts. There was a growing appreciation that there was more than one role in the systems development process. These were the roles of the computer operator, programmer and systems analyst. As seen in figure 2.1 (b), the operator controlled the running of the computer and the systems analyst acted in a role between the user and the programmer. Sometimes this distinction was developed further, as two types of analyst emerged: the business-orientated analyst and the technically-orientated systems analyst (figure 2.1 (c)). The former was concerned with understanding the organisational and business needs and communicating these to the technical systems analyst who was concerned mainly with the design of the technical systems which would meet these needs. The technical analyst would communicate this design to the programmers. However, in this text we use the term 'systems analyst' to cover the role of business analyst and technical analyst, and 'systems analysis' to cover both analysis and design aspects.

The historical perspective given above is certainly a distorted one. Many organisations did not follow the staged process suggested above. Many present-day organisations may follow any of the three types (and others) suggested in figure 2.1 and do so effectively. However, this perspective does provide an overview of the likely development of systems in organisations, highlighting some of the problems that occurred. An alternative pattern for organisations' use of information technology over the years is postulated by Nolan (1979) and revised by Galliers and Sutherland (1991).

Another reaction to the problems discussed in this section was the appearance of formalised methodologies to develop computer applications and we look at this change in the next section.

2.2 THE TRADITIONAL SYSTEMS DEVELOPMENT LIFE CYCLE (SDLC)

It was to answer the problems discussed in the previous section that methodologies similar to that recommended by the National Computing Centre (NCC) in the United Kingdom were adopted by many installations. This particular methodology was designed in the late 1960s. It is fully described in Daniels and Yeates (1971) and its revised form is described in Lee (1979). We shall concentrate on the earlier NCC version because it had a great impact on the data processing community and represented a typical methodology of the 1970s based on the SDLC. Most alternative approaches of the time (in North America as well as Europe) were of the general structure described below.

This methodology has the following steps:
- Feasibility study
- System investigation
- Systems analysis
- Systems design
- Implementation
- Review and maintenance.

These stages together are frequently referred to simply as 'conventional systems analysis', 'traditional systems analysis', 'the systems development life-cycle' or the 'waterfall model'. The term 'life-cycle' indicates the iterative nature of the process as by the time the review stage came, the system was frequently found to be inadequate and it was not long before the process started again to develop a new information system with a feasibility study.

1 Feasibility Study

In section 1.5 we suggested possible reasons why a computer system might be contemplated to replace a manual system. These reasons may also be applicable to computer systems replacing existing computer systems with the additional reason of advancing technology (which seems to be the only consistent aspect of computing as processing speeds increase, storage gets larger and costs, for the same computing power, decrease). The next stage is to look in more detail at the present system and then to determine the requirements of the new one.

The feasibility study looks at the present system, the requirements that it was intended to meet, problems in meeting these requirements, new requirements that have come to light since and briefly investigates alternative solutions. These must be within the terms of reference given to the analyst relating to the boundaries of the system and constraints, particularly those associated with resources. The alternatives suggested might include leaving things alone and improved manual as well as computer solutions.

For each of these, a description is given in terms of the technical, human, organisational and economic costs and benefits of developing and operating the system. Thus, any proposed system must be feasible:

- Legally (that is, it does not infringe any national or, if relevant, international company law)
- Organisationally and socially (that is, it is acceptable for the organisation and its staff, particularly if it involves major changes to the way in which the organisation presently carries out its processing)
- Technically (that is, it can be supported by the technology available and there is sufficient expertise to build it)
- Economically (that is, it is financially affordable and the expense justifiable).

Of the possible alternatives, a 'recommended solution' is proposed with an outline functional specification. This information is given to management as a formal report and often through an oral presentation by the systems analysts to management who will then decide whether to accept the recommendations of the analysts. This is one of the decision points in the SDLC as to whether to proceed or not.

2 Systems Investigation

If management has given its approval to proceed, the next stage is a detailed fact finding phase. This purports to be a thorough investigation of the application area. It will look at:

- The functional requirements of the existing system (if there is one) and whether these requirements are being achieved
- The requirements of the new system as there may be new situations or opportunities which suggest altered requirements
- Any constraints imposed
- The range of data types and volumes which have to be processed
- Exception conditions
- Problems of the present working methods.

This information will be much more detailed than that recorded in the feasibility report.

These facts are gained by interviewing personnel (both management and operational staff), by questionnaires, by direct observation of the application area of interest, by using techniques such as sampling, and by looking at records and other written material related to the application area:

- Observation can give a useful insight into the problems, work conditions, bottle-necks and methods of work
- Interviewing, which may be conducted with individuals and groups of users, is usually the most helpful technique for establishing and verifying information, and also provides an opportunity to meet the users and start to overcome possible resistance to change.
- Questionnaires are usually used where similar types of information needs to be obtained from a large number of respondents or from remote locations
- Searching through previous records and documentation may highlight problems, but the analyst has to be aware that the documentation may be out-of-date
- Sampling and other techniques may be used but often require specialist help from people with statistical and other skills.

A great deal of skill is required to use any of these effectively, and results need to be cross-checked by using a number of these approaches. It may be possible to find out about similar systems implemented elsewhere as the experience of others could be invaluable.

The NCC approach recommends a number of documentation aids to ensure that the investigation is thorough. These documentation aids include:

- Various flowcharts, which help the analyst to trace the flow of documents through a department. It is possible to include in the flowchart the processes that are applied to the document. These processes could include, for example, the checking and error correction procedures. Another flowchart could illustrate the system in outline following the functions as they are carried out in a number of departments
- An organisation chart, showing the reporting structure of people in a company or department
- Manual document specifications, giving details of documents used in the manual system
- Grid charts, showing how different components of the system, such as people and machines, interact with each other
- Discussion records on which the notes taken at interviews could be recorded.

Typical charts which were used in this approach are outlined in figure 2.2.

3 Systems Analysis

Armed with the facts, the systems analyst proceeds to the systems analysis phase and analyses the present system by asking such questions as:

- Why do the problems exist?
- Why were certain methods of work adopted?
- Are there alternative methods?
- What are the likely growth rates of data?

In other words, it is an attempt to understand all aspects of the present system and why it developed as it did, and eventually indicate how things might be improved by any new system.

4 Systems Design

This analysis phase leads to the design of the new system. Although usually modelled on the design suggested at the feasibility study stage, the new facts may lead to the analyst adopting a rather different design to that proposed at that time. Much will depend on the willingness to be thorough in the investigation phase and questioning in the analysis phase. Typically, however, the new design would be similar to the previous manual system, but using the power of computing equipment, and (hopefully) avoiding the problems that occurred with the old system and without including any new ones.

This stage involves the design of both the computer and manual parts of the system. The design documentation set will contain details of:

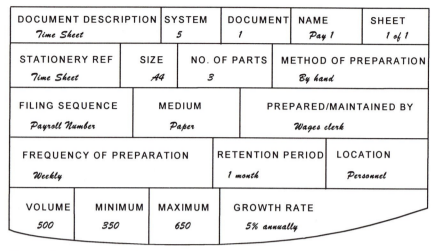

DOCUMENT DESCRIPTION	SYSTEM	DOCUMENT	NAME	SHEET
Time Sheet	*5*	*1*	*Pay 1*	*1 of 1*

STATIONERY REF	SIZE	NO. OF PARTS	METHOD OF PREPARATION
Time Sheet	*A4*	*3*	*By hand*

FILING SEQUENCE	MEDIUM	PREPARED/MAINTAINED BY
Payroll Number	*Paper*	*Wages clerk*

FREQUENCY OF PREPARATION	RETENTION PERIOD	LOCATION
Weekly	*1 month*	*Personnel*

VOLUME	MINIMUM	MAXIMUM	GROWTH RATE
500	*350*	*650*	*5% annually*

Manual Document Specification

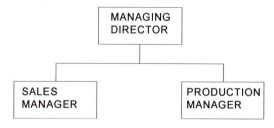

Organisation Chart

TITLE	SYSTEM	DOCUMENT	NAME	SHEET
Payroll system	*5*	*1*	*Pay 1*	*1 of 7*

PARTICIPANTS	DATE
Jim Smith, Payroll Manager; Tom Brown	*25/7/75*

OBJECTIVE/AGENDA	DURATION	LOCATION
Overview of manual system	*1 hour*	*Personnel Dept*

There are 17 staff involved, 5 part-time and ....
....

Discussion Record

Fig. 2.2: Various documentation for the systems investigation phase

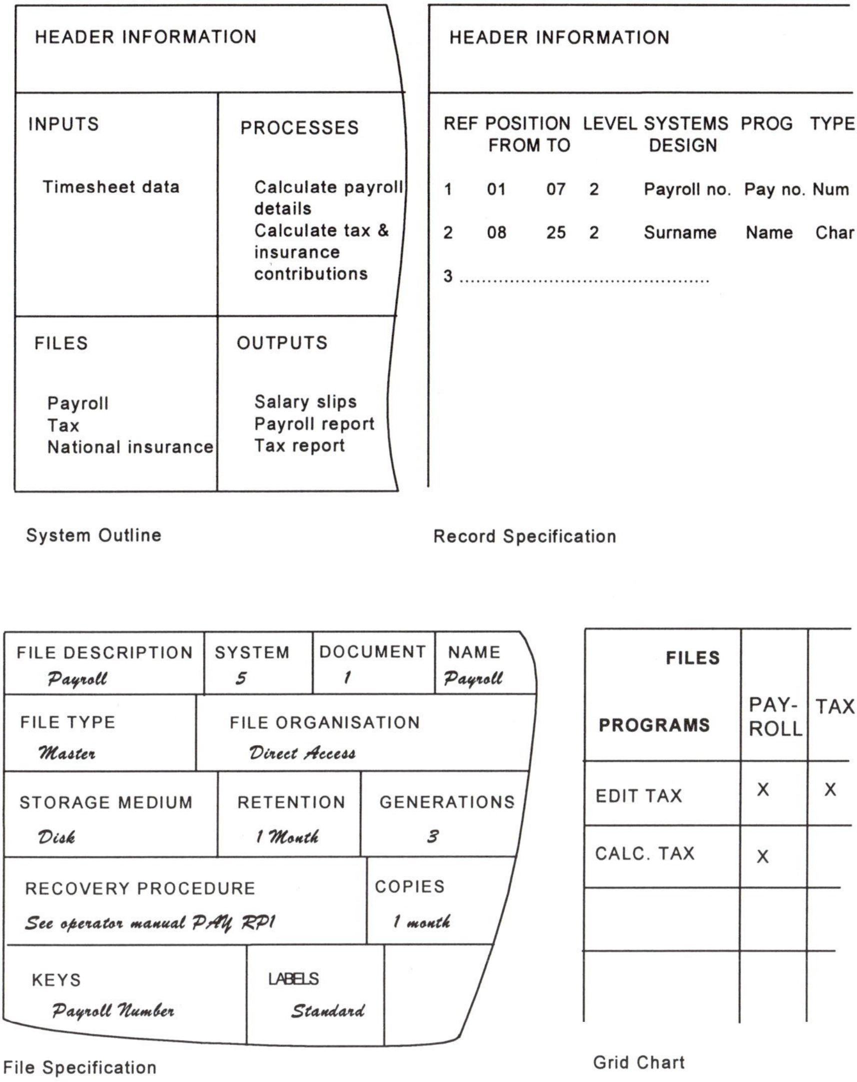

HEADER INFORMATION	
INPUTS	PROCESSES
Timesheet data	Calculate payroll details
	Calculate tax & insurance contributions
FILES	OUTPUTS
Payroll	Salary slips
Tax	Payroll report
National insurance	Tax report

System Outline

REF	POSITION FROM	TO	LEVEL	SYSTEMS DESIGN	PROG	TYPE
1	01	07	2	Payroll no.	Pay no.	Num
2	08	25	2	Surname	Name	Char
3	....					

Record Specification

FILE DESCRIPTION *Payroll*	SYSTEM 5	DOCUMENT 1	NAME *Payroll*
FILE TYPE *Master*	FILE ORGANISATION *Direct Access*		
STORAGE MEDIUM *Disk*	RETENTION *1 Month*	GENERATIONS *3*	
RECOVERY PROCEDURE *See operator manual PAY RP1*		COPIES *1 month*	
KEYS *Payroll Number*	LABELS *Standard*		

File Specification

FILES		
PROGRAMS	PAY-ROLL	TAX
EDIT TAX	X	X
CALC. TAX	X	

Grid Chart

Fig. 2.3: Documents used in the systems design phase

- Input data and how the data is to be captured (entered in the system)
- Outputs of the system
- Processes, many carried out by computer programs, involved in converting the inputs to the outputs
- Structure of the computer and manual files which might be referenced in the system

- Security and back-up provisions to be made
- Systems testing and implementation plans.

Again, there are documentation tools provided with which to detail the input, file and output formats, and to chart the procedures. An outline of some documents is shown in figure 2.3. The system outline provides a list of the inputs, outputs, processes and files to be used in the system. A file design specification includes details of:

- File medium (tape or disk)
- File organisation, frequently sequential or random access
- File size
- Back-up procedures.

The record specification describes all the data items in a record, including:

- Name and description
- Size
- Format
- Possible range of values.

Grid charts may be used to show which programs use which files, either for updating or reading. The flow of procedures, either manual or computer, are illustrated by flowcharts. The computer run flowchart for a payroll system is shown in figure 2.4. It shows the sequence of programs to be run and the inputs, outputs and files required at each stage in the computer run. Even this is simplified, as there would have been a number of sort programs run to ensure the data is always in the correct sequence prior to the update or print processes. Other flowcharts included the flow of procedures (including clerical procedures) and the overall system flowchart as processing progresses through the various departments. More rarely, decision tables, described in section 4.9, might be used.

5 Implementation

Following the systems design phase are the various procedures which lead to the implementation of the new system. If the design includes computer programs, these have to be written and tested. New hardware and software systems need to be purchased and installed if they are not available in the organisation at present. It is important that all aspects of the system are proven before cutover to the new system, otherwise failure will cause a lack of confidence in this and, possibly, future computer applications. The design and coding of the programs will normally be carried out by computer programmers. In this approach, the analysis and programming functions are considered as separate tasks carried out by different people.

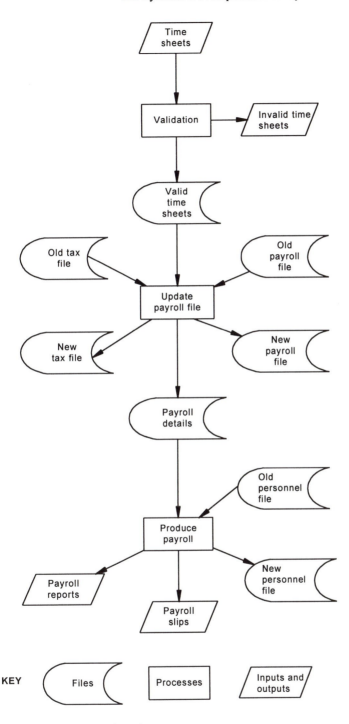

Fig. 2.4: Documenting the computer run

This represented an improvement over systems development previously when the two functions were combined.

A major aspect of this phase is that of quality control. The manual procedures, along with the hardware and software, need to be tested to the satisfaction of users as well as analysts. The users need to be comfortable with the new methods. The departmental staff need to practise using the system and any difficulties experienced should be ironed out. The education and training of user staff is therefore an important element of this phase. Without thorough training, users will be unfamiliar with the new system and unlikely to cope with the new approach (unless it is very similar indeed to the old system).

Documentation, such as the operations and user manuals, will be produced and approved and the live (real, rather than test) data will be collected and validated so that the master files can be set up. Security procedures need to be tested, so that there is no unauthorised access and recovery is possible in case of failure. Once all this has been carried out, the system can be operated and the old one discontinued. If cutover to the new system is done 'overnight' then there could be problems associated with the new system. It is usually too risky an approach to cutover. Frequently, therefore, there is a period of parallel running, where old and new systems are run together, until there is complete confidence in the new system. Alternatively, parts of the new system can be implemented in turn, forming a phased run. If one part of the system is implemented 'to test the water' before the rest of the system is implemented, this is referred to as a pilot run. Acceptance testing comes to an end when the users feel assured that the new system is running satisfactory and it is at this point that the new system becomes fully operational (or 'live').

6 Review and Maintenance

The final stage in the system development process occurs once the system is operational. There are bound to be some changes necessary and some data processing staff will be set aside for maintenance which aims to ensure the continued efficient running of the system. Some of the changes will be due to changes in the organisation or its environment; some to technological advances; and some to 'extras' added to the system at an agreed period following operational running. Inevitably, however, some maintenance work is associated with the correction of errors found since the system became operational.

At some stage there will also be a review of the system to ensure that it does conform to the requirements set out at the feasibility study stage, and the costs have not exceeded those predicted. A report should be produced.

The evaluation process might lead to an improvement in the way other systems are developed through the process of 'organisational learning'.

As commented earlier, because there is frequently a divergence between the operational system and the requirements laid out in the feasibility study, there is sometimes a decision made to look again at the application and enhance it or in the worst case develop yet another new system to replace it. This could also occur for another reason. Changes in the application area could be such that the operational system is no longer appropriate and should be replaced. The SDLC then finishes and starts at the point where there is a recognition that needs are not being met efficiently and effectively and the feasibility of a replacement system is then considered and the life cycle begins again.

2.3 POTENTIAL STRENGTHS OF THE TRADITIONAL SDLC

This conventional systems analysis methodology has a number of features to commend it. It has been well tried and tested. The use of documentation standards helps to ensure that the specifications are complete, and that they are communicated to systems development staff, the users in the department, and the computer operations staff. It also ensures that these people are trained to use the system. The education of users on subjects such as the general use of computers is also recommended, and helps to dispel fears about the effects of computers. Following this methodology also prevents, to some extent at least, missed cutover dates (the date when the system is due to become operational) and unexpectedly high costs and lower than expected benefits. At the end of each phase the technologists and the users have an opportunity to review progress. By dividing the development of a system into phases, each sub-divided into more manageable tasks, along with the improved training and the techniques of communication offered, the conventional approach gave much greater control over the development of computer applications than before. Indeed, it makes possible, due to the establishment of development phases, the use of project management techniques and tools (see section 5.2).

In other words, it was a methodology and this was an improvement on previous practice. It had all the attributes that we expect of a methodology:

- A series of phases starting from the feasibility study to review and maintenance. The phases are expected to be carried out as a sequential process. Each of these has sub-phases and the activities to be carried out along with the outputs (or 'deliverables') of each sub-phase are spelt out in some detail. Deliverables may include documents, plans or computer programs.

- A series of techniques which include ways to evaluate the costs and benefits of different solutions and methods to formulate the detailed design necessary to develop computer applications.
- A series of tools which might include project management tools, as control of the overall project is vital as a way of improving the likelihood of meeting deadlines.
- A training scheme so that all analysts and others new to their roles and responsibilities could adopt the standards suggested. Indeed, the authors of this book used to teach the approach in a six week full time course. Delegates on the course were usually experienced computer programmers wishing to become systems analysts.
- A philosophy, perhaps implied rather than stated, which might be that 'computer systems are usually good solutions to clerical problems'.

However, potentially at least, there are also serious limitations to this approach. This text addresses itself to methodologies, techniques and tools, which may offer improvements on this conventional approach or could be incorporated into this approach to make it more effective. It is therefore useful to discuss criticisms of the conventional approach and how it is used before considering alternatives to it. We would like to stress that the methodologies typified by the NCC approach of the 1970s were adequate, indeed enlightened perhaps, for their day and are still being used successfully today, but there have been a number of developments since then which make alternative approaches viable and potentially more effective.

2.4 POTENTIAL WEAKNESSES OF THE TRADITIONAL SDLC

The criticisms of the systems development approach to applications development, or to be more precise, of the way in which it was used, include:

- Failure to meet the needs of management
- Unambitious systems design
- Instability
- Inflexibility
- User dissatisfaction
- Problems with documentation
- Lack of control
- Incomplete systems

- Application backlog
- Maintenance workload
- Problems with the 'ideal' approach.

We will discuss each of these in turn.

- *Failure to meet the needs of management:* As can be seen in figure 2.5, although systems developed by this approach often successfully deal with such operational processing as payroll and the various accounting routines, middle management and top management were sometimes ignored by computer data processing. Management information needs, such as that required when making decisions as to where to locate a new factory, which products to stop selling, what sales or production targets to aim for and how sales can be increased, are neglected. Although some information may filter up to provide summary or exception reports, the computer is being used largely only for routine and repetitive tasks. Instead of meeting corporate objectives, computers are being used to help solve low-level operational tasks. Top managers and computers are not mixing, apart from the lip-service required to sanction the expenditure necessary to buy and develop mainframe computer systems. For many years managers have been demanding that computers help them more directly in their tasks.

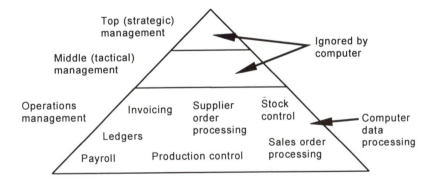

Fig. 2.5: Failure to meet all the needs of management

- *Unambitious systems design*: Computer systems usually replace manual systems, which have proved inadequate in changing circumstances. But apart from using a new technology, the computer system designs are often similar to those of the existing systems. Systems analysts talk of 'computerising' the manual system. It is therefore not surprising that the new designs are unambitious. More

radical, and potentially more beneficial, computer systems are not
being implemented.

- *Models of processes are unstable*: The conventional methodology
attempts to improve the way that the processes in businesses are
carried out. However, businesses do change, and processes need to
change frequently to adapt to new circumstances in the business
environment. Because computer systems model processes, they have
to be modified or rewritten frequently. It could be said therefore that
computer systems, which are 'models' of processes, are unstable
because the real world processes themselves are unstable.

- *Output driven design leads to inflexibility*: The outputs that the
system is meant to produce are usually decided very early in the
systems development process. Design is 'output driven' (see figure
2.6) in that once the output is agreed, the inputs required, the file
contents, and the processes involved to convert the inputs to the
outputs, can all be designed. However, changes to required outputs
are frequent and because the system has been designed from the
outputs backwards, changes in requirements usually necessitate a
very large change to the system design and therefore cause either a
delay in the implementation schedule or an unsatisfactory and
inappropriate system.

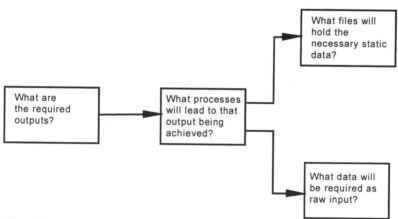

Fig. 2.6: Design is output-driven

- *User dissatisfaction*: This is often a feature of many computer
systems. Sometimes systems are rejected as soon as they are
implemented. It may well be only at this time that the users see the
repercussions of their 'decisions'. These decisions have been
assumed by the analysts to be firm ones and as computer systems

prove inflexible, it is normally very difficult to incorporate changes in requirements once the systems development is under way. Many companies expect users to 'sign off' their requirements at an early stage when they do not have the information to agree the exact requirements of the system. The users sign documentation completed by computer-orientated people. The documents are not designed for the users. On the contrary, they are designed for the systems analysts, operations staff and programming staff who are involved in developing the system. Users cannot be expected to be familiar with the technology and its full potential, and therefore find it difficult to contribute to the debate to produce better systems. The period between 'sign-off' and implementation tends to be one where the users are very uncertain about the outcome and they are not involved in the development process and therefore may lose their commitment to the system. They might have become disillusioned with computer systems and, as a consequence, fail to cooperate with the systems development staff. Systems people, as they see the situation, talk about user staff as being awkward or unable to make a decision. Users may first see the system only on implementation and find it inappropriate. This alienation between technologists and users has even been known to lead to users developing their own informal systems which are used alongside the computer system, ultimately causing the latter to be superfluous.

- *Problems with documentation*: In section 2.2, one of the benefits of the NCC approach discussed was the documentation that the methodology uses. Yet this is not ideal. The orientation of the documentation, towards the computer person and not the user, has already been mentioned as a source of problems. The main purpose of documentation is that of communication, and a technically-orientated document is not ideal. There was no parallel user-orientated documentation. A further problem which occurs is that the forms tend to be completed reluctantly by the programmer or analyst as a requirement of the computer or data processing department. It is rarely done well. Worse, frequently it is not changed when modifications to the system are implemented during development or maintenance. This makes the documentation useless because it cannot be relied upon as an accurate reflection of the actual system.

- *Lack of control:* Despite the methodological approach enabling estimates of time, people and other resources needed, these estimates proved to be unreliable because of the complexity of some phases and the inexperience of the estimators. Computer people were therefore seen as unreliable and some disenchantment ensued in relation to computer applications.

- *Incomplete systems*: Computers are particularly good at processing a large amount of data speedily. They excel where the processing is the same for all items: where the processing is structured, stable and routine. This often means that the unusual (exceptional conditions) are frequently ignored in the computer system. They are too expensive to cater for. If they are diagnosed in the system investigation stage, then manual staff are often assigned to deal with these exceptions. Frequently they are not diagnosed, the exceptions being ignored or forgotten, and the system is falsely believed to be complete. These exceptions cause problems early on in the operational life of the system. These problems mark a particular technical failure, but it also indicates a systems failure, a failure to analyse and design the system properly.

- *Application backlog*: A further problem is the application backlog found in data processing. There may well be a number of systems waiting to be developed. The users may have to wait some years before the development process can get under way for any proposed system. The process to develop an information system will itself take many months and frequently years from feasibility study to implementation. Users may also postpone requests for systems because they know it is not worth doing because of the backlog. This phenomenon is referred to as the invisible backlog.

- *Maintenance workload*: The temptation, then, is towards 'quick and dirty' solutions. The deadline for cutover may seem sacrosanct. It is politically expedient to patch over poor design rather than spend time on good design. This is one of the factors which has led to the great problem of keeping operational systems going. With many firms, the effort given to maintenance can be as high as 60-80 per cent of the total data processing workload. With so many resources being devoted to maintenance, which often take priority, it is understandable that there is a long queue of applications in the pipeline. The users are discouraged by such delay in developing and implementing 'their' applications.

- *Problems with the 'ideal' approach:* The SDLC model assumes a step-by-step, top-down process. This is somewhat naive. It is inevitably necessary to carry out a series of iterations when, for example, new requirements are discovered or sub-phases might prove unnecessary. In other words, information systems development is an iterative process, whatever text books say. The political dimension where, for example, people have their own 'agenda' transcends the rationale of any methodology. Users and analysts will need to interpret the methodology, and its techniques and tools, to be relevant to a particular problem situation. It also assumes a tailor-made rather than packaged solution, usually for a medium- or large-scale application using mainframe computers. With the widespread use of PC systems, such an approach may be inappropriate and unwieldy.

The above criticisms of the traditional SDLC should be regarded as 'potential' criticisms as many organisations using the approach did not fall into all or even most of the potential traps. However, this book concerns itself with improving information systems development by adapting the SDLC or using alternative approaches to developing information systems.

2.5 CONCLUSION

Before we consider approaches to systems analysis which represent advances on the traditional SDLC approach, the reader should stop and consider that many systems developed today are still done using the rule of thumb methods discussed in section 2.1, in other words, using no real methodology at all. This is particularly true in organisations using PCs as their first experience of computing, especially where there are no package solutions. Even when there is a package, it often needs adapting for the particular application. An alternative 'solution' to dealing with an inadequate package, but one thwart with dangers, is to adapt the business application to the needs of the package.

It is often said that businesses developing their own systems for PCs are making the same mistakes that systems analysts made when using the large computers of the 1960s. We do not wish to assert that the methodologies we have chosen to call traditional are of little value. Indeed, they are used successfully today and many of the alternative methods developed in the 1980s follow the SDLC in general, but address some of the potential weaknesses suggested above by improving one or more of its phases. This improvement might lie in the use of an alternative technique or a new software tool. However, it is the purpose of this text to discuss methodologies which improve the craft and engineering science of systems analysis and the effectiveness of information systems, and therefore we examine alternative approaches which either alter aspects of the traditional SDLC or offer a much more radical alternative approach.

In many respects, there is nothing intrinsically wrong with the traditional SDLC. Much depends on the way in which it is used. There must be sufficient resources assigned to the process; it needs to be well managed and controlled so that any deviation from the plan is noticed and dealt with; it should be seen not as a rigid process but as a flexible and iterative one; the feasibility study needs to be seen as a way of exploring alternatives rather than as a way of advancing a limited point of view; and systems development should not be seen as a purely technological process but one involving all users and developers, indeed the organisation as a whole.

We also do not wish to assert that any of the alternative methodologies represents a panacea. Major concerns in computing remain, for example:

- Meeting project deadlines
- Program maintenance
- Staff recruitment and retention
- General user dissatisfaction
- Rapidly changing requirements.

Notwithstanding these continuing pessimistic trends, there are many successful information systems, some developed using the conventional SDLC. However, the next chapter looks at the various themes in information systems development methodologies that are seen as alternatives to the conventional approach discussed in this chapter. These alternative approaches usually address one or more of the criticisms of the conventional approach discussed in the previous section.

FURTHER READING

Daniels, A. and Yeates D. A. (1971) *Basic Training in Systems Analysis.* 2nd ed., Pitman, London.

The classic description of the National Computing Centre (NCC) methodology which typifies the traditional systems development life cycle described in this chapter.

Ahituv, N., Neumann, S. and Riley, H. N. (1994) *Principles of Information Systems for Management.* 4th ed., Brown, Dubuque, Ia.

Mason, D. and Willcocks, L. (1994) *Systems Analysis, Systems Design.* McGraw-Hill, Maidenhead.

Shah, H. U. and Avison, D. E. (1995) *The Information Systems Development Cycle: A First Course in Information Systems.* McGraw-Hill, Maidenhead.

These texts provide thorough descriptions of an up-to-date view of the systems development life cycle.

Chapter 3
Themes in Information Systems Development

3.1 INTRODUCTION

In this chapter we explore some of the themes that have evolved in information systems development which address some of the weakness of the traditional SDLC approach discussed in Chapter 2. However, there is no panacea: no one approach solves all the problems.

The first themes address information systems for the organisation as a whole: their inter-connection, strategic use, re-engineering and planning. We look first at the systems approach which presents a holistic view of organisations and therefore addresses the piecemeal aspect of the traditional SDLC, which we criticised in Chapter 2, because the systems approach looks at the problem situation as a whole. In its 'pure' form it is somewhat theoretical, but the issues raised and principles established have been very influential in some practical information systems methodologies. Next, we look at strategic information systems which address the needs of top management. This could be said to be a 'head-on' attack at the emphasis placed on computer applications at the operational level of the firm in the traditional approach. This is followed by a section on business process re-engineering, which again is a re-examination of the present systems, including information systems of the organisation. Such re-examination leads to redesign within the context of both the other processes and the needs of the organisation as a whole. We then look at planning approaches which emphasise the way future information systems development can be organised and integrated so that strategic as well as the tactical and operational needs are included.

We then look at modelling and approaches which emphasise the modelling aspects either on the process side or on the data side. Modelling represents an important aspect of many methodologies. Following the section on modelling in general, we look at structured approaches, which centre around the techniques and tools for modelling processes. The next theme, data modelling, concentrates on understanding and classifying data and the relationships between data in an organisation and is therefore related to the development of databases. Object orientation has been a more recent theme in information systems, and in this approach we model objects of all kinds in the organisation: data, processes, people and software can all be modelled as objects. This is therefore a unifying approach within the overall theme of modelling.

Another general area is that of software. The first theme in this category is automation. Attempts have been made to automate all aspects of the systems development process and these are described in section 3.10. We look then at prototyping which uses system development tools in a bid to improve requirements analysis, that is, ensure 'what the users want is what they get'. Prototyping is also claimed to lead to more rapid application development. We then consider software engineering and formal methods, approaches which tend to concentrate on quality software which is, of course, an important part of any computerised information system.

The final major theme emphasises the role of people in developing and using an information system. The importance of the people using the information systems and other interested parties, frequently called stakeholders, is addressed in the theme of participation. Expert systems also address people's needs, and this is discussed as the final theme. In expert systems one of the important areas is capturing the knowledge of people who are considered experts in the particular application domain.

Most of the themes discussed in this chapter have their counterparts in Chapter 6: actual methodologies which are used in organisations. Others will have greatly influenced some of the methodologies discussed in that chapter. In a final section of this chapter, we outline some of the issues in choosing an information systems development methodology. These will include the techniques and tools which will be discussed in Chapters 4 and 5 respectively. We will come back to the topic of methodology choice in much more detail in Chapter 7.

3.2 THE SYSTEMS APPROACH

General systems theory attempts to understand the nature of systems which are large and complex. It stems in modern times from the work of Bertalanffy (see for example Bertalanffy, 1968). Although such works might not seem relevant to information systems methodologies at first reading, many of the principles have been taken up by the information systems community. A consideration of systemic activities is an integral feature of a number of information systems methodologies, in particular SSM and Multiview, which are discussed in sections 6.11 and 6.12 respectively.

In Chapter 1 we attempted to define the nature of systems. We saw how systems relate to each other and that they themselves consisted of subsystems. This gives rise to the definition of a system as a set of inter-related elements (Ackoff, 1971). A system will have a set of inputs going into it, a set of outputs going out of it, and a set of processes which convert the inputs to the outputs.

We define a boundary of a system when we describe it. This may not correspond to any physical or cultural division. A payroll system might include all the activities involved in the payment of staff in a business. These activities fall within the boundary of that system. Those systems outside it, with which it relates, are referred to as the environment. The systems approach concerns itself with interactions between the system and its environment, not so much with how the system works, which is considered as a 'black box'. The staff recruitment system and production systems within the firm will be part of the environment of the payroll system, as will the government's system to increase employment.

One of the bases of systems theory concerns Aristotle's dictum that the whole is greater than the sum of the parts. This would suggest that we must try to develop information systems for the widest possible context: an organisation as a whole rather than for functions in isolation. If we fail to follow this principle then a small part of the organisation may be operating to the detriment of the organisation as a whole. If we do break up a complex problem into smaller manageable units, we need to keep the whole in mind. Otherwise, this may be reductionist, the process distorting our understanding of the overall system. Users of many of the approaches discussed, in particular the structured approach, part of process modelling (described in section 3.7), may succumb to this danger unless they use the approach with care. Decomposing complex structures is the accepted approach in a scientific discipline, but information systems concern people as well as technology, and the interactions are such that in these human activity systems it is important to see the whole picture. The human components in particular may react differently when examined singly as when they play a role in the whole system.

Organisational systems are not predictable as they concern human beings. The outputs of computer programs may be predictable and capable of mathematical formalism. Human activity systems are less predictable because human beings may not follow instructions in the way a piece of software does, nor interpret instructions in the same way as other people do or in the way that they themselves might have done on previous occasions.

Another aspect of systems theory is that organisations are open systems. They are not closed and self-contained, and therefore the relationship between the organisation and its environment is important. They will exchange information with the environment, both influencing the environment and being influenced by it. The system which we call the organisation will be affected by, for example, policies of the government, competitors, suppliers and customers, and unless these are taken into account, predictions regarding the organisation will be incorrect. As organisations are complex systems this would suggest that we require a wide range of expertise and experience to understand their nature and how they react with the outside world. Multi-disciplinary teams might be needed to attempt to understand organisations and analyse and develop information systems.

Human activity systems should have a purpose and the inter-related elements interact to achieve this purpose. What then is the purpose of an information system? Depending on the application area, an information system may be constructed to help managers decide on where to build a new factory. It may provide information about customers so that decisions can be made about their credit rating. It might be to maximise the use of aircraft seats in an airline ticket reservation system. With this information provided, it is possible to control the environment rather than passively react to it.

Information systems usually have human and computer elements and both aspects of the system are inter-related. However, the computer aspects are closed and predictable, the human aspects are open and nondeterministic. Although not simple, the technological aspects are less complex than the human aspects in an information system, because the former are predictable in nature. However, many information systems methodologies only stress the technological aspects. This may lead to a solution which is not ideal because the methodologies underestimate the importance and complexity of the human element.

Systems theory has had widespread influence in information systems work. It would suggest, for example, that whatever methodology is adopted, the systems analyst ought to look at the organisation as a whole and also be aware of externalities beyond the obvious boundaries of the system. These include customers, competitors, governments and so on. Systems theory would also suggest that a multi-disciplinary team of analysts, not all computer-orientated, is much more likely to understand the organisation and suggest better solutions to problems. After all, specialisms are artificial and arbitrary divisions. Such an approach should prevent the automatic assumption that computer solutions are always appropriate as well as preventing a study of problem situations from only one narrow point of view. With this approach comes the acknowledgement that there may be a variety of possible solutions, none of them obviously 'best' perhaps, but each having some advantages. These solutions may involve structural, procedural, attitudinal or environmental change.

Checkland (1981), developed further in Checkland and Scholes (1990), has attempted to turn the tenets of systems theory into a practical methodology which is called Soft Systems Methodology (SSM). It is further developed, in the context of information systems, in Wilson (1990). Checkland argues that systems analysts apply their craft to problems which are not well defined and soft thinking attempts to understand the fuzzy world of complex organisations. By contrast, 'hard' approaches, such as the structured and data analysis methods, emphasise the certain and the precise in a particular domain and tend to look at the domain from one point of view, a major contrast compared to soft systems thinking.

Checkland's description of human activity systems also acknowledges the importance of people in organisations. It is relatively easy to model data and processes, but to understand organisations, it is essential to include people in the model. This is difficult because of their unpredictable nature. The claims of the proponents of this approach are that a better understanding of these complex problem situations is more likely to result when compared to using hard approaches alone.

We will now look at one final contrast in hard and soft systems viewpoints. Analysts following hard approaches think in terms of systems as though they exist as such and can be engineered. The soft systems viewpoint is that systems do not exist but represent a way of viewing, and therefore understanding, complex real-world activities. However, the implication of this is that different analysts will see the real world in different ways, depending on their background, experience and so on. The discussion between different analysts can therefore provide understanding of the real world as well as expose its complexity. It may lead to a completely different systems view of the organisation being studied.

We look at Soft Systems Methodology in section 6.11. Multiview, described in section 6.12, incorporates these ideas as a first stage towards developing a computer information system, and later stages are more influenced by a 'hard' systems viewpoint. There are other approaches which have used systems ideas in their design. Beer's Viable Systems Model, for example (see Beer, 1985, and Espejo and Harnden, 1989) provides a tool to study organisations holistically, analysing their structure and their information systems from many viewpoints.

3.3 STRATEGIC INFORMATION SYSTEMS

As we have seen in Chapter 2, the first business activities that were computerised tended to be the basic transaction processing systems such as payroll, sales order processing, stock control and invoicing. The approach used was simple and did not involve changing the nature of the business. The change usually only concerned the tool by which existing activities were undertaken. Computerisation aimed to promote efficiency, in particular, to reduce labour costs. The savings could be quantified, although the benefits were usually exaggerated and the costs of computer staff frequently underestimated or ignored. Where the labour displacement argument has also been used to justify the transfer from an existing computerised system to a new and improved computerised system, the case is usually more difficult to make and prove because there are fewer staff to displace. Overall the propensity for reducing traditional labour costs via information systems is declining. Nevertheless, we have more recently seen a number of justifications based on cost savings relating to a reduction in the managerial workforce, particularly middle management levels. It is this area that will probably provide the basis for most future labour saving cost displacement information systems projects.

Labour costs have formed the largest and easiest costs to displace using information technology (IT), but other costs have also been amenable to displacement, at least to some degree. Paper and postage costs have been displaced by Electronic Data Interchange (EDI) and electronic mail.

Property costs have been displaced, or reduced, where organisations have used IT to enable the move of back office functions away from town centres to cheaper areas (or even to home working). Inventory costs have been displaced by information systems-enabled Just In Time (JIT) systems. Such cost displacement savings can be relatively easily quantified in these systems which seek to perform essentially the same functions as before, but to perform them at less cost.

In many organisations the propensity for further information systems investments based on efficiency criteria are limited. First, the number of opportunities are reduced as more and more projects are implemented, and second, the returns are declining as the most clear-cut efficiency improving opportunities have already been addressed. The second problem is that certain information systems opportunities will never be able to demonstrate efficiency cost savings of the type discussed above and are thus very unlikely to pass the usual efficiency-based evaluation or justification process.

More recently, an additional role for information systems has emerged. This is the use of information systems and IT as a direct tool for obtaining competitive advantage. Information systems can be used to improve the business in the market place and in this way to help:

- Redefine the boundaries of particular industries
- Develop new products or services
- Change the relationships between suppliers and customers
- Establish barriers to deter new entrants to marketplaces.

The basic objectives of this type of information system is to identify better ways of doing things, leading to increased revenues, greater functionality, better products and service, improved presentation or image, an improvement to the organisation's competitive positioning, or whatever is required. This is usually referred to as using information systems for competitive advantage. We might argue that this makes more effective use of IT than the traditional efficiency argument.

Although making justifications based on effectiveness is not impossible, it is a much more difficult process than making a financial case on the basis of efficiency improvements. One of the reasons for this increased difficulty is because there is an extra stage of proof to be gone through. In efficiency projects, the benefit suggested is reduced cost, though it may not necessarily accrue in practice. For effectiveness projects, it is not just necessary to show that the benefit will be, for example, better service, but also that the recipients of that better service will recognise and value the improvement and change their behaviour in some positive way as a result. For example, existing customers may buy more services or not transfer their allegiances elsewhere.

One tangible result of this in some information systems development methodologies is to incorporate a benefits realisation programme to ensure that any measure is implemented in the information system and this leads to the hoped-for change in behaviour. This is long term and difficult to prove and it has often been the case that organisations have abandoned any attempt at justification and instead relied on management perception of the benefits, sometimes referred to as strategic insight, intuition or blind faith.

The use of information systems for effectiveness or competitive advantage has been shown to have the possibility of generating very significant rewards for the organisation in terms of income generation rather than cost savings. There are some 'classic' examples which purport to show how some organisations 'have seized the opportunity to use information systems to gain competitive advantage', in particular:

- American Hospital Supply Corporation's then 'revolutionary' 1978 order-entry system directly linked customers into their systems which made ordering by the customer very easy. Customers have access to the order-distribution system so that, for example, customers could perform various functions, such as inventory control, for themselves using the system. This helped reduce costs for both the company and their customers, and enabled American Hospital Supply to develop and manage pricing incentives to the customer across all product lines.

- Merrill Lynch and Company's 1977 cash management account system which combined a number of previously separate services, including credit card, cheque and deposit accounts, and brokerage services, into one conceptual account provides another example. It was a highly innovative idea and led to the concept of the 'financial services' industry by breaching the traditional barriers between the banking and securities industries. To achieve this a sophisticated information system was required to communicate between Merrill Lynch on the brokerage side and Bank One that handled the cheque and credit card side for them. It was a great success and it is argued that it created a new market only made possible by the use of information systems.

- There are a variety of other examples where information systems have been used to establish competitive advantage, such as DEC XCON, Xerox's customer support system, American Airlines Sabre reservation system (followed by Apollo), and the 'electronic' newspaper, USA Today. In the UK there are also some well known examples, such as Thompson's Tour Operator Reservation System, Sainsbury's laser scanning point of sale system and that used by First Direct bank.

It is from these kind of examples that the concept of using information systems to gain competitive advantage has evolved and it has to be conceded that some of the examples are compelling. However, there is a degree of scepticism voiced by some that seeks to cast doubts on some of the more simplistic approaches to IT for competitive advantage. For example, success of the information system may have been due to the fact that the product was good or that success may not be sustainable as competitors copy and improve on the information system. One response to the latter is to create barriers to entry.

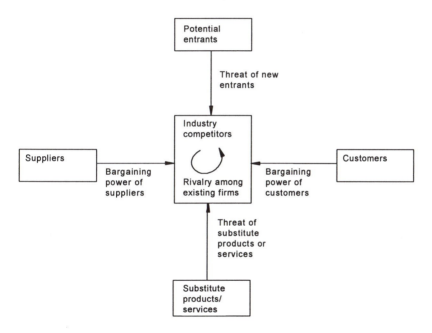

Fig. 3.1: A framework of competitive strategy (adapted from Porter, 1980)

Some industries may not lend themselves to such manoeuvres and Porter (1980) offers an industry analysis framework of competitive strategy to help identify the five competitive forces that any company needs to consider. These forces are illustrated diagramatically in figure 3.1 and Porter suggests that significant strategic advantage can be gained by diminishing supplier or customer power, holding off new entrants into the industry, lowering the possibility of substitution for its products, or gaining a competitive edge within the existing industry. This framework can be helpful in focusing the role of information systems to improve competitive positioning. Earl (1989), for example, identifies four ways in which Porter's model helps:

- It deals with the industry and competitive dynamics
- It highlights that competition is not simply concerned with the action of rivals
- It facilitates discussion and is based on sound principles of industrial economics
- It focuses on the few dominant forces necessary.

Competitive force	Potential of IT	Mechanism
New entrants	Barriers to entry	1. Erect 2. Demolish
Suppliers	Reduce bargaining power	1. Erode 2. Share
Customers	Lock in	1. Switching costs 2. Customer information
Substitute products/services	Innovation	1. New products 2. Add value
Rivalry	Change the relationship	1. Compete 2. Collaborate/alliances

Fig. 3.2: The strategic role of IT (adapted from Earl, 1989)

Figure 3.2 shows an extension to Porter's work by Earl to illustrate the strategic role that IT can play on each of Porter's dimensions. American Hospital Supply, for example, illustrates the use of IT to address the customer competitive force. IT was used to help lock in the customers using both mechanisms. Another aspect of this case is the role of a dedicated champion to push the ideas in the organisation, to overcome all the inevitable objections, to sustain the momentum over a fairly long period and to inspire people with the vision. This person is unlikely to come from the information systems domain, although he or she may be a 'hybrid manager', that is, a person having knowledge and experience in the IT and user domains. The role of people in successful information systems development is stressed further in the theme of participation, discussed in section 3.13.

Although many organisations now recognise the need to address effectiveness as well as efficiency in their use of IT and the resulting need for an IT strategy, the approach to developing such a strategy is by no means clear or universally agreed upon. Some organisations have adopted rather simplistic approaches as follows:

- *Technology driven model:* The reaction of some organisations has been implicitly to make the assumption that investment in IT will automatically result in business success and the achievement of competitive advantage. It is the embodiment of the view that if the technology exists it should be employed. It might be seen in the unthinking purchase of the latest IT product. This strategy is usually driven by technologists. A result of this approach is information technology that may not be appropriate for the needs of the organisation and a lack of control over IT budgets. As no business benefits are stated, there is no way of measuring whether any such benefits accrue. This strategy is obviously irrational although understandable in some ways due to the rapid pace of technological change exceeding the ability of many managers to keep up. Technology adopted in this way may cause disruption in the organisation and incompatibilities between information systems, rather than making for greater efficiency and effectiveness. For example, the introduction of the personal computer in many organisations was unplanned and could be described as technology-driven, and although this led to many individual benefits there were often other longer term problems of compatibility, support and integration. Strassmann (1990) concludes that there is little evidence that technology-based systems of information management have produced benefits that could be claimed to justify their costs. Even if the claimed (and unproved) individual gains are summed, they do not lead to an overall improved performance of the organisation. It would thus appear that an approach based purely on the technology-driven model is unlikely to prove adequate.

- *Competitor-driven model:* An alternative model or approach that some organisations have adopted is to react to their competitors by copying them. There is evidence that this happens in some sectors rather more than others. In the UK banking sector, for example, there seems to be a 'knee-jerk' reaction to copy each other in terms of technology and services. In manufacturing, there appears to be an assumption that anything Japanese is by definition the best approach to follow. The competitor-driven model is an approach based upon the fear that an organisation's competitors will use information technology to gain significant advantage over them. Therefore they must be 'tracked' and copied at each stage of their development. The fear is that companies that do not follow the same path in information technology will ultimately be squeezed out. The competitor model says that we will not allow this to happen by matching our competitors at every point. The problems with this approach are threefold. First, that by simply following competitors, the organisation will never innovate to its own particular strengths and advantages. Second, it may miss opportunities for being a leader itself. Third, it may still lose out by not itself being the first in the field, particularly in situations where being first in the field enables barriers to entry to be constructed.

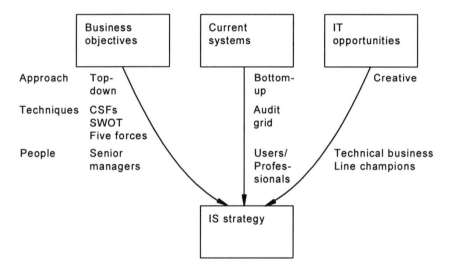

Fig. 3.3: Earl's multiple methodology (adapted from Earl, 1989)

A more thoughtful approach to the formulation of IT strategy and the alignment of IS development with the business needs, has been defined by Earl on the basis that no one-dimensional approach is likely to be successful. He therefore suggests an approach that combines a variety of different elements and techniques, including both top-down and bottom-up analysis. The individual elements are not necessarily new but they are combined into what is argued to be a comprehensive and effective approach. Figure 3.3 illustrates this 'multiple methodology'. The first element of the model is a top down analysis of the business and its goals and objectives leading to an identification of the potential role that IT might play in achieving these objectives. This is a top-down business-led activity in which IT people would play a role rather than the other way round and is best achieved by the use of established techniques such as critical success factors and SWOT analysis. It may also include a Porter 'five forces' type competitive positioning analysis and an analysis of competitors.

SWOT analysis is the identification of strengths, weaknesses, opportunities and threats to the business or organisation and the way in which information systems could enhance the strengths and opportunities and counter the weaknesses and threats. Critical success factors (CSF) (Rockart, 1979) are a 'limited number' of factors which are considered critical to the continued success of the business. The limited number is an important aspect of the analysis. If too many are identified, they are probably not all critical and at too low a level. The process thus includes an element of prioritisation of factors and a fundamental analysis of what is critical. A methodology based on critical success factors is described in Bullen and Rockart (1984). First, the business goals and objectives are analysed and then the factors critical to achieving each of those objectives are identified. Typically only about 4-6 CSFs would be identified for each objective. This is followed by an identification of the information and information systems required, if any, to support and monitor these critical success factors. According to Rockart, CSFs are likely to include:

- Factors critical to all organisations in the same industry
- Issues related to the particular organisation and to its position in the industry
- Environmental factors, such as legal, political, economic and social aspects
- Activities within the organisation that are proving to be short-term problems
- Monitoring and control procedures relating to the operations of the firm
- Factors which take into account the changes in the business environment

For each CSF, indicators or measures of performance must be established and trailed. The identified information systems will need to be developed or modified to ensure that the critical information is collected, analysed and distributed. In other words, it helps to ensure that the organisation's information systems support the agreed critical activities and thus the wider business objectives.

The second element of Earl's model is a bottom-up analysis of the organisation's current systems. This is a critical element of the model as in so many approaches to strategy formulation, the current systems are ignored and a 'green field' situation is assumed. This is usually quite unrealistic and leads to failure because the existing 'legacy' systems in an organisation have been ignored. The analysis consists of evaluating the strengths and weaknesses of the existing IT provision and existing systems on the basis of first, business contribution and value to users, and second, technical quality.

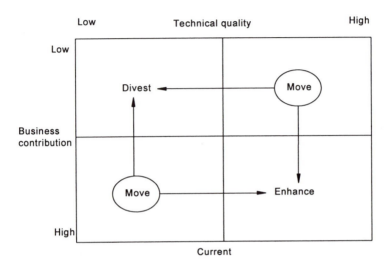

Fig. 3.4: Current systems audit grid

Figure 3.4 shows a typical systems audit grid that might be helpful in this analysis. One point of the analysis is to try to run down and exit from systems that perform poorly on both dimensions. It is frequently found that there are many such systems, particularly in organisations where IT has evolved in the organisation in a somewhat *ad-hoc* manner. These systems are often expensive to run and maintain, but are contributing little to the organisation in the direction that is required by the strategy. They should be phased out in order to free up valuable resources. On the other hand, those that are high on both dimensions should be enhanced and evolved. Earl makes the point that many strategic systems, or systems that have helped to provide competitive benefit, have in fact been based on enhancements and re-direction of existing systems rather than the construction of totally new ones. Where this can be achieved it obviously provides a head start and potentially a reduction in the cost and lead times for developing strategic systems. An example is the telephone banking system of First Direct in the UK which built the strategic telephone interface system on top of the existing traditional retail banking systems of Midland Bank. Those systems with combinations of highs and lows on the grid need to be carefully evaluated and moved into one of the other boxes. Probably the systems evaluated highly on the business contribution dimension would have more potential to be moved into the enhance and invest category rather than those systems that are simply high on technical quality.

The third element of the model is the identification of IT opportunities. This is not just an across-the-board look at the potential of the latest technological innovation, as in the technology driven model, but an attempt to assess the enabling effects of IT on the particular organisation and business, given its strategic direction. Of course in some cases the IT can help to shape the strategic direction and therefore the left and right legs of the multiple methodology model must be developed iteratively. Earl suggests that this element is a creative one and that many of the best ideas for applying IT in an organisation come from the people who understand the business on the ground, particularly at the customer interface, rather than the IT specialists, although the specialists can act as catalysts in this process. Thus a range of people need to be involved in this part of the process. The three elements of the model thus provide a multi-dimensional, integrated strategy formulation approach which, it is argued, results in a coherent information systems strategy which supports and reflects the overall business objectives, addresses the potential that IT might provide, and considers the reality of existing IS provision.

The emphasis on strategic information systems and a top management view has led to a number of implications for methodologies. Many, such as Information Engineering (section 6.4) have strategic planning as an important early phase in the development of an information system. In the next two sections we look at business process re-engineering and then to the planning of information systems development.

3.4 BUSINESS PROCESS RE-ENGINEERING

One approach to information systems development which takes into account strategic aspects is business process re-engineering (BPR). It has presented organisations with the opportunity to dispose of outdated rules and assumptions underlying business practices, and to re-engineer processes, including those performed by computer systems, which are responsible for under-performance. This re-engineering is enabled by technology. Although early initiatives have been found predominantly in the financial services sector, a wider variety of organisations are now instigating moves towards BPR.

The essence of business process re-engineering is a radical change in the way in which organisations perform business activities, thus it is defined by Hammer and Champy (1993) as:

the fundamental rethinking and radical redesign of business processes to achieve dramatic improvements in critical, contemporary measures of performance, such as cost, quality, service, and speed.

Re-engineering determines what an organisation should do, how it should do it, and what its concerns should be, as opposed to what they currently are. It is a radical approach and entails 'business reinvention - not business improvement, business enhancement, or business modification'. It is therefore not an incremental but a fundamental change. Emphasis is placed on the business processes (and therefore information systems that reflect them and enable the change), but it also encompasses managerial behaviour, work patterns and organisational structure.

The model of business process re-engineering created by Hammer and Champy describes the characteristics of re-engineered processes as follows. Several jobs are combined, performed by a 'case worker' responsible for a process. 'Case team' members are empowered to find innovative ways to improve service, quality, and reduce costs and cycle times. Due to process integration, fewer controls and checks are necessary, and defects are minimised by an entire process being followed through by those ultimately responsible for the finished product. Workers make decisions according to the requirements of the whole process. The steps in a process are performed in the order decided upon by those doing the work, rather than on the basis of fragmented and sequential tasks, and this enables the parallel processing of entire operations. Differing customer requirements may dictate that several versions of a product are created in one process.

Organisations re-engineer for four main reasons:

- They face commercial ruin and have no choice
- Competitive forces present problems unless the organisation takes radical steps to re-align business processes with strategic positioning
- Management in the organisation regard re-engineering as an opportunity to take a lead over the competition
- Publicity about BPR has prompted organisations to follow the lead established by others.

However, enthusiasm about BPR should be tempered with the reported failure rate of 50-70 per cent (Hammer and Champy, 1993). According to them, re-engineering projects fail primarily because senior managers lack the ambition for organisational change. Furthermore, many fail to comprehend the degree of change required, not only in business processes, but also in managerial behaviour and organisational structure. Piecemeal approaches mean that gains in individual processes fail to translate into improvements in the performance of the organisation as a whole. Top management often fails to define future operations clearly, and the extent to which competitive edge requires superior customer service, manufacturing efficiency or innovation. Re-engineering, in practice, often addresses non-critical business activities. Arguably, some BPR initiatives are being adopted either to gain publicity, or to further the careers of senior managers acting as the 'change agents'.

Management may confuse re-engineering with other, perhaps related, business improvement programmes, such as total quality management (TQM) and 'quality circles' concerned with the quality of products, in terms of specific standards. Consequently, processes may not be accurately identified, and organisations 'tinker' with aspects of business considered easier to change and from which performance improvements are easier to measure. Often aligned with standards such as BS5750 (ISO9000), process rationalisation and automation have not yielded the dramatic improvements organisations expect.

Short-run financial pressures on management might mitigate against the longer-term returns of re-engineering, profit and earnings per share taking precedence over market share and competitive positioning. Consequently, many BPR initiatives experience the problems of a lack of resources and senior management support (Moad, 1993) and, as a result, there is pressure to abandon the programme. The amount of senior management time or resistance to change may also lead to abandonment.

Re-engineering cannot normally be driven from the bottom, due to the need for a broad view of the organisation, and top-level guidance to promote cross-functional involvement. Without guidance and direction from the top, change is unlikely to be forthcoming. An exception might occur where re-engineering programmes are initiated by the IT department, and senior management is receptive to change and encourages bottom-up decision making. But this is not recommended. Re-engineering requires a consideration of jobs, structure, values, beliefs, and management and measurement systems, in addition to process redesign. Unless all the features are considered equally, re-engineering will not achieve change throughout the organisation. Similarly, a lack of focus on specific re-engineering projects distracts people from the main objectives of the organisation.

Senior managers, insulated from day-to-day operations, and unable to envisage the processes of change, are unlikely to be capable of 'championing' a re-engineering programme. Unless other managers can persuade their superiors of its value, organisations will not re-engineer.

BPR represents a major commitment and a radical approach to information systems development within the overall context of a re-examination of business processes as a whole within the organisation. It therefore has links with both the systems approach and the strategic information systems discussed in the previous two sections. It is unlike 'computerising existing systems' which was a feature of the traditional approach. BPR is a total approach, involving top management, total organisational restructuring, and a change in the way people think.

The practicalities behind Hammer and Champy's (1993) assertion that if the technology can be purchased by all organisations, 'a company will always be playing catch up with competitors who have already anticipated it' presents a problem for organisations who have invested heavily in large-scale information systems linked to functional activities which will be a result of traditional information systems development. But this can be argued about many of the themes discussed in this chapter.

Critical success factors could be used to evaluate the success of the re-engineering initiative, and might include, for example:

- Customer satisfaction
- Net profit and return on capital
- Expertise in credit and risk management
- Staff satisfaction
- Asset valuation.

During a re-engineering programme, IT should be regarded as an essential enabler, not a key driver of change. Information systems people should be involved in the early stages of planning, but normally should not be leaders of the change process. Other considerations could include:

- The way the company is organised
- The manner in which work is conducted
- The existing operational systems.

Failure to address all these concerns simultaneously is likely to lead to problems.

Following BPR, a cultural change is inevitable, and indicators of such change may include:

- A flatter organisational structure
- A focus on customers
- Greater teamwork
- Coaching rather than directing
- A facilitative style amongst teams, leading to a more widespread understanding of the roles of others.

To enable BPR, it may be necessary to recruit a re-engineering team consisting of, for example, strategists in information systems, business analysts with computer skills, organisational development specialists, and organisation and methods experts. Customer service teams may be used to maintain and develop the focus on future business and team 'facilitators' to coach team members, resolve conflicts and rectify operational problems.

Clearly, a balance must be struck between maintaining the authority of management and empowering the workforce to make decisions and become more self-sufficient and self-determining. In many cases, a substantial change in the senior management team has been necessary during re-engineering. Perceiving a personal threat to status and authority, many senior managers have balked at the extent of the changes necessitated by BPR. Conversely, however, successful BPR can significantly enhance the image of organisations and the personal images of the 'champions'.

According to Davenport and Short (1990), information systems and BPR have a recursive relationship. On the one hand, IT usage should be determined on the basis of how well it supports redesigned business processes. On the other, BPR should be considered within the realms offered by information systems (and the supporting information technology). The combination of information systems and BPR presents commercial organisations with the opportunity radically to change the way in which business is conducted. The increasing complexity of the environment has presented organisations with new threats and challenges. The need to maximise the performance of interrelated activities rather than individual business functions, combined with the opportunities offered by information systems, has meant that a new approach to the co-ordination of processes across organisations is necessary to achieve a sustainable competitive advantage. Consequently, organisations need to ensure the close alignment of information systems with business strategy, through strategic information systems, the latter being derived in the context of the changing commercial environment.

The following steps of process redesign are crucial to the success of BPR:

- Develop business vision and process objectives
- Identify processes to be redesigned
- Understand and measure the existing process, identify information systems levers which will help push the changes
- Design and build a prototype.

Following re-engineering, organisations have encountered further problems which require resolution to ensure the long-term impact of BPR which necessitate:

- Further process improvement
- Changes to the organisational structure
- New skill requirements as a result of changes in workforce activities and responsibilities
- Monitoring information systems.

Further, the future direction of information systems infrastructure should be ratified at the highest levels, to ensure that an adequate commitment of financial resources is made.

The long-term and radical nature of BPR has resulted in a diversity of experiences. The interpretation of BPR within each organisation determines the pace and impact of changes. BPR has been particularly successful in service and financial organisations, which have traditionally relied on IT. The role of information systems differs in accordance with the ambition of management to utilise its capabilities in the re-engineered processes, as well as the extent and sophistication of the information system prior to re-engineering. Organisations need to keep abreast of changes in technology supporting the information systems which could improve their ability to compete.

Whether BPR will have a significant impact on the prosperity of organisations will not be known for some time. Certainly, many businesses are questioning the 'rhetoric' associated with BPR. Analysis of organisations currently undertaking BPR provides material for future research, especially with respect to the availability and application of advanced technology in the development of information systems. In addition, re-engineered organisations may provide the opportunity to develop models and frameworks applicable to specific business domains.

There is not a consistent view of BPR at the time of writing:

- The long-term and radical nature of BPR has resulted in a diversity of experiences
- The role of information systems differs according to the ambition of management to utilise information systems capabilities in the re-engineered business processes, as well as the extent and sophistication of the applications prior to re-engineering.

We have seen that BPR is influenced by the systems approach and strategic considerations. It also has links with other approaches overviewed in this chapter. Automatic tools are useful in analysing the complex processes and enabling rapid applications development. Prototypes can be used to assess the new systems. The participation of people at all levels within the organisation is crucial. This includes the enthusiastic participation of information systems people in a supporting role as part of the BPR team. It also includes the involvement and leadership of strategic management, which links up with the planning approaches discussed in the next section. Surprisingly, perhaps, it has rather less to do with the structured approach because, although both emphasise processes, BPR is very much at a meta-level which crosses functional boundaries. Structured approaches concern the analysis and design of a single process. In section 6.13 we describe Process Innovation, which is Davenport's approach to BPR through the use of information technology.

3.5 PLANNING APPROACHES

Planning approaches stress the planning involved in developing information systems. Rather than look at individual applications and subsystems in detail, planning approaches involve the top (strategic) management of the organisation (the managing director, financial services manager and so on) in the analysis of the objectives of that organisation. Management should assess the possible ways in which these objectives might be achieved utilising the information resource. The approach therefore requires the involvement of strategic management in planning information systems.

Because of the requirement to develop an overall plan for information systems development in the organisation (within organisational planning as a whole), there are obvious links with the systems approach, and because of its emphasis on the role of strategic management, there are also obvious links to the section on strategic information systems. Further, because of this top management involvement, there are also links with the participative approach (section 3.13) as well as business process re-engineering.

Planning approaches are designed to counteract the possibility that information systems will be implemented in a piecemeal fashion, a criticism often made of the traditional approach. A narrow function-by-function approach could lead to the various subsystems failing to integrate satisfactorily. Further, it fails to align information systems with the business. Both top management and technologists should look at organisational needs in the early stages and they need to develop a strategic plan for information systems development as a whole so that information systems are integrated and compatible. This becomes a framework for more detailed plans. Individual information systems are then developed within the confines of these plans. Plans could be made at three levels:

1.   Long-term planning of information systems considers the objectives
      of the information systems function and provides rough estimates of
      resources required to meet these needs. It will normally involve
      producing a 'mission statement' for the information systems group
      which should reflect that of the organisation as a whole. The
      information systems plan at this stage will be an overview
      document, for example, providing only prospective project titles.

2.   Medium-term planning concerns itself with the ways in which the
      long-term plan can be put into effect. It considers the present
      information requirements of the organisation and the information
      systems that need to be developed or adapted to meet these needs.
      Information about each information system will be spelt out in
      detail. The ways in which the information systems are to be
      integrated will be detailed. Priorities for development will also be
      established and again these will reflect the long-term plan and
      mission statement. A document reflecting the plan will be produced
      which shows the current situation along with an action plan for
      future development.

3.   Short-term planning, perhaps covering the next twelve months, will
      again give a further level of detail. It concerns the schedule for
      change, assigning resources to effect the change and putting into
      place project control measures to ensure effectiveness. As well as
      detailing the resources required for each application in terms of
      personnel, hardware and budget, it will contain details of each stage
      in the development process as suggested in the systems development
      life cycle described in Chapter 2.

There is a need for information systems to address corporate objectives
directly. Planning approaches aim to ensure that management needs are
met by information systems. More radically, some information systems
are designed around management needs, a sort of 'top-down' approach,
sometimes ignoring operational needs which are assumed to be fulfilled
elsewhere. Managers may also set standards for information systems and
one of these requirements will be choosing the methodology for
developing information systems.

Many general information systems approaches discuss strategic
planning. Critical success factor analysis may be part of the approach. In
Avison (1992), business analysis is the first stage of the development of an
information system and involves the information systems strategy group in
the following tasks:

•   An assessment of the strategic goals of the organisation, which
      could be long-term survival, increasing market share, increasing

profits, increasing return on capital, increasing turnover or improving public image.

- An assessment of the medium-term objectives to be used as a basis for allocating resources, evaluating managers' performance, monitoring progress towards achieving long-term goals, establishing priorities.
- An appreciation of the activities in the organisation, such as sales, purchasing, research and development, personnel and finance.
- An appreciation of the environment of the organisation, that is, customers, suppliers, government, trade unions and financial institutions, whose actions will affect business performance.
- An appreciation of the organisational culture relating to values, networks and 'rites and rituals'.
- An appreciation of the managerial structure in terms of the layers of management or matrix structure, types of decision made, the key personnel and types of information needed to support the key personnel in their decision making.
- An analysis of the roles of key personnel in the organisation.

With this knowledge, it is possible to assess the type of information that an information system might provide. The first stage of Information Engineering (section 6.4) also concerns itself with planning aspects.

Addressing a more detailed later stage in the planning process, Lederer and Mendelow (1989) suggest a number of guidelines which should be considered when planning:

- *Develop a formal plan:* set objectives and policies in relation to the achievement of organisational goals and thereby enable the effective and efficient deployment of resources.
- *Link the information systems plan to the corporate plan:* provide an 'optimal project mix' which will be consistent with and link to the corporate plan, ideally over the same time period.
- *Plan for disaster:* ensure that dependencies are identified and damage likelihood identified.
- *Audit new systems:* evaluate present systems to identify mistakes and hence avoid their repetition and to identify areas where a small resource input might have led to a larger benefit.
- *Perform a cost-benefit analysis:* identify intangible and tangible benefits and costs before putting in the required resources.
- *Develop staff:* make use of and develop the skills of staff.
- *Be prepared to change:* as relationships, structures and processes change.

- *Ensure information systems development satisfies user needs:* understand the tasks and processes involved to establish the true user requirements.
- *Establish credibility through success:* build up user confidence through previous success, thereby promoting co-operation and lowering barriers.

Top management has in the past given a low priority to developing systems which attempt to seek out those opportunities that could give the business competitive advantage. It was not on the agenda of the followers of the traditional approach. Three approaches which do address these issues are BIAIT, BSP and ends/means analysis (Wetherbe and Davis, 1983 and Davis and Olsen, 1984).

In Business Information Analysis and Integration Technique (BIAIT) (see Carlson, 1979 and Burnstine, 1986), fundamental questions are asked which relate to the objectives of the business. In the approach these are reduced to seven basic questions. Each question is constructed in such a way that either there can only be two possible answers, 'yes' or 'no', or a choice is made from options provided. A grid or matrix of possible responses is formed in the model generated by BIAIT. This profile is used to suggest reports which analyse whether the information handling necessary and the set of objectives have been met. Only at this point does the method begin to look at possible information systems, computer-based or otherwise, which might support these requirements.

A European approach, Total Information Systems Management is found in Österle *et al.* (1993), but a more well-known approach is IBM's Business Systems Planning (BSP) (see IBM 1975). It also addresses the requirements of top management and the importance of ensuring that information systems development coincides with and supports the business plan. It attempts to develop an overall view of information systems planning and is sometimes referred to as information architecture. BSP follows three principles.

1. The need for an organisation-wide perspective. This factor alone separates the BSP approach from the traditional approach discussed in Chapter 2. Rather than address the information needs of any single area of the organisation, it takes the perspective of strategic management.

2. Analysis from top management downwards (but implementation from the detailed level upwards). It is strategic management who define organisational needs and priorities to help ensure that their perspective predominates in the initial system definition. The BSP approach takes on a bottom-up orientation during the design and

implementation phases. This is when the database is created and the processing requirements defined which are necessary to fulfil the organisational objectives.

3. The need for independence of the business plan from the computer application systems, including data storage aspects. This means that desirable organisation changes are not prevented from taking place because of restrictions of the computer systems. Similarly, the computer application systems can be modified, perhaps by using newer technology, without affecting the organisation.

The phases of Business Systems Planning are:

- Set up project team and gain commitment and support of top management
- Review objectives and schedules
- Identify key business activities
- Define information systems requirements related to each of these activities
- Analyse current systems support
- Design in overview the data structures, procedures and information systems
- Interview top management to verify information needs and priorities
- Construct the action plan that coincides with business objectives and the environment
- Design in detail
- Develop, test, install and operate information systems using an appropriate methodology.

The organisation-wide perspective is obtained by interviewing key executives and is structured to provide information relating to the organisational environment, organisational plans, planning processes, organisational structure, products, markets, geographical distribution, financial statistics, industry position, industry trends, and major problem areas. Information is also gathered about the current information systems and how these support the organisation. Following this analysis it is possible to show how information systems can better assist in improving management decision-making. These information systems should be part of an integrated plan for using the information system's resource most effectively to help top management both in the short and long terms.

3.6 MODELLING

In the previous four sections we have looked at various aspects of a major theme which has concerned information systems specialists responding to the weaknesses of the traditional systems development life cycle approach to information systems development. They seek to address the needs of the organisation as a whole, seek to meet the particular requirements of top management, to re-engineer systems so that they take into account these needs and planning information systems development, again, to meet these higher level requirements.

We now look at another over-riding theme in modern information systems development, that of modelling. A model is an abstraction, a representation of part of the real world. In the context of this book, it may concern a representation of one aspect of the present or proposed information system. Thus much emphasis in information systems has been placed in modelling process aspects and data aspects of the information system. Sections 3.7 and 3.8 look at process models in structured approaches and data models in data analysis approaches respectively. More recent interest has been placed on modelling all real world objects, which include processes, data, software and people. This object orientation is a theme discussed in section 3.9.

All these representations prove useful as an aid to understanding the real world and in analysing and designing the future information system. However, a model can only be a representation, it cannot be a perfectly accurate reflection of the real world. We saw in our discussion of the systems approach how the whole is greater than the sum of the parts. On the other hand, the real world is so complex, fuzzy and chaotic, that some would argue that it is necessary to simplify in order to make progress.

3.7 PROCESS MODELLING

Structured methodologies use many techniques of process modelling. They have as their unifying elements an emphasis on the processes and the basic technique of functional decomposition, that is, the breaking down of a complex problem in more and more detail in a disciplined way. At the lowest level the units are simple and manageable enough so that they could be reflected in a few lines of computer program code. An example of functional decomposition, which represents a simplification of a payroll system, is seen in figure 3.5. As well as enabling understanding of complex processes, it also enables people to view the processes at different levels. For example, the systems analyst may wish to view the system at a high level of abstraction and programmers at a lower level.

The possibilities of structured methodologies in systems analysis and design stemmed from the perceived benefits of software engineering (discussed in section 3.12). This concerns the improvement of software through a disciplined approach to software design. However, even if the software is of better quality, as shown by increased reliability, for example, this may not provide an overall improvement if the analysis and design that preceded it were poorly carried out. Further, even good analysis and design, if poorly documented, may lead to incorrect interpretation by programmers and even the best programming techniques cannot overcome such problems. Thus the question of quality analysis and design and its documentation has been addressed by structured approaches along with quality software.

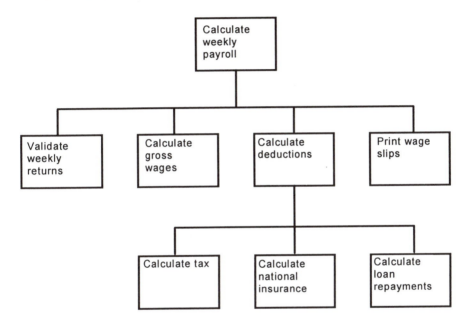

Fig. 3.5: Functional decomposition

Structured systems analysis and design has been associated with a number of consultancy houses and authors. These include Weinberg (1978), Yourdon and Constantine (1978), Gane and Sarson (1979) and DeMarco (1979). The most recent update of this school in published form is found in Yourdon (1989) and Yourdon Inc. (1993). In section 6.2, we look specifically at the Gane and Sarson methodology (STRADIS). This is a good example of the original structured approach. In section 6.3 we look at a later development of the approach, Yourdon Systems Methodology (YSM). Other methodologies discussed in Chapter 6, for example, SSADM and Merise, have been greatly influenced by this school.

Some of the techniques associated with structured systems analysis and design, specifically functional decomposition, decision trees, decision tables, data flow diagrams, data structure diagrams, and structured English, are looked at in Chapter 4, and tools used in structured approaches (and other methodologies), are discussed in Chapter 5. Many of the documentation and other techniques are easy to draw with computer diagramming and other packages

The techniques and tools to represent processes used in structured systems analysis and design prove to be a considerable improvement on the conventional techniques such as computer run and systems flowcharts discussed in section 2.2, both from the point of view of understanding the real-world processes and in communicating the knowledge acquired. Many of the representations are graphical and this encourages user involvement to some extent. Although the emphasis of the structured school is on processes, there is also a consideration of the structure of the data in some of the techniques and tools. For example, the normalisation of data and the data dictionary reflect the data in the application area rather than process aspects. These will also be described later.

Structured analysis documentation includes documents describing the logical (real-world) analysis of the processes and not just their physical (implementation) level designs. In other words, there is usually a clear distinction between any application logic (what a system is trying to achieve) and the computer representation of that logic (how the computer system achieves it). One of the major advantages of the approach is that the documentation is produced as part of the analysis and design process as 'deliverables' and not as an afterthought as can be the case with some of the documentation produced when using the traditional approach. Documentation can be cross-checked to ensure consistency in the analysis and design and hence enable quality control.

There is also a separation of data structures as seen by the user in the processing and its physical representation (a computer file or part of a database). This separation of logical and physical designs is an important element of a number of methodologies discussed in Chapter 6. It gives a level of 'data independence', in other words, the processes can change without necessarily changing the computer files. Similarly, the files can change without necessarily altering the user views of the data.

Two techniques that the analyst might use are data flow diagramming (section 4.7) and structured English (section 4.10). Data flow diagrams are a particularly useful aid in communicating the analyst's understanding of the system and are a feature of many commercial methodologies. Users, whether the operator of the system when it is operational or the manager of the department it is aimed to help, can see if the data flow diagram accurately represents the system and, once there is agreement, it can be converted to computer procedures. Structured English, which is another technique used in structured systems analysis and design, is very like a 'readable' computer program. It is not a programming language though it can be readily converted to a computer program, because it is a strict and logical form of English and the constructs reflect structured programming.

There are other variations on the 'English-like' languages used in these methodologies. Although 'English-like', they are not natural languages which tend to be ambiguous and long-winded. Structured English and its variants, such as pseudo code and tight English, are designed to express process logic simply, clearly and unambiguously. This is particularly important. In the past, systems analysts were first and foremost computer people. Now they are using terminology and techniques which can be understood by users. This might help in getting the willing participation of users. Much of the documentation aids are graphic representations of the subject matter. This is much easier to follow than text or computer-orientated documentation.

As well as improved communication tools, structured methodologies usually incorporate 'structured walkthroughs' which are used as another form of quality control. These are team meetings where the analysis and design specifications and other documentation (which should have been previously circulated) are exposed to review by the members of the team. Usually the group explores problems rather than looks for solutions (although some work this way) and the author of the documentation does not usually make comments (to avoid protracted discussions), although again this depends on the organisation. All comments are noted on 'action lists' so that they can be digested by the author and acted upon.

There are dangers however, even in this scenario, as unnecessarily long discussions and side-tracking must be avoided. They usually represent meetings of peers and 'management' are not involved. This will avoid the type of criticism which may have repercussions later on the team member's status or salary. A peer review is likely to reveal errors, omissions, ambiguities, inconsistencies and weaknesses in style, and also ensure a process of continuous training of the staff involved. Questions that the peer group might ask are:

- Can program designers use routines already available in the library?
- Is the user interface simple, understandable, and consistent?
- Does the design fulfil the specification fully?
- Will it work?

Having walkthrough sessions at each stage, requirements definition before analysis, analysis before design, and design before implementation, should avoid the late detection of errors and flaws in logic, and hence greatly reduce the risk of failure when the system is implemented. Early detection of errors usually implies cheaper correction. Some installations have frequent informal walkthroughs and more formal inspection meetings which take place after every major stage. There might be about four or five of these during the lifetime of the project.

Structured systems analysis and design can be considered as an alternative to the traditional approach. In fact this is a simplification. The authorities on structured analysis do not all view it in the same way. The techniques can be regarded as a useful alternative to many techniques used in the traditional SDLC, and therefore could be incorporated in that approach rather than replace it. Alternatively, a structured methodology may be seen to replace the traditional approach. Some writers emphasise the analysis aspect, such as DeMarco, while others, for example Gane and Sarson stress both analysis and design aspects. However, there is much more stress placed on analysis in these approaches than the traditional approach described in section 2.2. On the other hand, there is less stress on implementation aspects.

Most writers in this school stress the importance of techniques in systems analysis and design. Writers emphasise different techniques, although some are common to all structured methodologies, most notably, data flow diagrams. Indeed, so much emphasis is placed on the techniques in the approach that it is best described more fully in section 6.2 (in the context of the structured methodology STRADIS) following a discussion of the main techniques in Chapter 4.

3.8 DATA MODELLING

Whereas structured analysis and design emphasises processes, data analysis concentrates on understanding and documenting data which, it is argued, represents the 'fundamental building blocks of systems'. The data model, the result of data analysis, is orientated towards that part of the real world it represents (organisation, department, or whatever) and should be implementation independent. This means that the data model, and therefore the data analysis process that derives it, is suitable whether the physical model is a database, file or card index. There is therefore a separation of logical from physical data, that is, data independence. It also means that even if applications change, the data already collected may still be relevant to the new or revised systems and therefore need not be collected and validated again.

The interest in data analysis stems partly from the problem of 'graduating' from computer files to databases. What may be called 'informal data analysis', an *ad-hoc* approach to understanding the data structures in an organisation which is adequate for file processing systems, is not adequate for database applications where the data is shared between applications.

Early database experience did not always bring about the expected flexibility of computer applications, usually because the database was not a good reflection of the organisation it was supposed to represent. Modelling the organisation on a computer database is not a simple ambition. It is argued that data analysis has helped the likelihood of achieving a data model which is independent of any database package, accurate, unambiguous and complete enough for most applications and users. Its success comes in the systematic way by which it identifies the data in organisations and, more particularly, the relationships between these data elements, the 'data structure'.

Data analysis techniques (see Avison, 1992) attempt to identify the data elements and analyse the structure and meaning of data in the organisation. This is achieved by interviewing people in the organisation, studying documents, observation and so on, and then formalising the results through a process known as entity modelling (see section 4.5). A number of documentation aids, many of which are graphical, also help in the process of data analysis. This becomes the first step in abstracting aspects of the real world into a model that can be held on computer. The documentation produced in data analysis is more easily understood by users than the documentation for data representation used in the traditional NCC approach, such as the file and record specification forms which are very computer-orientated.

Data analysis does not necessarily precede the implementation of a database or computer system of any sort. It can be an end in itself, to help in understanding aspects of a complex organisation. Good models will be a fair representation of the 'real world' and can be used as a discussion document for understanding aspects of the organisation and the processes as well as improving the effectiveness of the role of data and information in an organisation.

There are a number of alternative approaches to data analysis. In the data collection approach, frequently referred to as document-driven analysis, the documents used in a department or organisation are analysed in a bottom-up manner. Such documents include reports, forms and enquiry formats. The analysis of each document in turn will lead to the formation, and then improvement, of a data model showing the data and the relationships between these data. The particular structures arrived at are relations (special kinds of tables), an example of which is shown in figure 3.6, and rules are applied to the set of relations to ensure that the model is flexible for future use, such as its mapping on to a database. These rules, which are described in section 4.6, are known as the rules of normalisation. The result of this type of data analysis is therefore a data model represented as a set of normalised relations.

PART NUMBER	COST	NAME	SUPPLIER
344	£10.00	Widgets	Smith
346	£12.00	Widgets	Jones
540	£10.00	Widget tops	Smith

Fig. 3.6: Part relation

The document-driven approach is a useful contribution, as documents in organisations are usually easy to analyse. But organisations are not always fully represented by the documents used in that organisation. Further, the number of documents is frequently such that they would take too long to analyse.

An alternative approach to data analysis, entity modelling, gains its information by interviewing people in the organisation, such as department managers and clerical staff. Entities, such as customers, suppliers, parts and finished goods, are identified and the relationships between these entities are also ascertained. The entities and their relationships can be expressed as a graphical model. This is a more common approach to data analysis and we look at this approach in detail in section 4.5.

Data analysis is stressed in many methodologies for a number of reasons, but particularly because data, it is argued, are more stable than are processes. Many of these arguments were made in section 2.4, but additional arguments for this approach are outlined below:

• The data model is not computer-orientated. It is not biased by any particular physical storage structure that may be used because of the logical/physical split. The model stays the same whether the storage structures are held on magnetic tape, disk, or main storage. If the target system is a computer system, it could be a mainframe, minicomputer or microcomputer. Further, it is not biased towards using any particular database management system (section 5.3).

- It is a model which is understandable by the technologists, but also by users, who can far more readily appreciate the meaning of the data relationships illustrated in graphical form than they could in the file and record specification forms of traditional systems analysis. It is not necessary to know anything about computers or computer file structures to understand the data model and to validate it.

- It does not show bias towards particular users or departmental views. The data model can reflect a variety of different views of the data.

- Although the model can represent the organisation as a whole, it can be adapted so that a particular part of the model can appear in a different form to different users.

- The data model is readily transformable into other models, such as relations, hierarchies or networks, which are often but not exclusively the ways in which database management systems require the data structures to be presented.

- The different data analysis techniques available allows a choice in situations where alternative methods are appropriate. The results of one technique can be cross-checked with another. There is therefore some expectancy that the data analysis does fairly represent that part of the real world modelled.

- Data modelling is rule-based, which means that the results of other analysts' work can be followed, assessed and, at least to a certain extent, proven. Further, the processes lend themselves to computerisation and many aspects of data analysis have been automated.

The approach does have critics. Some argue that although the documents and the techniques used are easier to understand by users when compared to some traditional methods, data analysis does not lend itself fully to user participation. Others have argued that the emphasis on data may be misplaced, people, rather than data, being the true 'lifeblood of the organisation'. Further, the assumption in data analysis that it is possible to model reality is questionable, see Kent (1978). The data model can only be *a* model and not *the* model of that part of the real world being investigated. It cannot reflect reality completely and accurately for all purposes. Even if data analysis has 'gone according to plan', the resultant data model cannot objectively represent the organisation. It is a subjective view distorted by the perceptive process. Having said this, however, the data model derived from data analysis usually proves in practice to be suitable for the purpose of building a database.

For many, data analysis is one-sided and although it is beneficial to develop a data model which can be mapped on to a database, there is still a parallel need to understand the functions that will be applied to the database when it is implemented. Indeed most data analysis-based methodologies now address process aspects as well. SSADM, section 6.5, is biased towards the data approach, on the other hand, Merise, section 6.6, is a more balanced approach of the data and process views of the computer-based information system development process.

3.9 OBJECT-ORIENTED SYSTEMS DEVELOPMENT

For some, object orientation has become the latest 'silver bullet' that is going to solve all the problems of development that have proved so tenacious and hard to overcome in the past (see Brooks, 1987). Jones (1990) suggests that in 'most environments, other than information systems, use of the object-oriented paradigm for developing new software is becoming the norm rather than the exception'. It is now beginning to make an influence in information systems development as well. It is introduced in this section and developed further in sections 4.13 and 6.8.

Booch (1991) suggests that the concept of object orientation emerged simultaneously in a number of different fields in the 1970s including computer architecture and operating systems, databases, cognitive science and artificial intelligence. However for most people it is associated with programming languages, in particular, Simula and Smalltalk.

Given that the concepts of object orientation have been around since the early 1970s it is surprising, perhaps, that they have only recently become so influential. This can be said of a number of 'new' themes discussed in this chapter. It may be that it takes about twenty years for such conceptual and theoretical advances to make their impact in practice. However, there are a number of explanatory factors in the case of object orientation, such as:

- The recent popularity of graphical interfaces
- The growing acceptance of the C++ programming language
- The focus of business on cost-cutting and the need to re-use software
- The increasing power of technology and the shift from mainframes to distributed computing and to PCs.

In the context of this book, perhaps, the most important development relates to object-oriented methodologies for the development of information systems.

The basic object-oriented concepts are quite different to traditional ones in systems analysis and design and this helps to explain why some people find it difficult to come to grips with the ideas. Coad and Yourdon (1991) suggest that the concepts are based on those we first learned as children, that is, objects and attributes, wholes and parts, classes and members. We learn by identifying particular objects, such as trees, and then identifying their component parts (that is, their attributes), for example, branches and leaves, and then to distinguish between different classes of objects, for example, those of trees and rocks. This implies that the concepts should be simple and indeed Smalltalk was originally developed for children to use. Nevertheless, many analysts find the concepts both different and difficult and we devote section 4.13 to the detailed description of the concepts in the next chapter.

There are a number of benefits of the approach.

- Object-oriented concepts unify many aspects of the information systems development process. For example, the analysis of the application area can be undertaken using object analysis and object modelling, the design of the new system can use object orientation as the design approach, applications can be developed using object programming languages, object-oriented CASE tools and 'data' (using a broad definition of the term to include text, sound and video) once collected can use object-oriented multi-media databases. There is no need to transform the objects into other representations. The object-oriented theme is relevant throughout. This contrasts with those methodologies which somehow have to blend and reconcile the results of different approaches, such as, that of the data and process views discussed in the previous two sections. For example, attempting to reconcile entity-relationship diagrams and data flow diagrams is not trivial because they represent different objects in different ways. The object-oriented approach represents data, processes, people and so on, all as objects.

- It facilitates the realistic re-use of software code and therefore makes application development quicker and more robust. In theory the organisation will develop a library of object classes that deal with all the basic activities that the organisation undertakes. Software development becomes the selection and connection of existing classes into relevant applications, and because those classes are well tried and tested as independent classes, when they are connected together they provide immediate industrial-strength applications that run correctly in a shorter period of time. Only completely new classes will need to be developed or purchased. Proponents of object orientation believe that eventually there will exist international libraries of object classes that developers will be able to browse to find the classes they require and then simply buy them. The classes in these libraries will be guaranteed to perform as specified, and so new applications are easily developed and existing (object-oriented) applications can be modified and extended in functionality just as easily. Software development is not only quicker and cheaper but the resulting applications are robust and error free.

Coad and Yourdon (1991) suggest a number of other motivations and benefits for object-oriented analysis, including:

1.   The ability to tackle more challenging problem situations because of the understanding that the approach brings to the problem situation.
2.   The improvement of analyst-user relationships, because the approach can be understood by both equally and because it is not computer-oriented.
3.   The improvement in the consistency of results, because it models all aspects of the problem in the same way.
4.   The ability to represent factors for change in the model so leading to a more resilient model.

These are ambitious claims, and in a later section these claims can be evaluated in the context of the Coad and Yourdon methodology (section 6.8) but more detail on the concepts and techniques of object-oriented analysis and design are provided first in section 4.13.

3.10 AUTOMATED TOOLS FOR INFORMATION SYSTEMS DEVELOPMENT

We now turn to a number of themes which relate to the overall theme of automation of the information systems development process and a greater emphasis on the software part of the information system. This section provides an historical perspective, Chapter 5 being devoted to a more detailed look at the facilities of modern tools. The hope of automating some aspects or even all of the information systems development process is an abiding one, and has intrigued developers for many years. It has often been recognised that certain aspects of the information systems development process are repetitive or rule-based and therefore susceptible to automation. Early examples include decision table software which could generate accurate code directly from a decision table (section 4.9), project control packages (section 5.2) to help organise and control the development process, and report generators (section 5.7) to help speed up some common programming tasks. On the analysis side, software has existed for many years to help in the construction of traditional programming flowcharts.

A variety of projects are well documented. ADS (Accurately Defined System), which was a system definition and specification technique developed by NCR in the 1960s. Five forms were provided which were used to specify outputs, inputs, computations, files and process logic in the form of decision tables. Efforts were made to automate some aspects of ADS by using programs which interpreted the data on the forms to produce information and for cross-referencing.

Systematics, see Grindley (1966 and 1968) attempted to incorporate, amongst other tools, automated decision table interpretation into an automated methodology of information systems design.

TAG (Time Automated Grid), (IBM, 1971), was another attempt to automate some aspects of the systems analysis process. Details of the required outputs of a system were fed into TAG and the package would work backwards to determine the inputs necessary to produce those outputs and in what sequence. Reports were produced which helped the analyst in the design process.

Perhaps the most important attempts in the 1960s and early 1970s were carried out at the University of Michigan which led to the ISDOS (Information System Design and Optimization System) project. The basis of ISDOS was PSL and PSA. Problem Statement Language (PSL) enables the analyst to state the requirements in formal terms (indeed in machine-readable form) which are then entered into a kind of database. The PSL statements are then analysed by the Problem Statement Analyzer (PSA) which has similarities with a data dictionary system and which interprets these statements, validates the data, stores it and analyses it, and then produces the output required. The PSL and PSA together produced a set of documents for the project including the systems analysts' formal specification and the users' requirements. These languages are described in Teichroew and Hershey (1977), Teichroew et al. (1979) and Welke (1987). The emphasis of the approach is on the documentation aspects, for they help the manual tasks of systems analysis rather than genuinely automating the processes of analysis and design.

ISDOS attempts to link all aspects of systems development, from a computerised problem statement into programming language statements. System requirements are input using PSL and PSA but ISDOS also incorporates an Optimizer, so that the code generated is as efficient as possible, and a file generation subsystem. ISDOS had as an aim the incorporation of all aspects of systems work in a complete automated package, but this proved over-ambitious at the time. Indeed, the ideal of converting a user specification in natural language automatically into verified software is still a long way off.

Whilst many of these attempts at automating various aspects of systems development were far sighted and valiant, they were not overly successful in terms of application. It has often been noted by users that data processing professionals have been very keen to automate everybody else but have shown a certain reticence to 'take their own medicine'. Apart from innate conservatism, there appear to have been a variety of reasons for this: some of the software was not very good or easy to use; it was often found that the benefits were outweighed by the effort and costs involved; and the technology was also a limiting factor.

Recently, some of these factors have changed. The technology is clearly more powerful, cheaper and widely available. Improved graphics facilities have also had an impact. The quality of the software has improved, and in general there is a growing climate of opinion that automated tools may be beneficial.

There are now a number of tools that support the analysis and design process. These are tools that help the user use the techniques described in Chapter 4, such as data flow diagrams, entity models and so on. They are sometimes described as documentation support tools, being designed to take the drudgery out of revising documents, because they make the implementation of changes very easy. In addition, they can contribute to the accuracy and consistency of diagrams. The diagrammer can, for example, cross-check that levels of data flow diagrams are accurate and that terminology is consistent. They can ensure that certain documentation standards are adhered to. Probably the greatest benefit is that analysts and designers are not reluctant to change diagrams, because the change process is simple. Manual re-drawing is not satisfactory, not just because of the effort involved but the potential of introducing errors in re-drawing. Many a small change required by a user was never incorporated into the system due to the effort required to re-draw all the documentation. These kinds of documentation tools have proved both practical and useful so that the change process is not now inhibited.

Tools supporting the use of single techniques have, however, also proved limited in the sense that much of the information required for a data flow diagram, for example, would also be required, in a slightly different form, for the process logic software, and so on. It was realised that it would make more sense to have a central repository of all the information required for the development project irrespective of its graphical representation. In fact, this is the data dictionary, or perhaps as it is more correctly known, because it contains information about processes as well as data, the systems dictionary, systems encyclopaedia or systems repository (section 5.5).

Once most of the information concerning a development project is on a data dictionary, it is in theory only a short step to the automation, or at least the automated support, of many of the stages of the development project. Further, one of the goals of a number of methodologies is to provide automated support for all of their stages. Some have the automatic generation of code as the end result of the automation of the information systems development process. Computer aided software engineering (CASE) tools, fourth generation systems and application workbenches are the modern equivalent of the ISDOS tools, but they are more widespread and effective.

The core CASE facilities include:
• Graphical facilities for modelling and design
• Data dictionary
• Automated documentation.

All CASE tools are likely to contain at least these facilities, whilst advanced CASE features also include:
• Code generation from systems specification or from the models designed using the tool
• Automatic audit trail of all changes
• Critical path scheduling with resource availability (that is, project control)
• Automatic enforcement of the standards of a chosen software development methodology.

It is important that these facilities are completely integrated so that they provide consistency in analysis and design. They are particularly effective in this regard if they are associated with a particular methodology. We devote much of Chapter 5 to a consideration of CASE support tools.

Perhaps it is appropriate to provide a few warning bells in relation to CASE tools. They are said to reduce the skill, complexity, time, error and maintenance associated with the development of information systems. However, most are modest in their facilities. Typically, they help people to draw diagrams, such as entity-relationship diagrams and data flow diagrams, and perform validation checks. Many of the more sophisticated tools are very costly in terms of price, but also in terms of training and support. They may also be appropriate for use with only one methodology, which ties the users to that approach and also may make updates (as the methodology is updated) both essential and expensive.

Before we leave the topic of automated analysis, many information systems are developed 'automatically' by purchasing and implementing a ready-made application package. This can be an information system that is bought off-the-shelf from a supplier. Application packages are usually available where the application is one which many organisations need.

Nevertheless, each business is likely to have certain 'quirks' which are not found elsewhere. For this reason, there still needs to be a full requirements definition, and an application package needs to be assessed on the basis of answers to the following questions:

1. Does it fit in with the information systems strategy?
2. Does it meet functional requirements?
3. Does it meet resource limits (for purchasing and running)?
4. Does it meet documentation standards?

These aspects are discussed fully in Avison (1990). However, because normally not all the requirements are met by any single package, it is not a perfect solution and rarely near perfect, so the use of application packages is limited and any purchased will have to be customised, unless the information systems application is particularly discrete, well defined, common to many organisations and stable. Further, it may be difficult to integrate the new software with existing information systems because they are likely to have been tailor-made or bought from other software vendors.

Should a package solution be feasible, it might be the recommended approach among the alternatives in a feasibility study. The requirements definition phase should be as thorough as in the standard systems development life cycle. The later phases will be somewhat modified, however. Instead of going on to design a tailor-made system, software suppliers are asked to show how they can meet the specification and at what price. The development phase is likely to be shorter, though some modifications of the application package to suit the organisation's needs are inevitable. This may be straightforward where all that is necessary is to decide between the options available. Of course, training, testing and other aspects of the implementation process must be at least as thorough.

3.11 PROTOTYPING

A prototype is an approximation of a type that exhibits the essential features of the final version of that type. Prototyping is common in many other areas. In engineering, mass production makes it imperative that the design has been tested thoroughly first. It is also found in areas where the final version is one-off, and it would be very expensive if the designers got it wrong, like a bridge or a building. Information systems tend also to be one-off, but the cost of building a prototype has in the past been a major proportion of final costs and therefore has been rarely included in information systems development, at least, if we discount rough drawings on paper. This has changed with the availability of software tools, in particular fourth generation systems, application generators and the like, which greatly reduce the costs associated with prototyping.

Prototyping addresses some of the problems of traditional systems analysis, in particular, the complaint that users only saw their information system at implementation time when it was too late to make changes. If a prototype version of the information system is not developed first, the systems analysts are experimenting on the user. The first version of the type is also the final version and this brings about an obvious risk of failure, including outright user rejection. Thus the prototyping approach is a response to the user dissatisfaction found when using the traditional approach to information systems development. Acceptance of the operational system is far more likely. By implementing a prototype first, the analyst can show the users something tangible, inputs, intermediary stages, and outputs, before finally committing the user to the new design. These prototypes are not diagrammatic approximations, which tend to be looked at as abstract things, but actual outputs on paper or on workstation screens. The formats can be changed quickly, as the users suggest changes, until the users are given a reasonable approximation of their requirements. Further, it may only be by using this approach that the users discover exactly what they want from the system, as well as what is feasible. It is also possible to try out a run using real data, perhaps generated by the users themselves.

Sometimes analysts may only use prototypes to examine areas where they are unsure of the user requirements or where they are unsure how to build the system. Some analysts recommend that only the most critical aspects of a new system should be prototyped. Alternatively, the prototype may be built up using the most straightforward aspects and add to the prototype as users and analysts understand the application area more fully.

Prototyping can therefore be seen as a much improved form of systems investigation and analysis, as well as an aid to design. It is particularly useful where:

- The application area is not well defined
- The organisation is not familiar with the technology (hardware, software, communications, designs and so on) required for the application

- The communications between analysts and users are not good
- The cost of rejection by users would be very high and it is essential to ensure that the final version has got users' needs right
- There is a requirement to assess the impact of prospective information systems.

It is also a way of encouraging a degree of user participation (see section 3.13).

A prototype is frequently built using special tools such as screen painters and report generators which facilitate the quick design of screens and reports. The user may be able to see what the outputs will look like quickly. Whereas a hand drawing of the screen layout will need to be drawn again for each iteration which leads to a satisfactory solution, the prototype is quickly re-drawn (or re-painted) using the tools available. As with word processing systems, the savings come in making changes, as only these need to be drawn. Thus the ease and speed with which prototypes can be modified are as important as the advantages gained from building the prototype in the first place. Iteration becomes a practical possibility.

Some earlier prototyping tools included the programming language APL, although this required some programming skill to use. The operating environment PICK also has facilities which can be used to develop prototypes. There are now PC versions of the popular mainframe operating environments. Sometimes a PC will be used for prototyping using a database package such as Access or Paradox. Once the user is satisfied with this version, an operational version might be built on the mainframe. However, there are now a number of tools available for prototyping, including the CASE tools discussed in Chapter 5.

Frequently a prototype system can be developed in a few days, and it may not take more than 10 per cent of the time and other resources necessary to develop the full operational system. This can be a good investment of time. Analysts can usually achieve rapid feedback from the users so that the iterative cycle can quickly work to a version acceptable to the users. Some systems teams use the prototype as the user sign-off. This is likely to be a much better basis for obtaining a user decision than the documentation of traditional systems analysis discussed in section 2.2.

One possible drawback of prototyping is that the users may question the long time required to develop an operational system when that taken to develop a prototype was so short. There is also the risk that the system requirements may change in the meantime. On the other hand, some users may argue that the time, effort and money used to develop a prototype is 'wasted'. It is sometimes difficult to persuade busy people that this effort does lead to improved information systems.

Prototyping may be more than just another tool available to the analyst. It could be used as a basis for a methodology of systems development in the organisation. This may have:

- An analysis phase, designed to understand the present system and suggest the functional requirements of an alternative system
- A prototyping phase to construct a prototype for evaluation by users
- A set of evaluation and prototype modification stages
- A phase to design and develop the target system using the prototype as part of the specification.

Prototyping as part of the systems development cycle is discussed in Dearnley and Mayhew (1983).

Many prototypes are intended to be discarded. They have not been designed to be used as operational systems as they are likely to be:

- Inefficient
- Incomplete, performing only some of the required tasks
- Poorly documented
- Unsuitable for integration with other operational systems
- Incapable of holding the number of records necessary
- Inadequate, being designed for one type of user only.

Here, they are used only as a development tool, as a learning vehicle. The prototypes will nevertheless enable users and analysts to 'act out' the future system. Prototypes may lack features which are essential in an operational system but inappropriate in a prototype, for example, security features, and this needs to be stressed to users who may expect the target system to be developed in the same time as the prototype.

An alternative approach is to use the final prototype as the basis for the operational system. In this scenario, the information system 'evolves' by an iterative process. Once the users are satisfied with the prototype, it 'becomes' the operational system. In this case the analysis and design needs to be thorough as prototyping is not a substitute for thorough analysis and design. Implementation compromises should not be made as these are likely to remain in the final system. This includes the documentation which might be neglected otherwise, as analysts argue 'it is only a prototype'. One of the necessary components of successful prototyping is management and control so as to ensure that compromises are not made and the process of repeated iteration does not go on too long (Avison and Wilson, 1991).

The information system may be implemented in stages. At each stage, the missing components are those which give the poorest ratio of benefits over costs. The analysts in this case will have to be aware of robust design and good documentation when the prototype is being developed. The prototype must be able to handle the quantities of live data that are unlikely to be incorporated when giving end users examples of the prototype's capabilities. Otherwise prototyping will not improve the quality of systems development. Therefore, although prototyping is frequently regarded as a 'quick and dirty' method, it need not be 'dirty'. If the prototype is well designed, the prototype can feature as part of a successful operational information system. The temptation, however, is for a quick and dirty solution because the tools can produce a quick result and analysts are tempted to move quickly on to the development phase before sufficient analysis has been carried out. The emphasis on controls in prototyping is therefore necessary. Corners should be cut only in situations where the information system will have a very short life span (or is set up for once only use).

Evolutionary information systems development is similar to the prototyping approach except that there is never a 'final version', there is always the possibility of iterative revision even when the system is operational. In this approach, the prototype may always develop in an evolutionary systems development process and may never 'die'. The prototype is changed to reflect new conditions. The tools that were used to build the prototype in the first place can be used for modifications. Here, maintenance is regarded as positive rather than negative (McCracken and Jackson, 1982), the likelihood of change is catered for, rather than being seen as a problem. This evolutionary approach therefore addresses the problem of dealing with change, an aspect which many other approaches do not address. In some environments continuous change is the norm, and evolutionary prototyping is the preferred approach to accommodate rapidly implemented change.

The term *rapid application development* (RAD) has more recently been introduced for information systems development using automated tools, reusable parts, small teams and user involvement, so that information systems are implemented at lower cost, higher quality, increased speed, meeting the needs of business better and leading to lower maintenance costs. Another essential ingredient is a methodology for the management of the whole project (Martin, 1991). The RAD methodology suggested by Martin is described in section 6.14 as an example of an approach that uses prototyping as part of the overall methodology.

3.12 SOFTWARE ENGINEERING AND FORMAL METHODS

Software engineering concerns the use of sound engineering principles, good management practice, applicable tools and methods for software development. This was thought to be the solution to the software crisis, that is, the ability to maintain programs, fulfil the growing demand for larger and more complex programs and the increased potential of hardware which has not been exploited fully by the software. The principles established in software engineering have now generally been accepted as a genuine advance and an improvement to programming practice, primarily by achieving better designed programs and hence making them easier to maintain and more reliable. This has led to improved software quality.

In the period before the advent of software engineering, the conventional way of developing computer programs was to pick up a pencil or to sit at a computer terminal and code the program without a thorough design phase. This *ad-hoc* process was also frequently used for larger programs, but a better method for these large programs was to develop a flowchart and code from this. However, even this method is not very satisfactory.

The time taken to develop a fully tested program will be far greater in the long run if effort is not spent on thorough design. Without this design, it is difficult to incorporate all the necessary features required of the program. However, these omissions will only be brought to light at the program testing stage, and it will be difficult to incorporate the changes required at such a late stage. Problem solving carried out by haphazard, trial-and-error methods is far less successful than where good analysis and design comes before coding.

As we have mentioned, one solution to better program design was the use of flowcharting methods, but although program flowcharting does discipline the programmer to design the program, the resultant design can prove inflexible, particularly for large programs. Flowcharting usually leads to programs which have a number of branches, effected in programming languages by GOTO commands. With this method, it is difficult to incorporate even the smallest amendments to the program in a way which does not have repercussions elsewhere. These repercussions are usually very difficult to predict and programmers find that after making a change to correct a program, some other part of the program begins to fail.

This problem does not stop at the testing phase, because once the system is operational, the amendments necessary in general maintenance will be difficult to incorporate. When they are made, the flowcharts, which represent the main form of program documentation, are frequently not updated. This means that the programs do not have good documentation. In turn, this makes future maintenance even more difficult.

Software engineering offers a more disciplined approach to programming which is likely to increase the time devoted to program design, but will greatly increase productivity through savings in testing and maintenance time. A good design is one achievement of the 'software engineering school' and a second is good documentation, which greatly enhances the program's 'maintainability'. A third is related to both these, that is, general control of the project, in particular, that obtained through software metrics (measurements).

One of the key elements of software engineering is functional decomposition. Here a complex process is broken down into increasingly smaller subsets. This was illustrated in figure 3.3. 'Calculate weekly payroll' at the top level can be broken down into first level boxes named 'validate weekly return', 'calculate gross wage', 'calculate deductions' and 'print wages slip'. Each of these boxes is a separate task and can be altered without affecting other boxes (provided output remains unchanged). Each of these can then be further broken down. For example, 'calculate deductions' can be broken down into its constituent parts, 'calculate tax', 'calculate national insurance' and 'calculate loan payments'. Eventually this top-down approach can lead to the level where each module can be represented as a few simple English statements or a small amount of programming code, the target base level. This hierarchy is sometimes described as a tree with the root at the top and the smallest leaves at the bottom. This is similar to the process referred to as stepwise (or successive) refinement (Wirth, 1971).

Figure 3.5 showed the way a process could be broken down into its constituent parts. Many programs can be broken down into the general structure shown as figure 3.7. The top level module controls the overall processing of the program. Separate modules at the lower level control the input and output routines (reading data and updating files), validation routines (checking that the data is correct), the processing routines, and the setting up of reports. Sometimes these structure charts include details of the data (called parameters) which are passed between modules. This information is written on the connecting lines in the diagram.

The alternative to a top-down approach is a bottom-up approach where the detail is specified first. Unfortunately, it is often difficult to make the 'chunks of detail' fit together at a later stage. Sometimes, particularly in a large software project, these chunks are designed and developed by different people and it is therefore likely that poor communication or vague specification will lead to problems when attempting to interface parts of the system. This may be discovered very late in the project, when delays will be particularly problematical and cause a loss of confidence in the system. With top-down development, these interfaces are designed first. At the eleventh hour it is far easier to correct a detail in an individual program than to correct a problem of fitting together programs or subsystems.

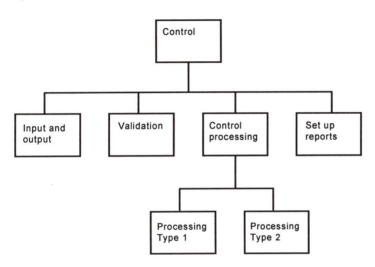

Fig. 3.7: Program design

So far, we have equated software engineering with good practice in programming design. But this is a narrow interpretation. The term is frequently used to cover the areas of requirements definition, testing, maintenance and control of software projects, as well as the design of software.

The requirements definition must be clear and unambiguous so that the software designer knows what the piece of software should do. The designer can then devise the way that these requirements will be achieved. Even the best design will not lead to a successful implementation if the requirements are not clear. The problem with natural language is that it is frequently ambiguous, although there is a series of techniques which can be used to clarify the requirements. We may here be going beyond the boundary of software engineering into information systems development. We take the view that the requirements of the software designer will be made clear by the systems analyst. Certainly where the software is associated with a part of the overall methodology for information systems development, the likelihood of success is increased.

Supporters of software engineering practice claim that the reliability of software designed in this way is improved. These improvements do not only derive directly from better design. There is an indirect benefit. If the software is better designed, it makes adequate testing of that software much easier. By designing the program so that it is split into smaller and separate elements, it is possible to construct a series of tests for each of these 'black boxes'. The number of tests for a larger program will be much greater. Perhaps simplistically (but it illustrates the point), a program split into four modules each containing five processing paths will require 5+5+5+5 (20) tests plus one for the program as a whole once the modules are tested, whereas testing one program will require 5x5x5x5 (625) tests, because for each of the first five paths there will be five of the second, and so on.

Maintenance is also important because, as we saw in the last chapter, it is an activity which takes up much of the analysts' and programmers' time. Maintenance can involve the update of programs whose requirements have changed, as well as the correction of programs which do not work properly. Maintenance will be made easier by the good documentation and good design which comes from the software engineering approach. The person making the changes will find that the software will be easier to understand because of the documentation. Further, as the program has been split into separate modules each performing a particular task, it will be easy to locate the relevant module and change it as appropriate. These changes will not affect other modules unexpectedly.

It is essential to use good control techniques to ensure that the implementation schedule for software is reasonable. Otherwise, the users and management will be unhappy about the system even if the requirements definition and software design work is first class, because expectations, which proved to be unrealistic, are dashed. A good methodology for the construction of software will be amenable to project control techniques. Breaking a system down into its parts makes estimating easier and more accurate.

A major aspect of control is the ability to measure software complexity, size of task and so on. In this way it will be possible to assess how long the software task should take and how many people should be involved and therefore be able to plan the whole process. Software metrics has become an important element of the software engineering school. Risk analysis, that is estimating the degree of uncertainty surrounding the metrics, has also become increasingly important.

Another aspect of control is ensuring the high quality of software. This implies that software is well tested; that standards, including those about documentation, have been applied; that it proves maintainable; that it is efficient; and that it meets the specification. There are now a number of official British, Australian and international standards relating to software, such as BS5750 (now, BS EN ISO 9000), AS3563 and ISO9001.

The term software engineering, therefore, is often used to cover much more than simply the design of computer programs. This is the view of many of the writers of texts on software. However, in this text the term is used to refer to the development of quality software. The wider 'environmental' issues are the concern of information systems.

In this text we are interested in producing quality information systems. We regard software engineering as a skill which improves program design and thereby makes software, an important element of computer information systems, more effective once implemented, and easier to maintain.

A software engineering phase is usually part (though sometimes not explicitly stated) of an information system development methodology, and is explicitly part of structured approaches. Areas such as the following lie outside the scope of software engineering:

- Understanding the problem area
- Understanding the needs of the user
- Looking at alternatives
- Deciding whether a computer system is needed.

If a computer system is recommended, then there will be a software engineering phase. However, the skills of software engineering should be separated from those required for information systems analysis and design. This is important as it was the lack of separation of the skills of the technologist from those of the analyst that led to major problems in the traditional approach to systems analysis. Further, it is important that the requirements of the analyst are expressed to the technologists in a way that the technologist can appreciate, and this provides one impetus in the movement towards the techniques of structured systems analysis and design which parallel, to some extent at least, those in software engineering.

We will now turn to *formal methods*, but before looking at formal methods, we should first clear up some confusion that may exist over the use of the term formal. The term is frequently used in the information systems community to mean 'clear structure' 'rule based' and 'technique orientated'. We prefer to use the term 'formalised' for this. In the context of this section the term 'formal' means expressed mathematically.

A methodology which incorporates formal methods uses mathematical precision in specification and design. For example, a formal specification will express in mathematically precise terms the user requirements which might originally have been expressed in natural language. Natural language can be criticised in that it is verbose and ambiguous. A formal design attempts to express these requirements (the what) concisely, unambiguously and completely and convert them into a design (the how) which reflect these requirements. The requirements statement drives the design.

Formal methods are often considered part of software engineering. They are certainly related because a software engineer may use formal methods to specify, develop and verify software through a formal, mathematical language which has no ambiguity in its expressiveness. The use of metrics may also help to detect any incompleteness and inconsistency in the specification.

One major advantage of this approach, its protagonists argue, is that the result can be proved to be correct (of particular importance in safety-critical applications). In other words, if the requirements are correct, and they can be expressed mathematically, then a solution can be derived which will work. The process of proving both the syntax and the semantics of a solution can be automated:

- Syntax refers to the logic of the requirements expressed in the grammar of the system
- Semantics refers to the set of expressions used which will achieve the results predicted.

Formal methods are particularly appropriate to producing quality software and can be usefully incorporated into a software engineering methodology. This, in turn, can be a subset of an information systems methodology. For example, Jackson Structured Programming (JSP) is incorporated into Jackson Systems Development (JSD), discussed in section 6.7. Other methodologies in formal methods include Vienna Development Method (VDM) (Andrews and Ince, 1991). These enable the unambiguous, precise and rigorous statement of requirements.

However, many aspects of user requirements in information systems, for example the design of the human-computer interface and other behavioural aspects, are perhaps outside the realm of 'being able to be expressed mathematically'. Thus, formal methods might be used to express some aspects of information systems requirements but not all. Similar comments can be made about aspects of software engineering, in particular, the inability to measure aspects which are not rule-based, formal and so on.

There are other problems associated with formal methods. Only some technologists have the skills necessary to use the language, and the symbolism of mathematics alienates many people. These techniques are unlikely to encourage user participation. Further, formal methods do not easily lend themselves to modularity, which is desirable for larger programs. Finally, striving for programs which can be proved correct can be an expensive way to develop software.

An area where formal methods are particularly useful is in process control applications, such as missile systems or factory production systems. They are less applicable in the information systems arena, because of its people-orientation and its applications tend not to be amenable to mathematical representation. We have chosen, therefore, not to discuss formal methods further.

3.13 PARTICIPATION

In this chapter we have covered themes within various movements in the information systems development arena. The first emphasised an organisational view and the needs of top management regarding strategic level decision making; the second concerned modelling, in particular, that of data and process modelling and object orientation; and the third emphasised software development and the use of software tools in the development process. We now turn to a final major theme, that of the people concerned in the development of an information system, particularly users and other stakeholders, that is, other people having a stake in the information systems project.

In the traditional systems analysis methodology, the importance of user involvement was frequently stressed. However, the computer professional was the person who was making the real decisions and driving the development process. Systems analysts were trained in, and knowledgeable of, the technological and economic aspects of computer applications but far more rarely on the human (or behavioural) aspects which are at least as important. The end user (the person who is going to use the system) frequently felt resentment, and top management did little more than pay lip-service to computing. The systems analyst may be happy with the system when it is implemented. It may conform to what the systems analysts understand are the requirements and does so efficiently. However, this is of little significance if the users, who are the customers, are not satisfied with it.

The strategic view of information systems highlighted the necessity for top management to play a role in information systems development. The approach discussed in this section highlights the role of all users who may control and take the lead in the development process. If the users are involved in the analysis, design and implementation of information systems relevant to their own work, particularly if this involvement has meant users being involved in the decision-making process, these users are more likely to be fully committed to the information system when operational. This will increase the likelihood of its success. Indeed, in some Scandinavian countries such a requirement may be embodied in law, with technological change needing the approval of trade unions and those who are to work with the new system.

Some information systems may 'work' in that they are technically viable, but fail because of 'people problems'. For example, users may feel that the new system will make their job more demanding, less secure, will change their relationship with others, or will lead to a loss of the independence that they previously enjoyed. As a result of these feelings, users may do their best to ensure that the computer system does not succeed. This aggression may show itself in attempts to 'beat the system', for example, by 'losing' documents or even by more obvious acts of sabotage. Frequently it manifests itself in people blaming the system for causing difficulties that may well be caused by other factors. This is sometimes referred to as projection, that is, they project their problems on to the system. Some people may just want to have 'nothing to do with the computer system', a kind of avoidance tactic. In this kind of situation, information systems are unlikely to be successful, or at the very least, fail to achieve their potential.

These reactions against a new computer system may stem from a number of factors, largely historical, but the proponents of participation would argue that they will have to be corrected if future computer applications are going to succeed and that it is important that the following views are addressed:

- Users may regard the IT department as having too much power and control over other departments through the use of technology
- Users may regard computer people as having too great a status in the organisation, and they may not seem to be governed by the same conditions of work as the rest of the organisation
- Users may consider the pay scales of computer staff to be higher than their own and that the poor track record of computer applications should have led to reduced salaries and status, not the opposite.

These are only three of the arguments. Some views are valid, others less so, but the poor communications between computer people and others in the organisation, symptomised by the prevalence of computer jargon, have not helped. Training and education for both users and computer people can help address the cultural clash between them. Somehow these barriers have to be broken down if computer applications are really going to succeed.

One way to help both the process of breaking down barriers and to achieve more successful information systems is to involve all those affected by computer systems in the process of developing them. This includes the top management of the organisation as well as operational level staff. Until recently, top management have avoided much direct contact with computer systems. Managers have probably sanctioned the purchase of computer hardware and software but have not involved themselves with their use. They preferred to keep themselves at a 'safe' distance from computers. This lack of leadership by example is unlikely to lead to successful implementation of computer systems: managers need to participate in the change and this will motivate their subordinates. With the implementation of executive information systems and the like, this attitude is becoming less prevalent.

Attitudes are also changing because managers can see that computer systems will directly help them in their decision making. The widespread use of PCs and the information about the technology available in newspapers and other sources has also diminished the 'mystique' that used to surround computing. Earlier computing concerned itself with the operational level of the firm; modern information systems concern themselves with decision support as well, and managers are demanding sophisticated computer applications and are wishing to play a leading role in their development and implementation.

Communications between computer specialists and others within the organisation also need to improve. This should establish a more mutually trusting and co-operative atmosphere. The training and educating of all staff affected by computers is therefore important. In turn, computer people should also be aware of the various operating areas of the business. This should bring down barriers caused by a lack of knowledge and technical jargon and encourage users to become involved in technological change.

Another useful way of encouraging user involvement is to improve the human-computer interface. There are a number of qualities that will help in this matter. These include visibility, simplicity, consistency and flexibility.

1. *Visibility:* This has two aspects. Firstly it means that the way that the system works is seen by the users. This aspect is related to participation, for, if users understand the system, they are more likely to be able to control it. Secondly, visibility means providing information on the current activity through messages to the users so that they know what is happening when the system is being run.

2. *Simplicity:* This means that the presentation of information to the users should be well structured, that the range of options at each point is well presented and that it is easy to decide on which option to choose.

3. *Consistency:* This means that the human-computer interface follows a similar pattern throughout the system. Indeed, wherever possible, all systems that are likely to be used by one set of users should follow this pattern.

4. *Flexibility:* This means that the users can adapt the interface to suit their own requirements.

User involvement should mean much more than agreeing to be interviewed by the analyst and working extra hours as the operational date for the new system nears. This is 'pseudo-participation' because users are not playing a very active role. If users participated more, even being responsible for the design, they are far more likely to be satisfied with, and committed to, the system once it is implemented. It is 'their baby' as well as that of the computer people. There is therefore every reason to suppose that the interests of the users and technologists might coincide. Both will look for the success of the new system. With a low level of participation, job satisfaction is likely to decrease, particularly if the new system reduces skilled work. The result may be absenteeism, low efficiency, a higher staff turnover, and failure of the information system.

The advocates of the participative approach (for example, Land and Hirschheim, 1983) recommend a working environment where the users and analysts work as a team rather than as expert and non-expert. Although the technologist might be more expert in computing matters, the user has the expertise in the application area. It can be argued that the latter is the more important when determining the success or failure of the system. An information system can make do with poor equipment, but not poor knowledge of the application. Where the users and technologist work hand in hand, there is less likely to be misunderstandings by the analyst which might result in an inappropriate system. The user will also know how the new system operates by the time it is implemented with the result that there are likely to be fewer teething troubles with the new system.

The role of systems analysts in this scenario may be more of facilitators, implementing the choices of users. This movement can be aided by the use of application packages which the users can try out and therefore choose what is best for them. Another possibility is the development of a prototype which users can use as a basis from which to agree final design.

Mumford (1983b) distinguishes between three levels of participation.

1.   Consultative participation is the lowest level of participation and leaves the main design tasks to the systems analysts, but tries to ensure that all staff in the user department are consulted about the change. The systems analysts are encouraged to provide opportunity for increasing job satisfaction when redesigning the system. It may be possible to organise the users into groups to discuss aspects of the new system and make suggestions to the analysts. Most advocates of the traditional approach to system development would probably accept that there is a need for this level of participation in the design process.

2.   Representative participation requires a higher level of involvement of user department staff. Here, the 'design group' consists of user representatives and systems analysts. No longer is it expected that the technologist dictates to the users the design of their work system. Users have an equal say in any decision. It is to be hoped that the representatives do indeed represent the interests of all the users affected by the design decisions.

3.   Consensus participation attempts to involve all user department staff throughout the design process, indeed this process is user-driven. It may be more difficult to make quick decisions, but it has the merit of making the design decisions those of the staff as a whole. Sometimes the sets of tasks in a system can be distinguished and those people involved in each task set make their own design decisions.

In section 6.10 we discuss the ETHICS methodology which has been designed around the principles of user participation.

Of course participation does have its problems. It might result in polarising or fragmenting user groups and there is a possibility of manipulating the process by selecting only those participants that are considered 'right' or by suggesting that users decide 'this....or there will be unhappy consequences'. Further, participation may cause resentment, either from analysts, who resent their own job being taken over by unskilled people, or by users, who feel that their job is accountancy, managing or whatever, and this is being cramped by demands to participate in the development of computer systems.

One reaction to lip service participation is the growth in *end user computing*, that is, users developing their own applications. With the increase in computer literacy, software tools being designed with end user computing in mind and low cost hardware, end user computing is indeed feasible. It is also one reaction to the application backlog as users see the only way that applications development will take place is to develop the applications themselves. It can be relatively unsophisticated, for example, users using menu-driven office systems, such as word processing, spreadsheet and database packages running on the PC with Windows, through to users writing their own software using programming languages. Such end-user computing gives users control over their applications. Potential weaknesses relate to the possible inefficiency, neglect of integrity and security issues, and the lack of an organisation-wide perspective of these application 'islands'. The growth of information centres, which are a source of advice to end users developing their own applications and cheaper hardware and software through bulk buying opportunities, is a response to some of the potential disadvantages. However, this is a centralising move which is resented by some end user departments. Some user departments might have their own specialist computer people who are members of the department, not that of the computing department or information centre, and their role might be to provide general advice through to acting as facilitators.

Another response to the excessive power of computing people in information systems development is joint requirements planning (JRP) and joint application design (JAD), (see section 6.14). Representatives from the user groups and computer people conduct workshops to progress the information system through the planning and design stages. The leader of the workshops has a particularly crucial role and should be trained and experienced. Executives are likely to be involved in the JRP workshops along with end users, but they are unlikely to play a leading role in the design phase. Computer people will also be more prevalent in JAD workshops, and the use of CASE tools and prototypes may help the process. In some organisations JAD workshops may take place in group decision support rooms consisting of linked computer systems with software, database and other support (Nunamaker *et al.,* 1991) but these can be dominated by the technology and the role of the human facilitator is again crucial.

Grundén (1986) suggests that participation implies even more fundamental changes in the organisation. Figure 3.8, adapted from that paper, gives some of the characteristics of this approach when compared to the traditional approach.

Conventional view   Human-oriented view

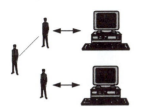   

Computer systems development   Information systems development
Hardware and software   Human beings
Cost effectiveness   Job satisfaction/people systems
Technologist-specified   User-developed systems
Narrow, specialist view   Systems view

Fig. 3.8: Comparison of conventional and human-oriented views of information systems development

The focus of the two approaches is very different. In conventional systems development, the emphasis is on the technology: computer systems, hardware and software. The technologist drives the system. Users are given rules to follow and departures from these norms are not tolerated.

The human-oriented view focuses on the people in the organisation. This may result in less complex, smaller systems which are not necessarily the most efficient from a technical point of view. Nevertheless, they are more manageable, less reliant on technology and on 'experts'. PCs are more frequently used as the technological base than mainframes. The traditional view emphasises the technology, whereas the human-oriented view is more interested in the organisation as a whole and the user as creator in that environment. The implication is that the conventional view is more common in traditional hierarchical, bureaucratic organisations; the human-oriented view is more common in democratic, growing and changing organisations.

However, in the 10 years since this publication, both views can now be seen as distortions. In theory and practice, organisational and business needs are seen as paramount. In practice, many if not most IT staff do take account of both these views.

3.14 EXPERT SYSTEMS

The term 'expert system' derives from the fuller, and more descriptive term, 'knowledge-based expert system' and these terms tend to be used synonymously. As is implied from the term 'knowledge-based', such systems have their roots in the field of artificial intelligence (AI). An expert system (ES) is a system which will simulate the role of an expert. It is distinguished from other applications because its usefulness is derived from the knowledge and reasoning ability of the expert system and not from number crunching (carrying out large and complex calculations) or the repetitive processing of data, which characterise most scientific and business computing applications respectively.

Feigenbaum (1982), one of the early pioneers in the area, defines an expert system as: 'an intelligent computer program that uses knowledge and inference procedures to solve problems that are difficult enough to require significant human expertise for their solution'. Somewhat more formally, the British Computer Society's Expert Systems Specialist Group defines an expert system as follows:

> An expert system is regarded as the embodiment within a computer of a knowledge-based component from an expert skill in such a form that the system can offer intelligent advice or take an intelligent decision about a processing function. A desirable additional characteristic, which many would consider fundamental, is the capability of the system, on demand, to justify its own line of reasoning in a manner directly intelligible to the enquirer. The style adopted to attain these characteristics is rule-based programming.

More recently, Buchanan and Smith (1993) have characterised expert systems programs as having five desirable properties:
1.   Reasons with domain specific knowledge
2.   Uses domain specific methods (these can be heuristic or algorithmic)
3.   Performs well in its problem area
4.   Explains or makes understandable both what it knows and the reasons for its answers
5.   Retains flexibility.

These properties suggest that expert systems are essentially artificial intelligence programs that contain (in some way) some of the knowledge that human specialists have. This does not imply that an expert system builds a psychological model of how the specialist thinks, but rather that it contain a model of the expert's model of the domain. The domain of the expert system may be a discipline or knowledge area, such as, geology or medicine. Alternatively, it may be a particular narrow subset, such as, 'risk factors and insurance premiums in California'.

The third property suggests that expert systems should achieve high standards when compared with human experts. In practice, they avoid a lot of the major challenges of a general purpose and all-encompassing expert by narrowing the scope of the problem to a practical and achievable domain. It is important that the user understands the expert system in the terminology of the expert specialism and also that the strengths and weaknesses of the expert system and the dynamics of reasoning are understood. The last property implies that the expert system should be capable of modification and change as new expertise and information is discovered.

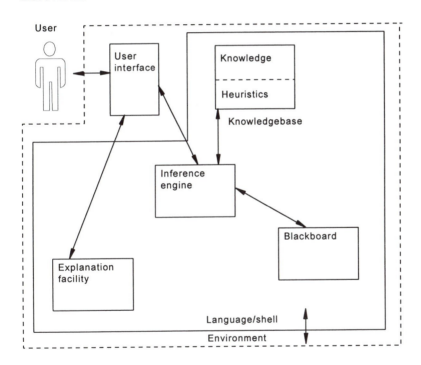

Fig. 3.9: Structure of an expert system (generic)

An expert system is basically an intelligent adviser concerning one or more areas or domains of knowledge. This knowledge exists in the minds of human experts, it has been developed and evolved over time by education and experience, and it has to be captured in some way by the expert system, usually in the form of sets of rules and groups of facts. The expert system is then informed about a particular situation of concern to a user, usually achieved by the user answering questions posed by the expert system. Then the expert system comes up with intelligent advice concerning that situation. Ideally, it is able to explain to the user how it arrived at the particular advice.

In practice, expert systems vary considerably, but there are a number of common components in a structure as illustrated in figure 3.9:

- *The knowledgebase:* This contains two elements. The first is the knowledge necessary for understanding the domain, essentially the facts of the problem area. The second element is the heuristics that indicate how to use the facts to solve specific problems, essentially the rules. These can be of various types. They might be absolute, scientific-type rules or they may be informal, judgmental, rules of thumb. The most common way of representing rules in expert systems is by IF-THEN structures as follows:

     If P1 and P2 and .....Pn,

     Then Q1 and Q2 and .....Qn

  which reads:

     If premises P1 and P2 ... Pn are true,

     Then perform actions Q1, Q2 and ... Qn.

  The premises are sometimes called 'conditions' and the Qs 'conclusions'. This type of IF-THEN rule is known as a production rule or production system. The conclusions can sometimes be drawn with certain degrees of confidence by the introduction of probabilities. For example, a rule might be that 'if X is a bird then conclude that X can fly'. This rule is not true in all circumstances because emus are birds but they cannot fly. We might then introduce another rule that states that 'if X is a bird and X is an emu then conclude that X cannot fly' or we might say that 'if X is a bird then conclude with a probability of .99 that X can fly'. The facts contained in the knowledgebase are simply assertions, for example, a robin is a bird.

  Although such rules are very common in expert systems, they are not the only way of representing knowledge and semantic nets. Frames, scripts, semantic graphs, amongst others, have also been used. Recently, rules have been complemented by frames in some expert systems. A frame is a representation of a group of related knowledge about a particular object. These are useful in situations where there are stereotypical objects. The characteristics of a car give an example (Turban, 1993, Giarratano and Riley, 1994). Most cars have stereotypical characteristics, for example, they all have engines, wheels, acceleration speeds and so on. All these characteristics are gathered together in a frame. The frame includes slots and facets. Slots are attributes of the object, for example, engine size. Facets are values, procedures or knowledge concerning the slot, for example, '2000cc' or 'if >1800cc, then insurance group = group + 1'. Frames are often structured into hierarchies and are good for structuring managerial and business knowledge. Frames have been described by Giarratano and Riley (1994) as being analogous to records in a traditional file, with slots and facets equivalent to fields and values.

- *An inference engine:* The inference engine controls the process of invoking the rules that pertain to the solution of the problems posed to the system. It is used for reasoning about the information in the knowledgebase and in the 'blackboard' (see below) and for formulating conclusions. It provides directions and organises and controls the processing of the knowledge and the rules to make inferences based on the problems, that is, the procedures for problem solving. Three elements of the inference engine can be identified. The first is the *rule interpreter* which effectively executes the rules. The second is the *scheduler* which controls when things should be performed and in what order. The third is the *consistency checker* which maintains a consistent form of the process and solution. The inference engine usually keeps track of the execution steps so that the user can backtrack through the interaction as necessary.

- *A language*: This is the language in which the expert system is written. A number of specific expert system programming languages have been developed, probably the best known being Prolog and LISP, but expert systems can be developed in any language, for example, C++ or even COBOL.

Probably the most common approach to expert systems development is to use a *shell*. A shell is an expert system with all the inference capability but without any domain specific knowledge. It is thus ready for users to input their own rules and knowledge to create their own specific expert systems application. Using a shell obviously saves time as all of the system except for the specific application knowledge is already programmed. Many shells were originally full expert systems applications that subsequently were converted to general purpose shells by removing their knowledge component. There are a variety of different types of shell, some are PC-based, whilst others are for large mainframe computers. They are continuing to evolve and some are even classed as end-user tools in which domain experts can build their own expert systems. Shells help focus the effort on the domain and knowledge acquisition rather than on the program development. They are not suitable for all types of application and can be limiting and inflexible. Examples of expert systems shells are VP Expert, Personal Consultant Plus and ESE.

- *An explanation generator or facility:* This is the part of the system that presents the reasoning behind how the system arrived at its conclusions.

- *A blackboard:* This is a temporary workspace that records intermediate hypotheses and decisions that the expert system makes as the problem and the solution evolve. It consists of three elements. The first is the plan and approach to the current problem; the second is the agenda, which records the potential actions awaiting execution, that is, the rules that appear to be relevant; and the third is the solution elements, which are the candidate hypotheses and decisions the system has generated so far. Blackboards do not exist in all expert systems, but they are particularly relevant in systems which work by generating a variety of solutions to the problem posed. They are also important in systems that integrate the knowledge of several different specialists.

- *A user interface:* This varies widely, but is a very important element in the architecture of an expert system. There will be an interface for the user and also one for entering and updating the knowledgebase.

- *The environment:* This is the variety of hardware and software that surrounds the expert system. An expert system may be a stand-alone system or it may be embedded within other systems.

Expert systems emerged as an identifiable element of AI in the late 1960s and early 1970s when, according to Buchanan and Smith (1993), it was realised that in certain subject areas, advances could be better achieved by providing a program with substantial subject-specific knowledge and relatively simple inference capabilities rather than following the previous approach, which was to make deductions based on the general axioms of a subject area.

An early successful example was the DENDRAL program which represented many specific facts about organic chemistry and used those facts in rather simple, but nevertheless powerful, inferences. DENDRAL provided Feigenbaum with what has come to be known as the first principle of expert systems building which is that 'in knowledge lies power'. Turban (1993) suggests that DENDRAL also led to a number of significant conclusions for expert systems as follows:

- General problem solvers are too weak
- Human problem solvers are good only if they operate in a very narrow domain
- Expert systems need to be updated frequently and rule-based representations enable this to be achieved efficiently
- Complex problems require substantial knowledge of the problem area.

Another well known system also credited with establishing a degree of credibility for expert systems was MYCIN. This is an expert system developed in the 1970s in collaboration with the Infectious Diseases Group at Stanford University to carry out medical diagnosis in the domain of blood infections and meningitis. The early diagnosis and correct treatment of meningitis is critical, and MYCIN helped doctors with diagnosis and treatment in the crucial 24-48 hour period after the detection of symptoms, when little information is typically available. MYCIN is based on a set of about 500 IF-THEN inference rules, many including probability concepts that allowed MYCIN to make conclusions based on uncertain evidence. The system is easy to use and doctors would enter an interactive dialogue concerning a specific patient. When the doctor keyed in 'RULE', MYCIN would provide an explanation of the last rule applied, and if 'WHY' was keyed in, an explanation of the inference process would be provided. The system is said to equal the performance of top human experts.

Expert systems have proved to be of value in certain scientific applications, particularly medical, but have been slower to establish themselves in commercial or business areas. Many commentators thought that business applications should be amenable to expert systems as many areas of business rely on the expertise and knowledge of specialists. Areas of finance, such as tax, credit and risk assessment and portfolio management provide examples. A number of business-based expert systems were developed in the mid 1980s including XCON and XSEL from Digital Equipment Corporation (DEC), ExpertTax by Coopers and Lybrand, Credit Clearinghouse by Dun & Bradstreet and Authorizer's Assistant by American Express. However, although there were some successful business applications of expert systems, they were relatively few, and in general they have not lived up to their expectations. A study by Gill (1995) of 97 commercial expert systems built in the US before 1988 revealed that by 1992 the majority were no longer in use, having been abandoned or fallen into disuse, and that only one third were in widespread use. A similar result has been found in the UK (DTI, 1992).

The reasons for the lack of successful business applications of expert system are varied, but certainly one reason is the rather ill defined and uncertain nature of most business domains. Although experts do exist, their expertise is not of the same nature as that of a chemist or a doctor. In business, there are very few hard and fast rules that everybody will agree on. One expert's views may be totally different to another. This has meant that a consensus on definitions of rules (more where probabilities are included) has proved very difficult to achieve. Most business-based expert systems have restricted themselves to areas where the domain is relatively narrow and where there exists a basic set of rules and standards. One example is the application of the rules of particular methodologies in CASE and other tools (see Chapter 5). This helps explain why in the areas of strategy and high-level business planning, where some people originally hoped that expert systems would make their biggest impact, we see little in the way of expert systems.

Apart from the problem of finding the right applications, the process of knowledge acquisition or knowledge elicitation has proved problematical. This is essentially the process of obtaining the expertise from human experts, although it may also involve the obtaining of knowledge from books, papers, files, systems, manuals and so on. There are also a variety of different types of knowledge. This variety can be expressed in many different ways, but we categorise it in two types. Descriptive knowledge consists of the facts and descriptions, including the associations and relationships between them. Procedural knowledge is information about processes, procedures and constraints involved in applying the information, for example, for decision making.

In designing expert systems, the particular level of knowledge is important. The first level is the domain knowledge, this is information specific to the domain under consideration, for example, 'knowledge of corporation tax'. A second level is also domain specific, but at a higher level, for example, the national taxation system, in which corporation tax is a part. A final level might be the environment at large within which the taxation system operates, and relevant knowledge here might be cultural, or philosophical or it might relate to the belief and value systems. All these are potentially important to an expert system concerned with corporation tax as a way of explaining and understanding the context, constraints, ethics, history and so on. However, the difficulties of capturing and representing the knowledge becomes progressively more difficult as the levels get higher. There are many potential levels, but in practice, expert systems concern themselves mainly with the first level and then the rest as one general level.

The formulation of decision rules is no easy task, as we have seen above, but it is not just because business knowledge is not scientific and that the views of experts may differ considerably. It is also to do with the fact that experts are not always good at structuring or organising their knowledge and decision-making criteria in any formal way. Experts cannot always explain why they know something or the basis for their decisions. The definition of probabilities is also extremely difficult, as is achieving consensus. Even when experts are in agreement, the knowledge still has to be formulated in ways that enable the expert systems to make use of this knowledge. The representation of knowledge, sometimes known as the construction of a knowledge map, is extremely problematical. It is these problems of knowledge acquisition and representation that are the real bottleneck in the development of expert systems for business applications.

There are also a number of other problems concerning business expert systems, one concerning the difficulties of testing and validating. It is difficult to prove that the expertise has been captured correctly and that the rules and inferences will lead to good and effective results when applied. Some companies have found this such a problem that they have abandoned their expert systems. In the United States, in particular, the fear of litigation for proffering incorrect commercial advice is very high.

Yoon *et al.* (1995) identify a number of other characteristics that were found to be important for successful expert systems. The study was based on a study of 69 projects. The factors considered important included:

- High levels of management support for the project
- High user participation
- Skilled expert system developers
- Use of a systems approach to analyse the business problems
- Management of unrealistic expectations
- Use of good domain experts
- Understanding of the impact on end-user jobs.

Students of non-expert systems development will not be surprised at these findings. This raises the question of the relevance of expert systems to methodologies. Expert systems can be used for handling and applying the rules of a systems development methodology in a CASE tool. Another area relates to the use of methodologies, not just to develop standard information systems, but to develop expert systems. The approach to their development has been very *ad hoc*, and there is no such thing as a standard expert systems development methodology. Indeed, up until recently, there has been little interest in methodological issues in the expert systems community. The prevailing approach to the development of expert systems may be characterised as prototyping or evolutionary. This has evolved from the trial and error approach of the early expert systems where the developers would code up a few rules and then try out the program on the users, find it was inadequate, and change or add some more rules, try it out again and so on.

More recently, there has been an increasing focus on approaches which first attempt to acquire and structure the knowledge and then build the system. Most expert systems developments have typically been separate from information system development methodologies, in the sense that different people are involved and the focus tends to be on solving the technical problems rather than on the process of development. For some people, expert systems have particular characteristics that mean that information systems development approaches are not relevant. They argue that the knowledge domain is more complex and that the knowledge acquisition process and the representation of rules are the key issues in expert systems development. Mockler (1992) suggests that this complexity implies that to develop expert systems requires somebody to interface between the domain experts and the technical expert systems developers. This person is known as a knowledge engineer and typically has good cognitive and interpersonal skills.

Whether developing expert systems is radically different from developing information systems is a matter of debate but if the above mentioned findings of Yoon *et al.* are true, then information system development methodologies may have much to offer the developers of expert systems, although there is not much evidence that information systems development methodologies are evolving to address knowledge acquisition and representation issues. On the other hand, some argue that the knowledge acquisition task is not that significantly different from the standard systems analysis task. It may be that current methodologies may not need too much adapting to handle knowledge acquisition and representation as well as the acquisition and representation of data and processes. A third area of interest relates to whether any of the approaches to expert systems development have anything to offer information system development methodologies. Some of these issues are further discussed in Chapter 6 where an expert systems development methodology (KADS, section 6.15) is described.

3.15 CONCLUSION

Over this chapter we have looked at a number of general themes of information systems methodologies. Each of the actual methodologies discussed in Chapter 6 is likely to have as its base one of these themes, but may have elements of others. Some, for example Multiview, have attempted to incorporate features of many of these themes or 'views'. The techniques and tools in different approaches are not mutually exclusive, and Multiview is also representative of those methodologies that are referred to as contingency approaches, where the techniques and tools available are chosen and adjusted according to the particular problem situation.

It is sometimes argued that professional systems analysts should not rely too heavily on one approach. These approaches can be suitable in some organisations and not in others. Each organisation is unique and radical solutions may only work, for example, if the political climate in the organisation is conducive to change. There is, however, a contrary argument. This is that the use of one methodology imposes good standards for the organisation and many of the advantages of using information systems methodologies will be sacrificed by too much flexibility.

It is impossible to prove that any solution is optimal. In fact analysts using the same method may well come up with very different solutions. This is not necessarily a criticism of the method, but it is a criticism of the view that there is only one appropriate solution. In any case, a sub-optimal technical solution may well be appropriate if this gains user acceptance.

Some methodologies emphasise techniques and tools. These can be very useful to understand and communicate, but methodologies are usually much more. They are also about people, tasks, skills, control and evaluation. But even human-oriented approaches use techniques and tools.

When each of the methodologies is discussed in Chapter 6, it might be useful to see how the methodology answers some of the criticisms aimed at the conventional approach to systems analysis which were described in section 2.3. Each of the themes discussed above have their critics, for example:

- 'Data analysis may not solve underlying problems that the organisation might have. Indeed it may have captured the existing problems into the data model, and made them even more difficult to solve in the future.'
- 'In breaking down a system into manageable units and then even more manageable units, structured analysis offers a simplistic view of a complex system and fails to identify fully the importance of the links between systems.'
- 'Participation leads to inefficient systems designed by those who are good managers, clerks, or salespeople, but poor, and unwilling,

systems analysts, and further, it is demotivating to people trained and experienced as 'true' systems analysts.'
- 'Systems theory represents an idealistic academics' position and is not relevant to the practitioner.'
- 'Prototyping is only concerned with the user interface and does not address the fundamental problems of systems analysis, it simply makes poor systems palatable to users'.
- 'All methodologies are designed to provide information systems, but do not solve the fundamental problems of management'.

We will return to some of these topics in the final chapter.

FURTHER READING

Checkland P. B. (1981) *Systems Thinking, Systems Practice.* Wiley, Chichester.
    This book provides an excellent discussion of systems thinking as well as soft systems methodology.

Earl, M. J. (1989) *Management Strategies for Information Technology.* Prentice Hall, Englewood Cliffs, New Jersey.
Porter, M. E. (1980) *Competitive Strategy.* Free Press, New York.
Strassman, P. (1990) *The Business Value of Computers.* The Information Economics Press, New Canaan, Conn.
    Useful books on the use of strategic information systems for competitive advantage.

Hammer, M. and Champy, J. (1993) *Reengineering the Corporation: A Manifesto for Business Revolution.* Harper Business, New York.
    A follow-up to Hammer's classic paper on BPR in the Harvard Business Review and a thorough, if optimistic, overview of BPR.

Martin, J. and Leben, J. (1989) *Strategic Information Systems Planning Methodologies.* Prentice Hall, Englewood Cliffs, New Jersey.
    A thorough overview of planning approaches including Business Systems Planning.

Yourdon, E. (1989) *Modern Structured Analysis.* Prentice Hall, Englewood Cliffs, New Jersey.
    The most recent thorough coverage of the structured approach.

Avison, D. E. (1992) *Information Systems Development: A Database Approach.* 2nd ed., McGraw-Hill, Maidenhead.
Covers a database approach to information systems development.

Coad, P. and Yourdon, E. (1991) *Object Oriented Analysis.* 2nd ed., Prentice Hall, Englewood Cliffs, New Jersey.
A good exposition of the object-oriented approach.

Martin, J. (1991) *Rapid Application Development.* Prentice Hall, Englewood Cliffs, New Jersey.
This book covers tools and prototyping.

Pressman, R. S. (1992) *Software Engineering: A Practitioner's Approach.* 3rd ed., McGraw-Hill, New York.
A thorough text on software engineering and formal methods.

Land, F. and Hirschheim, R. (1983) Participative systems design: rationale, tools and techniques. *Journal of Applied Systems Analysis*, 10.
A good review paper on the participative approach.

Turban, E. (1993) *Decision Support and Expert Systems.* 3rd ed., Macmillan Publishing Co., New York.
A standard text on expert systems.

Chapter 4
Techniques

4.1 INTRODUCTION

We have chosen to describe techniques in a separate chapter for two reasons: first, most are common to more than one methodology and therefore to leave them to the methodologies chapter would lead to repetition there, and second, so that the principles contained in the methodologies (Chapter 6) can be described without going into the techniques used in too much detail.

Although the techniques described are used in a number of methodologies, this does not mean that they are interchangeable because, as used in any particular methodology, they could address different parts of the development process, be used for different purposes or be applicable to different objects. More obviously, but less fundamentally, they often use different diagrammatic conventions to show the same things.

The first three techniques, rich pictures, root definitions and conceptual models, help to understand the problem situation being investigated by the analysts. They originated in Checkland's Soft Systems Methodology, but have been incorporated into other approaches, for example, Multiview.

Increasingly, these techniques are being used by analysts who may be following a methodology which does not include them 'officially' as part of that approach, but nevertheless find them helpful, particularly in the early stage of a project. Rich pictures prove particularly useful as a way of understanding the problem situation in general at the beginning of a project; root definitions help the analyst to identify the human activity systems they are to deal with; and conceptual models show how the various activities in the human activity system relate to each other. The three sections describe each of the three techniques, but many analysts using one of these techniques will use them all and therefore the reader will notice much cross-referencing in the descriptions of each.

The next three techniques, entity modelling, normalisation and data flow diagrams, are fundamental and common to many methodologies and are described at length. Entity modelling and normalisation are techniques for analysing data. Many of the other techniques described in this chapter, which can be categorised as process logic, analyse processes in some respect, but of these, the most fundamental and well used are data flow diagrams. Although entity modelling and normalisation are data-oriented and data flow diagrams process-oriented, there is sometimes overlapping. For example, when constructing data flow diagrams it is necessary to know what data is required to support a process, and this data is depicted in data stores. Nevertheless, the technique is primarily one for analysing processes, the data identified is simply a by-product. Similarly, entity modelling is primarily a data analysis technique despite the fact that it cannot be carried out without knowing something of the processes that exist.

Following a full description of data flow diagramming, the process logic techniques described are decision trees, decision tables, structured English and action diagrams.

The next technique described is the entity life cycle. The technique of entity life cycle analysis is also common to a number of methodologies. The entity life cycle is not, despite its name, a technique of data analysis, but more a technique of process analysis. On the other hand, the following technique described, object orientation, can represent data and processes, indeed the same technique can represent just about everything to do with the information system. It is not surprising, therefore, that the approach has attracted so much interest.

The final two techniques described in this chapter, structure diagrams and matrices, are used in all walks of life as well as in many information systems development methodologies and frequently at different stages in a methodology. Structure diagrams show hierarchical structures, be it a computer program; that to represent relationships between people in a department; or the structure of processing logic. Matrices show the relationship between two things, for example, entities and processes, departments and documents or roles of people and processes.

STAGE	OVERALL	DATA	PROCESSES
STRATEGY	Rich pictures		
ANALYSIS	Rich pictures Objects Matrices Structure diagrams	Entity modelling	Data flow diagrams* Entity life cycle* Process logic Root definitions* Conceptual models*
LOGICAL DESIGN	Objects Matrices Structure diagrams	Normalisation Entity modelling*	Process logic
PROGRAM/ DATABASE DESIGN	Objects Matrices Structure diagrams	Normalisation*	Process logic*

Fig. 4.1: A classification of techniques

Figure 4.1 shows the position or stage in the development process where any particular technique is utilised and whether it is primarily regarded as general, or data- or process-orientated. We have divided the life cycle into four overall stages for this purpose: a strategic or planning phase; an investigation and analysis phase; a design phase; and an implementation phase. The asterisk indicates the stage or stages where the technique is most commonly utilised. For example, entity modelling is used at two stages, but its use at the logical design stage is most common. We have grouped decision trees, decision tables, structured English and action diagrams under the heading process logic.

The description of the same techniques sometimes seem to differ considerably from methodology to methodology although in most cases the principle of the technique is common and it is the notations and conventions that differ. We describe each technique once only but note the variations that exist. Sometimes the use of different conventions, particularly in the diagrammatic techniques, can make the result look radically different. It is a good discipline to look beyond the conventions of a technique and try to identify the underlying principles involved.

Many of these techniques described, particularly the diagrammatic ones, relate to the documentation of the processes or activities involved in developing an information system. These techniques of documentation can be used to communicate the results achieved to other analysts, users, managers and programmers.

The techniques can also be used to help in the process of analysis and design and to verify that all the steps in the methodology have been carried out. One of the generally acknowledged advances in information systems development is the improvement that has resulted from the use of these techniques.

In addition, most of the techniques, in particular those operating at a high level, embody the principles of abstraction and generalisation. This means that the issues or concepts of importance are brought out in the first level diagram and that the detail is ignored at that stage. For this to be effective, it is necessary that the right objects are depicted in the technique. In entity modelling, for example, the entities and their relationships are thought to be important and the attributes are regarded as a point of detail. Second, the emergence of the detail at the later stage must not invalidate the results of the earlier use of the technique. This is not to say that iteration must not take place, iteration helps obtain the correct results, it means that the techniques must lead us to the things of importance amid the confusion of all the detail. It is useful to consider techniques in this light when exploring them in this chapter.

4.2 RICH PICTURES

The analysis of such factors as interfaces, boundaries, subsystems, the control of resources, organisational structure, roles of personnel, organisational goals, employee needs, issues, problems and concerns are not all contained in other techniques but are of interest to the analyst. Understanding political aspects is essential for successful information systems. A high percentage of failure is due to ignoring these issues. When constructing a rich picture diagram such issues are taken into consideration.

An understanding of what the organisation is 'about' need not take a diagrammatic form. It could be a mental map of some sort. We describe a possible diagrammatic form for rich pictures but we use the term rich picture rather than rich picture diagrams.

The technique stems from Checkland's Soft Systems Methodology (section 6.11) and the description here is based on that used in Multiview (Avison and Wood-Harper, 1990). A rich picture diagram is a pictorial caricature of an organisation and is an invaluable tool for helping to explain what the organisation is 'about'. The rich picture should be self-explanatory and easy to understand.

One may start to construct a rich picture by looking for elements of structure in the problem area. This includes things like departmental boundaries, activity types, physical or geographical layout and product types. Having looked for elements of structure, the next stage is to look for elements of process, that is, 'what is going on'. These include the fast-changing aspects of the situation: the information flow, the flow of goods and so on.

The relationship between structure and process represents the 'climate' of the situation. Very often an organisational problem can be tracked down to a mismatch between an established structure and new processes formed in response to new events and pressures.

The rich picture should include all the important hard 'facts' of the organisational situation, and the examples given have been of this nature. However, these are not the only important facts. There are many soft or subjective 'facts' which should also be represented, and the process of creating the rich picture serves to tease out the concerns of the people in the situation. These soft facts include the sorts of things that the people in the problem area are worried about, the social roles which the people within the situation think are important, and the sort of behaviour which is expected of people in these roles.

Representing the situation in terms of 'information systems needed' should be discouraged at this stage. These should come once the analysis has been carried out. Again, the question is not 'what systems does the manager think exist?', but rather 'what systems can be described in the situation?'. A 'system' in this sense is not about hardware and software but is a perceived grouping of people, objects and activities which it is meaningful to talk about together.

Typically, a rich picture is constructed first by putting the name of the organisation that is the concern of the analyst into a large 'bubble', perhaps at the centre of the page. Other symbols are sketched to represent the people and things that inter-relate within and outside that organisation. Arrows are included to show these relationships. Other important aspects of the human activity system can be incorporated. Any symbols can be used which are appropriate to the specific situation. We use crossed-swords to indicate conflict and the 'think' bubbles indicate the worries of the major characters.

In some situations it is not possible to represent the organisation in one rich picture. In this case, further detail can be shown on separate sheets. The perceived relative importance of people and things could be reflected by the size of the symbols on any one rich picture.

Figure 4.2 represents a rich picture for a professional association. We will use this case study in the description of rich pictures and the following two techniques described. The work concerning the case had three phases: an initial study, a full requirements analysis and finally the development of some of the computer applications, including a computer system which handled some of the association's examinations.

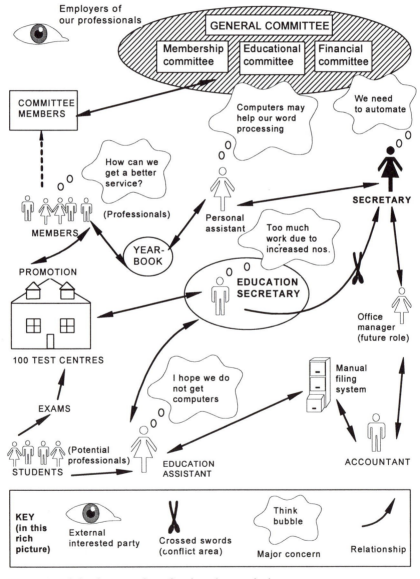

Fig. 4.2: Rich picture of professional association

The initial study was requested in correspondence from the secretary of the association. This was a top post in the organisation. She felt that many of the systems ought to be computerised and she wished to know the type of computerisation that would be appropriate for the situation, and whether the association should establish its own computer system or go to a service bureau. The professional association is a professional body initiated for people working in or attempting to enter a particular profession. The current administrative system was purely manual. All the functions were under the control of the secretary. The education subsystem was administered by an education secretary.

The sorts of application included membership administration, examination administration, and tuition administration, requiring information about subjects, tutors and fees. It was found that the work load at peak times of the year was becoming too demanding, membership was growing rapidly and the administration and accounts occupied much of the time of the senior management, particularly that of the secretary.

Figure 4.2 represents an early draft of part of the rich picture of this human activity. If it has been well drawn, you should get a good idea of who and what is central to the organisation and what are the important relationships. Bear in mind that there is no such thing as a 'correct' rich picture. Drawing the rich picture is a subjective process.

The act of drawing a rich picture is useful in itself because:

- Lack of space on the paper forces decisions on what is really important (and what are side issues or points of detail for further layers of rich pictures)
- It helps people to visualise and discuss their own role in the organisation
- It can be used to define the aspects of the organisation which are intended to be covered by the information system
- It can be used to show up the worries of individuals, potential conflicts and political issues.

Differences of opinion can be exposed, and sometimes resolved, by pointing at the picture and trying to get it changed so that it more accurately reflects people's perceptions of the organisation and their roles in it.

Once the rich picture has been drawn, it is useful in identifying two main aspects of the human activity system. The first is to identify the primary tasks. These are the tasks that the organisation was created to perform, or which it must perform if it is to survive. Searching for primary tasks is a way of posing and answering the question 'What is really central to the problem situation?'. For example, it could be argued that the association aims to give an excellent service to its members (which might be implied by the diagram) or increase standards in its profession (which is not implied by the diagram). Everything else is carried out to achieve that end. Primary tasks are central to the creation of information systems, because the information system is normally set up to achieve or support that primary task.

The second way that the rich picture is of particular value is in identifying issues. These are topics or matters which are of concern. They may be the subject of dispute. They represent the (often unstated) question marks hanging over the situation. In the association, they might include 'what do we hope to achieve by installing a computer system?'. This was a major issue when the systems analysis was done. This process of identifying the issues will lead to some debate on possible changes. It might be possible for these issues to be resolved at this stage, but it is essential that they are understood. Issues are important features, as the behaviour resulting from them could cause the formal information system to fail. Unless at least some of them have been resolved, the information system will have little chance of success. In some situations, the issues can be more important than the tasks.

The analyst starts by looking at an unstructured problem situation. This emphasis on the 'problem situation' as opposed to the 'problem' is important. By looking at the problem, rather than the whole situation, it will be difficult to be able to tell whether the diagnosis of the problem is correct. All too often a client will say 'I am having a problem with X', when the problem is actually being caused by something else and X is a symptom of the over-riding problem. A problem with stock control may be caused by a weak stock records system or a lack of time in the shop to update records as items are sold or by the fact that there is a lot of pilfering. If the analysts are limited to the official statement of the problem, then these 'real' causes may never be uncovered. It is therefore necessary for analysts to keep eyes, ears and minds open and avoid jumping to early conclusions about the 'problem'.

There are a number of ways in which analysts get drawn into the problem situation and a number of roles that they may be called on to play. It is important that their role and their relationship with other people in the problem situation have been well defined and well explained. Whether an external consultant, a member of the data processing department, a representative of the supplier, or a friend giving advice, it is important to think about the roles of the client, problem owner and problem solver.

The client, sometimes referred to as the customer, is often the person who is employing the analyst and might also be the person most affected by the activities. The client may be the problem owner, though frequently the client is a senior manager who has called in the analyst to look at a problem in one area of the overall domain. In this case the problem owner will be the manager of that work area. Sometimes there is no one obvious problem owner: two departments could have developed a partial response to a changing situation and have different ideas about how to tackle it. The problem solver is normally the analyst, but frequently the problem owner is also trying to solve the problem. In this case it is important that the two roles are differentiated.

The rich picture can help the owners of the problem sort out the fundamentals of the situation, both to clarify their own thinking and decision making and also to explain these fundamentals to all the interested parties. The rich picture becomes a summary of all that is known about the situation. An analysis of the rich picture will help in the process of moving from 'thinking about the problem situation' to 'thinking about what can be done about the problem situation'.

In the example, the problem owner in the association shown in the rich picture is the general committee of the association. The analysts found that there was conflict between the secretary and the committee for whom she acted, on how best to serve the organisation in its overall goal of improving the standards of the profession. The secretary, who was the client, was expected to become the system user.

As the study was further under way, the role of the education secretary became more important. He was responsible for the professional examinations, and this system became the focal point of later analysis. There was a real conflict between the secretary and the education secretary regarding how and what to automate (hence the crossed swords in the rich picture). As the education secretary became a more powerful stakeholder and the analysts homed in on the examination system, he became the client. Hence he is central in the rich picture shown as Figure 4.2. At the time of the first part of the investigation, it was the secretary who was central.

The building shape in the left of the picture represents the one hundred test centres. It is important to draw attention to the difficulties of handling the examinations. The role of 'students', on the bottom left of the picture, changes from wanting to be professionals to 'members' who ask 'how can we get better service' once they have passed the examination at the test centre and have entered the profession.

Other stakeholders are also included in the rich picture. Developing further the theme of computerisation, some actors were less positive about the prospect. The education assistant, an important actor in the examination system, had not used computers before and the think bubble contains 'I hope we don't get computers'. To jump a few steps, the actual system that was implemented in this situation was not a complete success, and in retrospect, more attention should have been paid to her views. As one of the main persons involved, her misapprehension about the 'system' should have been considered more fully.

We have included the accounting system in our rich picture. The accountant, bottom right of the picture, was satisfied with this manual filing system at the time of the investigation, but it would be looked at in the future.

In drawing the rich picture, some things have been left out which are understood by the participants but which would not be assumed by an outsider. One of these is the social roles of the people in the association and the sort of behaviour which is expected of people in these roles. Sometimes footnotes are useful to describe or list these. A second aspect is in the level of detail. The complexity of the marking system and other regulations for admission to the association, with which the stakeholders were familiar and well understood, could not be gleaned by an outsider looking at the rich picture. A second rich picture, drawn at a greater level of detail, would help here. Rich pictures can be 'decomposed' into others of greater detail. On the other hand, alternative techniques might be more appropriate at this level of detail.

With the analysts coming in as outside consultants, it was important that these 'assumptions' are drawn out. The approach adopted here was for the analyst to ask the users for a detailed explanation of a complex situation. Where a team of analysts is available, they can be divided so that different sorts of questions can be asked: those relating to management strategy, those relating to data and those relating to people's roles. The rich picture, once drawn up, proved a very useful communication tool in this situation and was refined according to new information. No thought is given at this stage to possible solutions. One of the purposes of drawing a rich picture diagram is to avoid 'design before analysis'.

The simplicity of the final rich picture is achieved by pruning the answers so that there is as much agreement as possible and so that the final picture really does represent the important people, activities and issues of the problem situation.

Although this technique has been used by many systems analysts, it is certainly not as well-used as, for example, entity-relationship diagrams or data flow diagrams. It might be regarded as a 'joke' technique as it does not seem as formal as others and therefore may not have the credibility with managers who may also wish to avoid their political issues being disclosed and debated. Proponents of rich pictures might argue, however, that this is not because of any weakness of the approach but partly because of a lack of knowledge about them and partly because systems analysts are not prepared to spend enough time on analysis and rush to the design and development phases.

4.3 ROOT DEFINITIONS

The second technique originating from SSM, root definitions, can be used to define two things that are otherwise both vague and difficult. These are problems and systems. It is essential for the systems analyst to know precisely what human activity system is to be dealt with and what problem is to be tackled. The technique also originated from SSM and descriptions here are based on the Multiview approach.

The root definition is a concise verbal description of the system which captures its essential nature. Each description will derive from a particular view of reality. To ensure that each root definition is well-formed, it is checked for the presence of six characteristics. Put into plain English, these are *who* is doing *what* for *whom*, and to whom are they *answerable*, what *assumptions* are being made, and in what *environment* is this happening? If these questions are answered carefully, they should tell us all we need to know.

There are technical terms for each of the six parts, the first letter of each forming the mnemonic CATWOE. We will change the order in which they appeared in our explanation to fit this mnemonic:

- *Client* is the 'whom' (the beneficiary, or victim, affected by the activities)
- *Actor* is the 'who' (the agent of change, who carries out the transformation process)
- *Transformation* is the 'what' (the change taking place, the 'core of the root definition' (Smyth and Checkland, 1976))
- *Weltanschauung* (or world view) is the 'assumptions' (the outlook which makes the root definition meaningful)
- *Owner* is the 'answerable' (the sponsor or controller)
- *Environment* remains the 'environment' (the wider system of which the problem situation is a part).

The word *Weltanschauung* may be new to many readers. It is a German word that has no real English equivalent. It refers to 'all the things that you take for granted' and is related to our values.

The first stage of creating the definition is to write down headings for each of the six CATWOE categories and try to fill them in. This is not always easy because we often get caught up in activities without thinking about who is really supposed to benefit or who is actually 'calling the tune'. We may question our assumptions and look around the environment even more rarely.

Even so, the difficulty for the individual in creating a root definition is less than the difficulty in getting all the individuals involved to agree on a usable root definition. Only experience of such an exercise can reveal how different are the views of individuals about the situation in which they are working together.

In trying to create the root definition for the professional association's examination system as part of the case study, the following process was followed. First, what were thought of as the issues and primary tasks were identified. These represented the things that the users were concerned about:

- Efficient administration and management of the examinations system
- Choosing a solution which would not mitigate against the association's other systems, such as membership records management and accounting
- Building up a good reputation for the association.

Within this were identified three major components which are called the relevant systems. In the case study, the issues and primary tasks could largely be resolved by the following relevant systems:

- Administration and management system
- Communication and motivation system
- Information provision system.

These relevant systems are subsystems to support a higher system which is to maintain and improve the reputation of the profession by ensuring high standards of entry into the profession.

The working root definition was:

> A system owned and operated by the professional association to administer the examinations by registering, supervising, recording and notifying students.

In the case, when it was necessary to write the root definition, there was particular difficulty about the client. At first the obvious client was the secretary, but on further analysis, the view was that the real client was the education secretary. Yet, as a computer solution became very likely, the person exercising power proved to be the treasurer, a member of the financial committee who would only give his consent to purchase a computer system if it was a particular brand. This happened to be that which he was experienced at using and one which the analysts felt later was inappropriate to the examinations system. There was nothing that the analysts could do about this political in-fighting, but at least they were aware of the problem.

It is sometimes difficult to produce a rigorous root definition because of these political or other problems. Sometimes it is impossible to resolve differences. However, unless they are resolved, they may be a source of difficulties later.

The CATWOE criteria were used to check and revise the above root definition as follows:

- *Customers:* members of the association, the secretary, education secretary and treasurer.
- *Actors:* the association, its members, students attempting to join, and its full time staff
- *Transformation:* to provide examinations which will ensure entry at the right level to the profession
- *Weltanschauung:* the view that computer systems would be efficient and effective if they were used in this domain
- *Owners:* the general committee of the association (representing members of the association)
- *Environment:* the particular profession.

Thus the first use for the root definition is to clarify the situation. People involved in an enterprise have very different views about that enterprise. These views are frequently at cross-purposes. Not everyone, for example, thought that computer systems would be efficient and effective. This holds true even when the same words are used to describe things. This is because the differences are usually in the unstated assumptions or different perceptions of the environment. More significantly, there are sharp differences of opinion about whose problem the analysts are trying to solve, that is, who is the owner and who is the customer. It may not be possible to resolve differences of opinion and one root definition - a preferred root definition - might be chosen from the alternatives and used as a basis to further develop the information system.

Root definitions are particularly useful in exposing different views. We will look at an information system for a hospital to illustrate this. The different people involved in a hospital will look at the system from contrasting positions. Furthermore, these viewpoints in this problem situation are very emotive as they have moral and political overtones. In some situations this can lead to deliberate fudging of issues so as to avoid controversy. This is likely to cause problems in the future. Even if the differences cannot be resolved, it is useful to expose them.

Here are three different root definitions of a hospital system. They all represent extreme positions. In practice, anyone trying to start such a definition would make some attempt to encompass one or more of the other viewpoints, but any one of these could be used as the starting point for the analysis of the requirements of an information system in a hospital.

We will first look at the problem situation from the point of view of the patient, presenting a possible CATWOE and root definition.

THE PATIENT	
CLIENT	Me
ACTOR	The doctor
TRANSFORMATION	Treatment
WELTANSCHAUUNG	I've paid my taxes so I'm entitled to it
OWNER	'The system' or maybe 'the taxpayer'
ENVIRONMENT	The hospital

A hospital is a place that I go to in order to get treated by a doctor. I'm entitled to this because I am a taxpayer, and the system is there to make sure that taxpayers get the treatment they need.

The perception of the doctors will be different.

THE DOCTOR	
CLIENT	Patients
ACTOR	Me
TRANSFORMATION	Treatment (probably by specialised equipment, services or nursing care)
WELTANSCHAUUNG	It is important to treat as many people as possible within a working week.
OWNER	Hospital administrators
ENVIRONMENT	National Health Service (NHS) versus private practice. My work versus my private life.

A hospital is a system designed to enable me to treat as many patients as possible with the aid of specialised equipment, nursing care, etc. Organisational decisions are made by the hospital administrators (who ought to try treating patients without the proper facilities) against a background of NHS politics and my visions of a lucrative private practice and regular weekends off with my family.

The views of the hospital administrators are likely to be different still:

THE HOSPITAL ADMINISTRATOR	
CLIENT	Doctors
ACTOR	Me
TRANSFORMATION	To enable doctors to reduce waiting lists
WELTANSCHAUUNG	Create a bigger hospital within cash limits
OWNER	The government department of health
ENVIRONMENT	Politics

A hospital is an institution in which doctors (and other less expensive staff) are enabled by administrators to provide a service which balances the need to avoid long waiting lists with that to avoid excessive government spending. Ultimate responsibility rests with the government and the environment is very political.

We could therefore develop three sets of very different information systems depending on the view taken. The patient would have the system centred around patients' health records; the doctor would have the system designed around clinic sessions; and the administrator around the accounts.

These definitions have been deliberately controversial, but they attempt to show the private views of the participants as well as their publicly stated positions. There is no reason why definitions need to be formal and cold. Wilson (1990) carries out a similar exercise concerning the prison system. Dependent on the view taken, amongst other possibilities, the prison system could be seen as a:

- Punishment system
- Society protection system
- Behavioural experiment system
- A criminal training system
- Mail-bag production system
- People storage system
- An exclusive storage system.

These contribute to the eventual primary task definition:

a system for the receipt, storage and despatch of prisoners.

The alternative root definitions (briefly expressed above) indicate the difficulty of reconciling different viewpoints, and yet if one is not agreed, it will be even more difficult to agree on final information systems needs. However, without looking at these wider views, information systems might be developed on the basis of a single (client's) view of the problem situation. Information systems are designed to serve the needs of people, and analysts are always brought directly into contact with power struggles between individuals and between viewpoints. Analysts have to make decisions, consciously or unconsciously, about which particular view of the situation or combination of views to work from. One option is to attempt to be 'scientifically detached', but this is only one of the options and is difficult to achieve in reality. In any case, such an aim would seem to be in conflict with the 'philosophy' of root definitions. We cannot be 'objective'.

In some Scandinavian countries there are laws or public agreements which state that the views of the workers or their representatives have to be sought and clearly represented at all stages in the analysis, design and implementation of computer-based systems. In the United States, the analysts usually focus on the opinion of the people who have commissioned them or are the senior people in that situation. Many analysts argue that they are making 'objective' decisions, innocent of any prejudice, but these may be based on personal and political assumptions that are never made explicit. The process described in this section should help to avoid this pitfall.

The manager of a small firm contemplating a PC system may find this process rather long-winded and unnecessary. The system of communications is likely to be easier in smaller firms. Nevertheless, the undercurrent in a small business can be just as political.

4.4 CONCEPTUAL MODELS

Rich picture diagrams and root definitions, and the investigation and analysis preceding their construction, give an overall view of the organisation whose information processing may be computerised to some extent at least. It also provides some key definitions of the purposes to be furthered by the information system. To complete the analysis of the human activity system (following SSM and Multiview), we need to build a model which shows how the various activities are related to each other, or at least how they ought logically to be arranged and connected. This is called a conceptual model. (Unfortunately the term conceptual model is used in some other methodologies to refer to entity modelling).

If the analysis of the human activity system is to be helpful to the organisation, then it will show any discrepancies between what is happening in the real world and what ought to be happening. This may lead to changes in the organisation of human activities. The purpose of introducing a computer-based information system is to improve things, not just to 'automate the status quo', although, as we saw in Chapter 2, many computer information systems do little more than this. Thus, once conceptual models are constructed, we will have a model of the required activities which will serve as the foundation for the information model and a set of recommendations for an improved human activity system.

What do we mean by a conceptual model and what is it for? Perhaps these questions are best explained by analogy. When architects design a building they must produce two things: first, a set of artist's impressions and a scale model to show the client what is proposed and second, a set of plans for the builder. Together these constitute the model. They will enable the builder to say how much it will cost and how long it will take and will represent all that needs to be created for the parties concerned to decide whether to go ahead with that design, modify it or to choose an alternative.

The model serves three purposes:

- It is an essential element in the architects' design activities
- It is a medium of communication between architects and clients to enable the right design to be selected
- It is a set of instructions to the builders.

In computing, we also try to create models which will serve these three purposes, but the process is not so well known, or so well tried and tested, as it is in construction. As we will see in this chapter and on looking at the various methodologies in Chapter 6, there are probably in use almost as many ways of describing a proposed system as there are design teams. This creates problems for users and designers as they try to understand what is being proposed.

In information systems development there is no clear-cut distinction between artist's impressions and the engineer's blueprints. There is not one version of the model for the user and another version for the computer programmer. Some may argue that this would be a valid goal, but furthering our analogy, artist's impressions are notoriously optimistic and vague about difficulties, and engineer's blueprints are very difficult to interpret by all but the trained. It is not satisfactory for the untrained to have to accept the statement: 'trust us, we're the experts'.

This means that the users and the builder of the information system must both understand the conceptual model. Of course the information represented on a model can be complex, but the real world it represents is also complex.

Returning to the case study used for illustrative purposes in the previous two sections, the main activities of the examinations system, and consequently the information to support these activities prior to computerisation, is shown in figure 4.3.

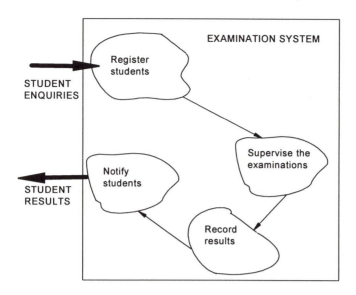

Fig. 4.3: Level 1 conceptual model for the professional association

The conceptual model is formed from the chosen root definition as follows:

1.   Form an impression of the system to carry out a physical or abstract transformation from the root definition.
2.   Assemble a small number of verbs which describe the most fundamental activities in the defined system.
3.   Develop this by deciding on what the system has to do, how it would accomplish the requirement and how it would be monitored and controlled.
4.   Structure similar activities in groups together.
5.   Use arrows to join the activities which are logically connected to each other by information, energy, material or other dependency
6.   Verify the model by comparing it against the perceived reality of the problem situation.

The conceptual model shown as figure 4.3 was derived from the root definition which was, for the professional association:

A system owned and operated by the professional association to administer the examinations by registering, supervising, recording and notifying students.

We start by taking significant aspects from the root definition and naming subsystems which will enable us to achieve what we require.

These are the subsystems to register students, supervise the examinations, record the results and notify the students.

So as to ensure that the most useful subsystems have been identified and understood, they are described in more detail in words and diagrammatically. In other words, there is a second layer in the conceptual model set which looks the same as the top layer, but is at a more detailed level. In other words, the technique lends itself to functional decomposition (as do many of the techniques described in this chapter).

In order to get agreement between problem solver and problem owner on these systems, it is important to ensure that there is a mutual understanding of the real world meaning of the terms. It is necessary, for example, for the analyst to get to know what is involved in registering students (and vetting enquires and selecting potential students for registration, which is at the level 2 conceptual model shown as figure 4.4) in order to understand that subsystem. As we have said above, the analyst is only concerned with 'what is conceptually necessary'. It does not matter, for example, how the enquiries are received; how the forms are sent out to be completed; or which member of staff deals with them.

The conceptual model is derived from the root definition. It is a model of the human activity system. Its elements are therefore activities and these can be found by extracting from the root definition all the verbs that are implied by it. The list of active verbs should then be arranged in a logically coherent order.

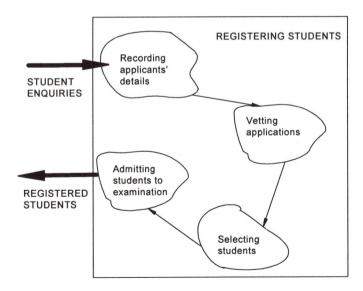

Fig. 4.4: Level 2 conceptual model

We would expect the number of activities to be somewhere in the range of five to nine. Activities should be grouped to avoid a longer list, as a long list is too complicated and messy to deal with. A shorter list suggests that the root definition is too broad to be useful.

Having listed the major activities, some of these may imply secondary activities. These should also be listed and arranged in logical order around their primary activities and will form level 2 conceptual models.

Conceptual modelling is an abstract process. The purpose of going into this abstract world of systems thinking is to develop an alternative view of the problem situation. When this alternative view has been developed, we can return to the real world and test the model. It is constructed in terms of what must go into the system, and it can therefore be set alongside the real world. We are not concerned here with how the system will be implemented. Other techniques will be used to determine this aspect.

The conceptual model illustrates what ought to be happening to achieve the objectives specified in the root definition. There is normally more than one way of doing something and so choices will have to be made about the structure of the conceptual model. Many systems analysts come from a scientific intellectual background where they are encouraged to believe that answers are either right or wrong. School science subjects tend to instil this view and not all people are able to take a different view. An inexperienced analyst may put a flowchart in front of the user and ask 'is this right?' Users who may have had many years of coping with 'messy' reality may well be reluctant to answer so positively. They might wish to respond 'Yes, but ...' or 'No, but ...'. Unfortunately the politics of the situation may be such that they may be forced to say 'yes' if they recognise some resemblance between the flowchart and reality. Alternatively they may be left to 'pick at the details'. One of the problems that the user faces is the inappropriateness of the flowcharts used in conventional systems analysis, and a strength of the conceptual model is its usefulness as a communication tool.

This conceptual model needs to be compared with reality to see whether improvements should be made to the way in which activities are organised. For example, in the second or third level detail of conceptual models, they might highlight bottlenecks, such as too many small decisions waiting for the manager or too many assistants waiting to use the same price catalogue. They may also show up circuitous routes for transferring information.

In small organisations, information handling is very informal, everyone sees what is happening or works alongside the people who know. As work diversifies and more staff are taken on, information flow is based around the experienced staff who become 'walking databases'. Such an arrangement can then ossify and become increasingly dysfunctional to new functions and new personalities. Many apparently efficient offices are thrown into disarray by the loss of the one person 'who seems to know everything'.

In comparing the conceptual model with reality the analyst will ask the question 'Does the information flow smoothly?'. There are two extreme forms of organisation: where one person sees a job through from beginning to end or where each person handles a specialised part of the work. Of course, most organisations have aspects of each, and both have different implications for information flow.

Many factors must be taken into account when matching functions to staff. These include the capabilities and aspirations of staff, the demands of different aspects of the business and the need for management to keep control of what is going on. A number of these might change if a computer system were to be introduced. The conceptual model can be used as a technique for thinking about how subsystems should be organised in order to achieve the purposes set out in the root definition. Questions can be asked about which subsystems should be linked together and whether they can be handled more efficiently if they are kept separate.

The conceptual model can also be used in the design of new human activity systems, such as the setting up of a new company or department, because it shows what activities should be carried out and how they should be related to each other.

The techniques of rich pictures, root definitions and conceptual models will be explored further in the context of SSM and Multiview (section 6.11 and 6.12).

4.5 ENTITY MODELLING

Traditional systems analysis procedures were applied to single applications that were the first to be computerised in the organisation. As we saw in the last chapter, more emphasis is now placed on applications being developed as part of an integrated information system and the techniques described in section 2.2 prove inadequate. The most obvious situation which requires a different approach is the development of a database. In a database environment, many applications share the same data. The database is looked upon as a common asset.

The theme of data modelling has been discussed in section 3.8. Data modelling concentrates on the analysis of data in organisations and entity modelling is the main technique used to achieve this in many methodologies, including SSADM, Merise and Information Engineering. Data analysis techniques were largely developed to cater for the implementation of database systems, although that does not mean that they cannot be applied to non-database situations. Entity modelling may even be carried out as a step in conventional file applications. It can also be of interest to management as a data-oriented way (and potentially an information-oriented way) of perceiving their organisation.

Just as an accountant might use a financial model, the analyst can develop an entity model. The entity model is just another view of the organisation, but it is a particular perception of the organisation which emphasises data aspects. Systems analysis in general, and data analysis is a branch of systems analysis, is an art or craft, not an exact science. There can be a number of ways to derive a reasonable model and there are a number of useful data models (there are of course an infinite number of inadequate models).

A model represents something, usually in simplified form which highlights aspects which are of particular interest to the user, and is built so that it can be used for a specific purpose, for example, communication and testing. As we saw in Chapter 3, many types of model are used in information systems work. A model is a representation of real-world objects and events, and good entity models will reflect certain aspects of the 'real world'. The entity model is an abstract representation of the data within the organisation. It can be looked on as a discussion document and its ability to reflect the real world can be verified in discussions with the various users. However, the analyst should be aware that variances between the model and a particular user's view could be due to the narrow perception of that user. The model should be a global view, not an application or function-oriented view. The 'globe' could be a department, a number of departments, a company or an organisation, such as a branch of government.

An entity-relationship model views the organisation as a set of data elements, known as entities, which are the things of interest to the organisation, and relationships between these entities. This model helps the computer specialist to design appropriate computer systems for the organisation, but it also provides management with a unique tool for perceiving aspects of the business. The essence of problem solving is to be able to perceive the complex, 'messy', real world in such a manner that the solution to any problem may be easier. This model is 'simple' in that it is fairly easy to understand and to use.

Each entity can be represented diagrammatically by soft boxes (rectangles with rounded corners). Relationships between the entities are shown by lines between the soft boxes. A first approach to an entity model for an academic department of computer science is given in figure 4.5. The entity types are STUDENT, ACADEMIC STAFF, COURSE and NON-ACADEMIC STAFF. The entity type STUDENT participates in a relationship with ACADEMIC STAFF and COURSE. The relationships are not named in figure 4.5, but it might be that STUDENT *takes a* COURSE and that STUDENT *has as tutor* ACADEMIC STAFF. The reader will soon detect a number of important things of interest that have been omitted (room, examination, research and so on). As the analysts find out more about the organisation, entity types and relationships will be added to the model.

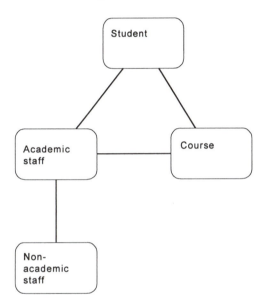

Fig. 4.5: Entity modelling - a first approach

A mistake frequently made at this stage is to define the entities to reflect the processes of the business, such as stock control, credit control or sales order processing. This could be a valid model of the business but it is not an entity model and cannot be used to produce the flexible database for the organisation that may be one of the 'deliverables' when using a methodology. A database so created would be satisfactory for some specific applications, but would not be adequate for many applications that might access a database. Data analysis differs from conventional systems analysis in that it separates the data structures from the applications which use them. The objective of data analysis is to produce a flexible model which can be easily adapted as the requirements of the organisation change. Although the applications will need to be changed, this will not necessarily be true of the data.

The entity model is sometimes referred to as the conceptual schema or conceptual model. However, in order to avoid confusion with Checkland's conceptual models described earlier in this chapter (an entirely different model) we will use the terms entity or data model in this book.

Probably the most widespread technique of data analysis is that proposed by Chen (1976). In Chen's entity-relationship (E-R) model, the real world information is represented by entities and by relationships between entities. In a typical business, the entities could include jobs, customers, departments, and suppliers. The analyst identifies the entities and relationships between them before being immersed in the detail, in particular, the work of identifying the attributes which define the properties of entities.

Figure 4.6 relates to part of a hospital. The entities described are DOCTOR, PATIENT and CLINICAL SESSION. The relationships between the entities are also described. That between DOCTOR and PATIENT and between DOCTOR and CLINICAL SESSION are one-to-many relationships. In other words, one DOCTOR can have many PATIENTS, but a PATIENT is only assigned one DOCTOR at a particular point in time. Further, a DOCTOR can be responsible for many CLINICAL SESSIONS, but a CLINICAL SESSION is the responsibility of only one DOCTOR. The other relationship is many-to-many. In other words, a PATIENT can attend a number of CLINICAL SESSIONS and one CLINICAL SESSION can be attended by a number of PATIENTS. A one-to-one relationship would be shown by a line without any 'crow's feet'.

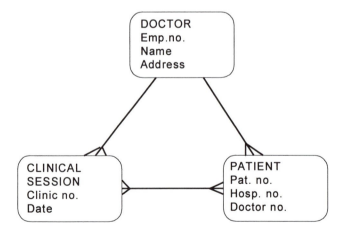

Fig. 4.6: An entity set relating to part of a hospital

The diagram also shows a few attributes of the entities. The particular attribute or group of attributes that uniquely identify an entity occurrence is known as the key attribute or attribute(s). The 'employee number' (Emp. no.) is the key attribute of the entity called DOCTOR.

The technique attempts to separate the data structure from the functions for which the data may be used. This separation is a useful distinction, although it is often difficult to make in practice. In any case, it is sometimes useful to bear in mind the functions of the data analysed. A DOCTOR and a PATIENT are both people, but it is their role, that is what they do, that distinguishes the entities. The distinction, formed because of a knowledge of functions, is a useful one to make. However, too much regard to functions will produce a model biased towards particular applications or users.

Another practical problem is that organisation-wide data analysis may be so costly and time-consuming that it is often preferable to carry out entity analysis at a 'local' level, such as in the marketing or personnel areas. If a local entity analysis is carried out, the model can be mapped on to a database and applications applied to it before another local data analysis is started. This is far more likely to gain management approval because managers can see the expensive exercise paying dividends in a reasonable time scale. An important preliminary step is therefore to define the area for analysis and break this up into distinct sub-areas which can be implemented on a database and merged later. Local data analysis should also be carried out in phases. The first phase is an overview which leads to the identification of the major things of interest in that area. At the end of the overview phase, it is possible to draw up a second interview plan and the next, longer, phase aims to fill in the detail.

Although it is relatively easy to illustrate the process of modelling in a book or at a lecture, in real life there are problems in deciding how far one should go and what level of detail is appropriate. The level of detail must serve two purposes:

1. It must be capable of explaining that part of the organisation being examined.
2. It must be capable of being translated into a physical model, usually for mapping onto a computer database.

It is important to realise that there is no logical or natural point at which the level of detail stops. This is a pragmatic decision. Certainly design teams can put too much effort into the development of the model. Some decisions are based on the way in which the data is used. An example of this could be entity occurrences of persons who are female, where they relate to:

- Patients in a hospital
- Students at university
- Readers in a library.

In the patient example, the fact that the person occurrence is female is important, so important that the patient entity may be split into two separate entities, male patients and female patients. In the student example, the fact that the person is female may not be of great significance and therefore there could be an attribute 'sex' of the person entity. In the reader example, the fact that the person is female may be of such insignificance that it is not even included as an attribute. There is a danger here, however, as the analyst must ensure that it will not be significant in all applications in the library. Otherwise the data model will not be as useful.

An entity is a thing of interest in the organisation, in other words it is anything about which we want to hold data. It could include all the resources of the business, including the people of interest such as EMPLOYEE, and it can be extended to cover such things as SALES-ORDER, INVOICE and PROFIT-CENTRE. Some entities are obvious physical things, like customers or stock. Others are transactions, like orders, sales and hospital admissions. Some entities are more or less artificial. These are rather like catalogue entries in the library: the only reason to have them is to help people find books which would otherwise be difficult to locate. It covers concepts as well as objects. A SCHEDULE or a PLAN are concepts which can be defined as entities. An entity is not data itself, but something about which data should be kept. It is something that can have an independent existence, that is, can be distinctly identified.

In creating an entity model, the aim should be to define entities that enable the analyst to describe the organisation. Such entities as STOCK, SALES-ORDER and CUSTOMER are appropriate because they are quantifiable, whereas 'stock control', 'order processing' and 'credit control' are not appropriate because they are functions: what the organisation does, and not things of interest which participate in functions. Entities will normally be displayed in small capitals in this book. Entities can also be quantified - it is reasonable to ask 'how many customers?' or 'how many orders per day?', but not 'how many credit controls?'. An entity occurrence is a particular instance of an entity which can be uniquely identified. It will have a value, for example, 'Jim Smith & Son' and this will be a particular occurrence of the entity CUSTOMER. There will be other occurrences, such as, 'Plowmans PLC' and 'Tebbetts & Co.'.

An attribute is a descriptive value associated with an entity. It is a property of an entity. At a certain stage in the analysis it becomes necessary not only to define each entity but also to record the relevant attributes of each entity. A CUSTOMER entity may be defined and it will have a number of attributes associated with it, such as 'number', 'name', 'address', 'credit-limit', 'balance' and so on. Attributes will normally be displayed within single inverted commas in this book. The values of a set of attributes will distinguish one entity occurrence from another. Attributes are frequently identified during data analysis when entities are being identified, but most come later, particularly in detailed interviews with staff and in the analysis of documents. Many are discovered when checking the entity model with users.

An entity must be uniquely identified by one or more of its attributes, the key attribute(s). A <u>customer number</u> may identify an occurrence of the entity CUSTOMER. A <u>customer number</u> and a <u>product number</u> may together form the key of entity SALES-ORDER. The key attribute functionally determines other attributes, because once we know the customer number we know the name, address and other attributes of that customer. Key attributes will normally be underlined in this book.

There often arises the problem of distinguishing between an entity and an attribute. In many cases, things that can be defined as entities could also be defined as attributes, and vice versa. We have discussed one example relating to the sex of people. The entity should have importance in the context of the organisation, otherwise it is an attribute.

In practice, the problem is not as important as it may seem. Most of these ambiguities are settled in the process of normalisation (section 4.6) and this often happens in database design. In any case, the analyst can change the model at a later stage, even when mapping the model onto a database, though the earlier the analyst gets it right the better. Entities are used by functions of the organisation and the attributes are those data elements that are required to support the functions. The best rule of thumb is to ask whether the data has information about it, in other words, does it have attributes? Entities and attributes are further distinguished by their role in events (discussed below).

A relationship in an entity model normally represents an association between two entities. A SUPPLIER entity has a relationship with the PRODUCT entity through the relationship supplies, that is, a SUPPLIER *supplies* PRODUCT. There may be more than one relationship between two entities, for example, PRODUCT *is assembled by* SUPPLIER. Relationships will normally be displayed in italics in this book.

A relationship normally arises because of:

1.   Association, for example CUSTOMER *places* ORDER
2.   Structure, for example ORDER *consists of* ORDER-LINE.

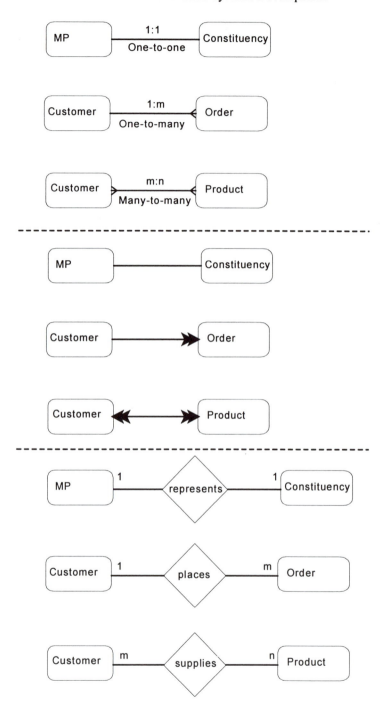

Fig. 4.7: Cardinality of a relationship - three diagramming conventions

The association between entities has to be meaningful in the context of the organisation. The relationship has information content, for example, CUSTOMER *places* ORDER. The action *places* describes the relationship between CUSTOMER and ORDER. The name given to the relationship also helps to make the model understandable. As will be seen, the relationship itself can have attributes.

The cardinality of the relationship could be one-to-one, one-to-many, or many-to-many. At any one time, a MEMBER-OF-PARLIAMENT can only represent one constituency, and one CONSTITUENCY can have only one MEMBER-OF-PARLIAMENT. A MEMBER-OF-PARLIAMENT *represents a* CONSTITUENCY. This is an example of a one-to-one (1:1) relationship. Figure 4.7 shows different conventions of representing relationships found in methodologies. Very often, a one-to-one relationship can be better expressed as a single entity, with one of the old entities forming attributes of the more significant entity. For example, the entity above could be MEMBER-OF-PARLIAMENT, with CONSTITUENCY as one of the attributes.

The relationship between an entity CUSTOMER and another entity ORDER is usually of a degree one-to-many (1:m). Each CUSTOMER can have a number of ORDERS, but an ORDER can refer to only one CUSTOMER: CUSTOMER *places* ORDER.

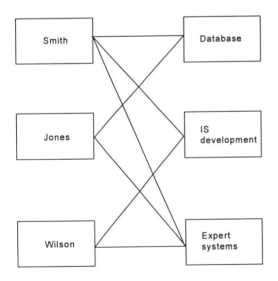

Fig. 4.8: Many-to-many relationship between STUDENT and MODULE

With a many-to-many (m:n) relationship, each entity can be related to one or more occurrences of the partner entity. A STUDENT can take many

MODULES; and one MODULE could be taken by a number of STUDENTS (MODULE *is taken by* STUDENT; STUDENT *takes* MODULE).

In this last example (of a many-to-many relationship), entity occurrences of the STUDENT entity could be 'Smith', 'Jones' and 'Wilson', and they could take a number of modules each. For example, Smith might take database, IS development and expert systems; Jones might take database and IS development; and Wilson IS development and expert systems. This is shown in figure 4.8.

Frequently there is useful information associated with many-to-many relationships and it is better to split these into two one-to-many relationships, with a third intermediate entity created to link these together. Again, this should only be done if the new entity has some meaning in itself. The relationship between COURSE and LECTURER is many-to-many, that is, one LECTURER *lectures on* many COURSES and a COURSE *is given by* many LECTURERS. But a new entity, MODULE can be described as used in the previous example which may only be given by one LECTURER and is part of only one COURSE. Thus a LECTURER *gives* a number of MODULES and a COURSE consists of a number of MODULES. But one MODULE *is given by* only one LECTURER and one MODULE *is offered to* only one COURSE (if these are the restrictions). This is shown in figure 4.9.

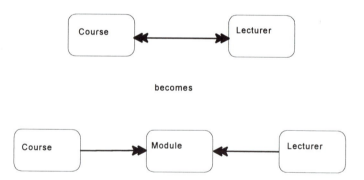

Fig. 4.9: A many-to-many relationship represented as two one-to-many relationships

There are other distinctions and sophistications which are often included in the model. Sometimes a 1:m or an m:n relationship is a fixed relationship. The many-to-many relationship between the entity PARENT and the entity CHILD is 2:m (that is, each child has two parents); but a PARENT *can beget* more than one CHILD.

Whilst some relationships are mandatory, that is, each entity occurrence must participate in the relationship, others are optional. An entity MALE and an entity FEMALE may be joined together by the optional relationship *married to*. Mandatory and optional relationships may be represented as shown in figure 4.10.

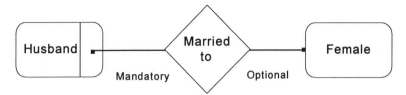

Fig. 4.10: Representation of mandatory and optional relationships

Other structures include exclusivity, where participation in one relationship excludes participation in another, or inclusivity, where participation in one relationship automatically includes participation in another.

A relationship may also be involuted where entity occurrences relate to other occurrences of the same entity. For an EMPLOYEE entity, for example, an EMPLOYEE entity occurrence who happens to be a manager *manages* other occurrences of the entity EMPLOYEE. This can be shown diagrammatically by an involuted loop, as in figure 4.11. Some approaches suggest that these should be eliminated by creating two entities (MANAGER and EMPLOYEE in this case).

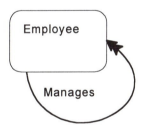

Fig. 4.11: An involuted relationship

Any relationship is necessarily linked to at least one entity. We have already looked at the involuted relationship. Where a relationship is linked to two entities (as in the case of the examples in figure 4.7), it is said to be binary. If a relationship is linked to three entities, as in figure 4.12, it is said to be ternary. In this example, EMPLOYEE fulfils a ROLE, EMPLOYEE fulfils a CONTRACT and ROLE fulfils a CONTRACT. Otherwise it is n-ary, with the value of 'n' equalling the number of entities.

A good model is one that is a good representation of the organisation, department or whatever is being depicted. The process of entity modelling is an iterative process and slowly the model will improve as a representation of the perceived reality. The entity model can be looked on as a discussion document and its coincidence with the real world is verified in discussions with the various users. However, the analyst should be aware that variances between the model and a particular user's view could be due to the narrow perception of that user. If a global entity model is built for a whole organisation it is usual for entities to be grouped into important clusters.

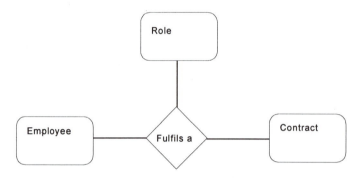

Fig. 4.12: Example of a ternary relationship

Up to now, we have considered the data-oriented aspects of data analysis, but in practice it is useful for functional considerations to be made in order to check the model. These relate to events and operations. Entities have to support the events that occur in the enterprise. Entities will take part in events and in the operations that follow events. Attributes are those elements which supply data to support events.

'Tom' is an occurrence of the entity EMPLOYEE. Tom's pay rise or his leaving the company are events, and attributes of the entity EMPLOYEE will be referred to following these events. Attributes such as 'pay-to-date', 'tax-to-date', 'employment status', and 'salary' will be referred to.

Operations on attributes will be necessary following the event: an event triggers an operation or a series of operations. An operation will change the state of the data. The event 'Tom gets salary increase of 10 per cent' will require access to the entity occurrence 'Tom' (or EMPLOYEE-NUMBER '756') and augmenting the attribute 'salary' by 10 per cent. Figure 4.13 shows the entity EMPLOYEE expressed as a relation with attributes. We have to check that the relation supports all the operations that follow the event mentioned.

Emp. no.	Name	Status	Pay-to-date	Tax-to-date	Salary
756	Tom	Full	734.30	156.00	14000

Does the entity support the operations following events?
e.g. employee leaves the company
     employee gets a pay rise

Fig. 4.13: Event driven (functional) analysis

Some methodologies, for example, Merise, also define the synchronisation of an operation (figure 4.14). This is a condition affecting the events which trigger the operation and will enable the triggering of that operation. This condition can relate to the value of the properties carried by the events and to the number of occurrences of the events. For example, the operation 'production of pay slips' may be triggered by the event 'date' when it equals '28th day of the month'.

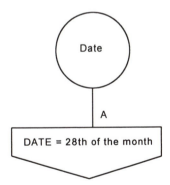

Fig. 4.14: Synchronisation of an operation

Some readers may be confused by the discussion of events (sometimes called transactions), which are function-oriented concepts, when data analysis is supposed to be function-independent. The consideration of events and operations is of interest as a checking mechanism. They are used to ensure that the entity model will support the functions that may use the data model. This consideration of the events and operations may lead to a tuning of the model, an adjustment of the entities and the attributes.

We will now look at the stages of entity analysis as a whole which are:

- Define the area for analysis
- Define the entities and the relationships between them
- Establish the key attribute(s) for each entity
- Complete each entity with all the attributes
- Normalise all the entities
- Ensure all events and operations are supported by the model.

We have looked at all these elements apart from normalisation which is discussed in section 4.6.

The first stage of entity analysis requires the definition of the area for analysis. This is frequently referred to as the universe of discourse. Sometimes this will be the organisation, but this is usually too ambitious for detailed study, and as we have seen the organisation will normally be divided into local areas for separate analysis.

Then we have the stages of entity-relationship modelling. It is a top-down approach in that the entities are identified first, followed by the relationships between them, and then more detail is filled in as the attributes and key attribute(s) of each entity are identified.

For each local area, then, the entities are defined. The obvious and major entities will be identified first. The analyst will attempt to name the fundamental things of interest to the organisation. As the analyst is gathering these entities, the relationships between the entities can also be determined and named. Their cardinality can be one-to-one, one-to-many or many-to-many. It may be possible to identify fixed relationships and those which are optional or mandatory. The analyst will be able to begin to assemble the entity-relationship diagram. The diagram will be rather sketchy, somewhat like a 'doodle', in the beginning, but it will soon be useful as a communication tool. There are computer software tools which can help draw up these diagrams and make alterations easily (see Chapter 5). The key of each entity will also be determined. The key attributes will uniquely identify any entity occurrence. There may be alternative keys, in which case the most natural or concise is normally chosen.

The analyst has now constructed the model in outline and is in a position to fill in the detail. This means establishing the attributes for each entity. Each attribute will say something about the entity. The analyst has to ensure that any synonyms and homonyms are detected. A product could be called a part, product or finished product depending on the department. These are all synonyms for 'product'. On the other hand, the term product may mean different things (homonyms), depending on the department. It could mean a final saleable item in the marketing department or a sub-assembly in the production department. These differences must be reconciled and recorded in the data dictionary (section 5.4). The process of identifying attributes may itself reveal entities that have not been identified. Any data element in the organisation must be defined as an entity, an attribute or a relationship and recorded in the data dictionary. Entities and relationships will also be recorded in the entity-relationship diagram.

Each entity must be normalised once the entity occurrences have been added to the model. This process is described fully in section 4.6. Briefly, the rules of normalisation require that all entries in the entity must be completed (first normal form), all attributes of the entity must be dependent on all the key (second normal form), and all non-key attributes must be independent of one another (third normal form). The normalisation process may well lead to an increase in the number of entities in the model.

The final stage of entity analysis will be to look at all the events within the area and the operations that need to be performed following an event, and ensure that the model supports these events and operations. Events are frequently referred to as transactions. For this part of the methodology, the analyst will identify the events associated with the organisation and examine the operations necessary on the trail of each of the events.

Events in many organisations could include 'customer makes an order', 'raw materials are purchased from supplier' and 'employee joins firm'. If, say, a customer makes an order, this event will be followed by a number of operations. The operations will be carried out so that it is possible to find out how much the order will cost, whether the product is in stock, and whether the customer's credit limit is OK. The entities such as PRODUCT (to look at the value of the attribute 'stock') and CUSTOMER (to look at the value of the attribute 'credit limit') must be examined. These attribute values will need to be adjusted following the event. You may notice that the 'product price' is not in either entity. To support the event, therefore, 'product price' should be included in the PRODUCT entity, or in another entity which is brought into the model.

Entity modelling has documentation aids like other methods of systems analysis. It is possible to obtain forms on which to specify all the elements of the data analysis process. The separate documents will enable the specification of entities, attributes, relationships, events and operations.

Typical forms are shown as figures 4.15, 4.16, 4.17, 4.18 and 4.19. These forms can be pre-drawn using software tools and their contents automatically added to the data dictionary.

ENTITY TYPE SPECIFICATION FORM

Entity name   *The standard name for the entity*

Description   *A brief description of the entity type*

Synonyms   *Other names by which the entity is known*

Indentifier(s)   *Name of the key attribute(s) which uniquely identify the entity occurrences*

Date specified

Minimum occurrences *expected*   Maximum occurrences *expected*

Average occurrences *expected*   Growth rate % *over time*

Create authority   *The names of the users who are allowed to create the entity*

Delete authority   *The names of the users who are allowed to delete the entity*

Access authority   *The names of the users who are allowed to read the entity*

Relationships involved (cross reference)
   As shown in the entity-relationship diagram

Attributes involved (cross reference)
   Attributes which are found in other entities (to cross reference) different entities for access

Functions involved (cross reference)
   Applications which require data contained in these and other entities

Comments

Fig. 4.15: Entity documentation

ATTRIBUTE TYPE SPECIFICATION FORM

Attribute name   *The standard name for the attribute*

Description   *A brief description of the attribute*

Synonyms   *Other names by which the attribute is known*

Date specified

Entity cross reference

Entities that include the attribute, including those where the attribute is a key or part of a key

Create authority   *The names of the users who are allowed to create the attribute*

Delete authority   *The names of the users who are allowed to delete the attribute*

Access authority   *The names of the users who are allowed to read the attribute*

Functions involved (cross reference)

Uses of the attribute in the functions of the organisation

Format

Attributes which are found in other entities (to cross reference) different entities for access

Values

The values that the attribute may have

Comments

Fig. 4.16: Attribute documentation

RELATIONSHIP TYPE SPECIFICATION FORM

Relationship name   *The standard name for the relationship*

Description   *A brief description of the relationship*

Synonyms   *Other names by which the relationship is known*

Date specified

Entities involved (owner)
            (member)   *The owner and member entities*

Occurrences

The numbers of each entity type involved in an occurrence of a relationship and the occurrence of that relationship

Cardinality   Optional   *Condition governing the existence*
(1:1; 1:m; m:n)   Mandatory   *of a relationship*

If exclusive   *State paired relationship name*

If inclusive   *State paired relationship name and first existence relationship name*

Create authority   *The names of the users who are allowed to create the relationship*

Delete authority   *The names of the users who are allowed to delete the relationship*

Access authority   *The names of the users who are allowed to read the relationship*

Comments

Fig. 4.17: Relationship documentation

EVENT TYPE SPECIFICATION FORM

Event name   *The standard name for the event*

Description   *A brief description of the event*

Frequency   *of the event*

Date specified

Operations following event

The procedures that are triggered by the event

Synchronisation

The condition of the event that affects the trigger

Pathway following event

A diagrammatic representation of the processes following the event through the entity types accessed

Create authority   *The names of the users who are allowed to create the event*

Delete authority   *The names of the users who are allowed to delete the event*

Access authority   *The names of the users who are allowed to read the event*

Comments

Fig. 4.18: Event documentation

OPERATION TYPE SPECIFICATION FORM

Operation name   *The standard name for the operation*

Description   *A brief description of the operation*

Access key

Date specified

Entities involved

   Entity names and processing carried out on those entities

Events proceeding operation

   Events caused by operation

Response time required

Frequency

Privacy level

Create authority   *The names of the users who are allowed to create the operation*

Delete authority   *The names of the users who are allowed to delete the operation*

Access authority   *The names of the users who are allowed to read the operation*

Comments

Fig. 4.19: Operation documentation

As we have already stated, it may be possible to use completed documents directly as input to a data dictionary system so that the data is held in a readily accessible computer format as well as on paper forms. Entity modelling can be used as an aid to communication as well as a technique for finding out information. The forms discussed also help as an aid to memory, that is, communication with oneself. The entity-relationship diagrams, which are particularly useful in the initial analysis and as an overview of the data model, can prove a good basis for communication with managers and users. They are much more understandable to non-computer people than the documents used in traditional data processing, although they are also a good communications aid between computer people. They provide a graphical description of the business in outline, showing what the business is, not what it does. Managers and users can give 'user feedback' to the analysts and this will also help to tune the model and ensure its accuracy. A user may point out that an attribute is missing from an entity, or that a relationship between entities is one-to-many and not one-to-one as implied by the entity-relationship diagram. The manager may not use this terminology, but the analyst will be able to interpret the comments made.

Data analysis is an iterative process: the final model will not be obtained until after a number of tries and this should not be seen as slowness, but care for accuracy. If the entity model is inaccurate so will be the database and the applications that use it. On the other hand, the process should not be too long or 'diminishing returns' will set in.

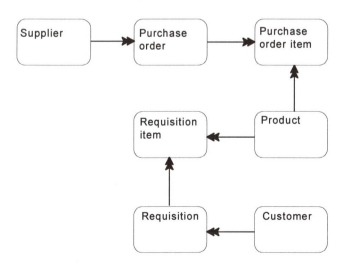

Fig. 4.20: Entity-relationship diagram - a first approach for a wholesaler

The entity-relationship diagram given in figure 4.20 shows the entities for part of a firm of wholesalers. The attributes of the entities might be as follows:

SUPPLIER	<u>Supp-no</u>, Supp-name, Address, Amount-owed
PURCHASE ORDER	<u>P-O-no</u>, P-O-date
PURCHASE ORDER ITEM	<u>P-O-item-no</u>, Quantity
REQUISITION	<u>Req-no</u>, Req-date
REQUISITION ITEM	<u>Req-item-no</u>, Quantity
CUSTOMER	<u>Cust-no</u>, Cust-name, Address, Amount, Credit
PRODUCT	<u>Product-no</u>, description

The key attributes are underlined. Perhaps you would like to verify that you can understand something of the organisation using this form of documentation. It is a first sketch of the business, and you may also verify the relationships, add entities and relationships to the model or attributes to the entities, so that the model is more appropriate for a typical firm of wholesalers. For example, we have not included payments in this interim model.

The entity-modelling approach to data analysis is interview-driven, that is, most information is obtained through interviewing members of staff. It is also top-down, in that the entities are identified first and then more and more detail filled in. It has proved very useful and is included in many information systems development methodologies, as we shall see in Chapter 6. Methodologies usually have entity modelling preceding the process of normalisation. This technique is described in the next section.

4.6 NORMALISATION

In 1970 Ted Codd published his influential paper 'A Relational Model of Data for Large Shared Data Banks'. This has had a profound effect on systems analysis and database design. Most of the early work in data processing had been done pragmatically with developments taking place in the data processing applications environment rather than in the research and academic environments. Little or no formal techniques were being used or developed: problems were solved by rule of thumb and guess-work.

However, by the 1970s the early pioneering days of computer data processing were over and more sophisticated integrated systems were being developed. Large integrated files were needed. Customer files, for example, were being developed which included data relevant to more than one application system. These could include sales order, sales ledger and bad debt processing. Later came databases. Codd's research was directed at a very real problem, that is the deletion, insertion and update of data on these very large files.

Being a mathematician, his approach to the problem was mathematical. In describing the work in this text, however, we avoid a formal mathematical treatment.

SALES-ORDER

CUSTOMER NAME	PART NUMBER	QUANTITY ORDERED
Lee	25	12
Deene	38	18
Smith	38	9
Williams	87	100

Fig. 4.21: Sales-order relation

As seen in figure 4.21, a relation is a table or flat file. This relation is called SALES-ORDER and it could show that Lee ordered 12 of 'part number' 25, Deene and Smith ordered 18 and 9 respectively of 'part number' 38, and Williams ordered 100 of 'part number' 87.

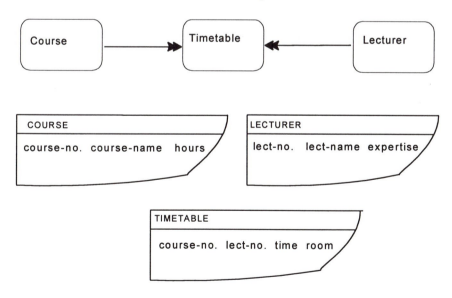

Fig. 4.22: Entities expressed as relations

The entities and relationships identified in the entity-relationship model can both be represented as relations in the relational model. In figure 4.22, there are three entities in the entity model and they are expressed as three relations: COURSE, LECTURER and TIMETABLE.

We will now introduce some of the terminology associated with the relational model. Each row in a relation is called a tuple. The order of tuples is immaterial (although they will normally be shown in the text in a logical sequence so that it is easier to follow their contents). No two tuples can be identical in the model. A tuple or column will have a number of attributes, and in the SALES-ORDER relation of figure 4.21, 'name', 'part' and 'quantity' are attributes. All items in a column come from the same domain, that is, the domain is the set of values from which valid attributes can be drawn. There are circumstances where the contents from two or more columns come from the same domain. The relation ELECTION-RESULT (figure 4.23) illustrates this possibility. Two attributes come from the same domain of political parties. The number of attributes in a relation is called the degree of the relation. The number of tuples in a relation define its cardinality.

ELECTION-RESULT

ELECTION YEAR	FIRST PARTY	SECOND PARTY
1974	Labour	Conservative
1974	Labour	Conservative
1979	Conservative	Labour
1983	Conservative	Labour
1987	Conservative	Labour
1992	Conservative	Labour

Fig. 4.23: Election-result relation

In the SALES-ORDER relation, the key is 'customer name'. It might be better to allocate numbers to customers in case there are duplicate names. If the customer may make orders for a number of parts, then 'part' must also be a key attribute as there will be several tuples with the same 'name'. In this case, the attributes 'customer name' and 'part number' will make up the composite key of the SALES-ORDER relation.

What is the key for the ELECTION-RESULT relation? On first sight, 'election year' might seem appropriate, but in 1974 there were two elections in a year, and even if all three attributes formed the composite key, there are still duplicate relations. It is necessary to add another attribute which is unique, such as, an election number which is incremented by one following each election. Alternatively, it would be possible to replace 'year' by 'election date' to make each tuple unique (there will not be two of these elections on the same day). Another alternative would be the composite key of 'election year' and a new attribute 'election number in year'.

There may be more than one possible key. These are known as candidate keys, that is, they are candidates for the primary key. 'Customer name' and 'customer number' could be candidate keys in a CUSTOMER relation. In this circumstance one of these is chosen as the primary key. An attribute which is a primary key in one relation and not a primary key, but included in another relation, is called a foreign key in that second relation.

The structure of a relation is conventionally expressed as in the following examples:

SALES-ORDER (name, part, quantity)

ELECTION-RESULT (elect-year, elect-number, first-party, second-party)

The process of normalisation is the application of a number of rules to the relational model which will simplify the relations. This set of rules proves to be useful because we are dealing with large sets of data and the relations formed by the normalisation process will make the data easier to understand and manipulate. The model so formed is appropriate for the further stages in most database methodologies and the database will be shareable, a fundamental justification for the database approach.

Normalisation is a technique which is used in a variety of methodologies. For example, it is used in Gane and Sarson (STRADIS), Information Engineering, SSADM, Merise and Multiview which are all discussed in Chapter 6. The technique of normalisation is applicable irrespective of whether a relational database is envisaged or not. It is often used in its own right as an analysis technique for the structuring of data, it can be used on its own or as a means of cross-checking or validating other models, particularly an entity model. Even in structured systems methodologies which stress processes rather than data, for example Gane and Sarson's STRADIS, it is used to consolidate all the various data stores that have been identified in a data flow diagram into a coherent data structure.

Normalisation is the process of transforming data into well formed or natural groupings such that one fact is in one place and that the correct relationships between facts exist. As well as simplifying the relations, normalisation also reduces anomalies which may otherwise occur when manipulating the relations in a relational database. In this simplifying process no data is lost or added to that provided in the original set of unnormalised relations.

Normalised data is stable and a good foundation for any future growth. It is a mechanical process, indeed the technique has been automated, but the difficult part of it lies in understanding the meaning, that is, the semantics, of the data, and this is only discovered by extensive and careful data analysis.

Codd originally developed three levels of normalisation, and the third and final stage is known as third normal form (TNF). It is this level of normalisation that is usually used as the basis for the design of the data model, as an end result of data analysis, and for mapping onto a database. There are a few instances, however, when even TNF needs further simplification, and these are also looked at later in this section. TNF is usually satisfactory in practice.

There are three basic stages of normalisation:

- First normal form: Ensure that all the attributes are atomic (that is, in the smallest possible components). This means that there is only one possible value for each attribute and not a set of values. This is often expressed as the fact that relations must not contain repeating groups.
- Second normal form: Ensure that all non-key attributes are functionally dependent on (give facts about) all of the key. If this is not the case, split off into a separate relation those attributes that are dependent on only part of the key together with the key.
- Third normal form: Ensure that all non-key attributes are functionally independent of each other. If this is not the case, create new relations which do not show any non-key dependence.

A rather flippant, but more memorable, definition of normalisation can be given as 'the attributes in a relation must depend on the key, the whole key, and nothing but the key'. This is an oversimplification, but it is essentially true and could be kept in mind as the normalisation process is developed.

A more detailed description of normalisation is now given. A key concept is functional dependency, which is often referred to as determinacy. This is defined by Cardenas (1985) as follows:

> Given a relation R, the attribute B is said to be functionally dependent on attribute A if at every instant of time each value of A has no more than one value of B associated with it in the relation R.

Thus, if we know a 'customer-number', we can determine the associated 'customer-name', 'customer-address' and so on, if they are functionally dependent on 'customer-number'. Functional dependency is frequently illustrated by an arrow. The arrow will point from A to B in the functional dependency illustrated in the definition. Thus, the value of A uniquely determines the value of B.

(a) COURSE-DETAIL (Unnormalised)

CRSE	COURSE-NAME	LEVEL	MOD	MOD-NAME	STATUS	UNIT-POINTS
B74	Comp Sci	BSc	B741	Program 1	Basic	8
			B742	Hardware 1		
			B743	Data Proc 1		
			B744	Program 2	Intermed	11
			B745	Hardware 2		
B94	Comp Apps	MSc	B951	Information	Advanced	15
			B952	Microproc		
			B741	Program 1	Basic	8

(b) COURSE-DETAIL

CRSE	COURSE-NAME	LEVEL	MOD	MOD-NAME	STATUS	UNIT-POINTS
B74	Comp Sci	BSc	B741	Program 1	Basic	8
B74	Comp Sci	BSc	B742	Hardware 1	Basic	8
B74	Comp Sci	BSc	B743	Data Proc 1	Basic	8
B74	Comp Sci	BSc	B744	Program 2	Intermed	11
B74	Comp Sci	BSc	B745	Hardware 2	Intermed	11
B94	Comp Apps	MSc	B951	Information	Advanced	15
B94	Comp Apps	MSc	B952	Microproc	Advanced	15
B94	Comp Apps	MSc	B741	Program 1	Basic	8

Fig. 4.24: First normal form

Figure 4.24 (a) is a non-normalised relation COURSE-DETAIL. Before normalising the relation, it is necessary to analyse its meaning. Knowledge of the application area gained from entity modelling will provide this information. It is possible to make assumptions about the inter-relationships between the data, but it is obviously better to base these assumptions on thorough analysis. In the relation COURSE-DETAIL, there are two occurrences of course ('crse'), one numbered B74 called computer science at the BSc. level and the other B94 called computer applications at the MSc. level. Each of these course occurrences has a number of module ('mod') occurrences associated with it. Each 'mod' is given a 'mod-name', 'status' and 'unit-points' (which are allocated according to the status of the 'module').

First Normal Form includes the filling in of details, ensuring all attributes are in their smallest possible components. This is seen in the example in figure 4.24(a) and is a trivial task. You may note that in figure 4.24(a), the order of the tuples in the unnormalised relation is significant. Otherwise the content of the attributes not completed cannot be known. As we have already stated, one of the principles of the relational model is that the order of the tuples is not significant. The tuples seen in figure 4.24(b) could be in any order in this relation. First normal form essentially converts unnormalised data or traditional file structures into fully completed relations or tables.

The key of the relation of figure 4.24(b) is 'crse' and 'mod' together (a composite key) and the key attributes have been underlined. A composite key is necessary because no single attribute will uniquely identify a tuple of this relation. There were in fact a number of possible candidate keys, for example, 'mod-name' and 'course-name', but we chose the primary key as above because they are numeric and unique.

Further work would have been necessary if the following was presented as the unnormalised relation:

CRSE, COURSE-NAME, LEVEL, MODULE-DETAILS

'Module-details' has to be defined as a set of atomic attributes, not as a group item, thus it has to be broken down into its constituents of 'mod-name', 'status' and 'unit-points':

CRSE, COURSE-NAME, LEVEL, MOD-NAME, STATUS, UNIT-POINTS

Second Normal Form is achieved if the relations are in first normal form and all non-key attributes are fully functionally dependent on all the key. The relation COURSE-DETAIL shown in figure 4.24(b) is in first normal form. However, the attributes 'mod-name', 'status' and 'unit-points' are functionally dependent on 'mod'. In other words, they represent facts about 'mod', which is not the whole key which is 'crse' and 'mod'. This is known as partial dependency. We may say that if the value of the module is known, we can determine the value of 'status', 'name', and 'unit-points'. For example, if 'mod' is B743, then 'status' is basic, 'mod-name' is data proc 1, and 'unit points' is 8. They are not dependent on the other part of the key, 'crse'. So as to comply with the requirements of second normal form, two relations will be formed from the relation and this is shown as figure 4.25. But this is only a partial advance to second normal form, there are elements in the first normal form relation not in this model.

(a) COURSE-MODULE

CRSE	MOD	COURSE-NAME	LEVEL
B74	B741	Comp Sci	BSc
B74	B742	Comp Sci	BSc
B74	B743	Comp Sci	BSc
B74	B744	Comp Sci	BSc
B74	B745	Comp Sci	BSc
B94	B951	Comp Apps	MSc
B94	B952	Comp Apps	MSc
B94	B741	Comp Apps	MSc

(b) MODULE

MOD	NAME	STATUS	UNIT-POINTS
B741	Program 1	Basic	8
B742	Hardware 1	Basic	8
B743	Data Proc 1	Basic	8
B744	Program 2	Intermed	11
B745	Hardware 2	Intermed	11
B951	Information	Advanced	15
B952	Microproc	Advanced	15

Fig. 4.25: First step towards second normal form

The relation COURSE-MODULE is still not in second normal form because the attributes 'course-name' and 'level' are functionally dependent on 'crse' only, and not on the whole of the key. A separate COURSE relation has been created in figure 4.26. The COURSE relation has only two tuples (there are only two courses), and all duplicates are removed. Notice that we maintain the relation COURSE-MODULE. This relation is all key, and there is nothing incorrect in this. Attributes may possibly be added later which relate specifically to the *course-module* relationship, for example the teacher or text. The relation is required because information will be lost by not including it, that is, the modules which are included in a particular course and the courses which include specific modules. The relations are now in second normal form.

Third Normal Form (TNF) is necessary because second normal form may cause problems where non-key attributes are functionally dependent on each other (that is, a non-key attribute is dependent on another non-key attribute). In the relation MODULE, the attribute 'unit-points' is functionally dependent on the 'status' (or level) of the course, that is, given 'status', we know the value of 'unit-points'. So 'unit-points' is determined by 'status' which is not a key. We therefore create a new relation STATUS and delete 'unit-points' from the relation MODULE. We check each non-key attribute and find that there are no more such dependencies. The third normal form is given in figure 4.27.

(a) COURSE-MODULE

CRSE	MOD
B74	B741
B74	B742
B74	B743
B74	B744
B74	B745
B94	B951
B94	B952
B94	B741

(b) COURSE

CRSE	COURSE-NAME	LEVEL
B74	Comp Sci	BSc
B94	Comp Apps	MSc

(c) MODULE

MOD	MOD-NAME	STATUS	UNIT-POINTS
B741	Program 1	Basic	8
B742	Hardware 1	Basic	8
B743	Data Proc 1	Basic	8
B744	Program 2	Intermed	11
B745	Hardware 2	Intermed	11
B951	Information	Advanced	15
B952	Microproc	Advanced	15

Fig. 4.26: Second normal form

Sometimes the term transitive dependency is used in this context. The dependency of the attribute 'unit-points' is transitive (via 'status') and not wholly dependent on the key attribute 'module'. This transitive dependency should not exist in third normal form.

The attribute 'status' is the primary key of the STATUS relation. It is included as an attribute in the MODULE relation, but it is not a key. This provides an example of a foreign key, that is, a non-key attribute of one relation which is a primary key of another. This will be useful when processing the relations as 'status' can be used to join the STATUS and MODULE relations to form a larger composite relation if this joint information is required by the user. The user requirements, which might include reports, are likely to contain data coming from the joining of a number of relations.

(a) COURSE-MODULE

CRSE	MOD
B74	B741
B74	B742
B74	B743
B74	B744
B74	B745
B94	B951
B94	B952
B94	B741

(b) COURSE

CRSE	COURSE-NAME	LEVEL
B74	Comp Sci	BSc
B94	Comp Apps	MSc

(c) MODULE

MOD	MOD-NAME	STATUS
B741	Program 1	Basic
B742	Hardware 1	Basic
B743	Data Proc 1	Basic
B744	Program 2	Intermed
B745	Hardware 2	Intermed
B951	Information	Advanced
B952	Microproc	Advanced
B741	Program 1	Basic

(d) STATUS

STATUS	UNIT-POINTS
Basic	8
Intermed	11
Advanced	15

Fig. 4.27: Third normal form

We will now consider the reasons why we normalise the relations in the first place. Unnormalised relations would have been formed by the analysts using information gained from interviews, for example, and are rough first-cut tabular representations of the data structures.

Relations are normalised because unnormalised relations prove difficult to use. This can be illustrated if we try to insert, delete, and update information from the relations not in TNF. Say we have a new 'module' numbered B985 called Artificial Intelligence and which has a 'status' in the intermediate category. Looking at figure 4.24, we cannot add this information in COURSE-DETAIL because there has been no allocation of this 'module' occurrence to any 'crse'. Looking at figure 4.25(b), it could be added to the MODULE relation, if we knew that the 'status' intermediate carried 11 unit-points. This information is not necessary in the MODULE relation seen in figure 4.27(c), the TNF version of this relation. We simply add to the MODULE relation in figure 4.27 (c), the tuple B985, artificial intelligence, intermediate. The TNF model is therefore much more convenient for adding this new information.

If we decided to introduce a new category in the 'status' attribute, called coursework, having a 'unit-points' attached of 10, we cannot add it to the unnormalised relation MODULE (figure 4.25(b)) because we have not decided which 'module' or modules to attach it to. But we can include this information in the TNF model by adding a tuple to the STATUS relation (figure 4.27(d)) which is coursework, 10.

Another problem occurs when updating. Let us say that we decide to change the 'unit-points' allocated to the Basic category of 'status' in the modules from 8 to 6, it becomes a simple matter in the TNF relation. The single occurrence of the tuple with the key Basic, needs to be changed from (Basic 8) to (Basic 6) in figure 4.27(d). With the unnormalised, first normal or second normal form relations, there will be a number of tuples to change. It means searching through every tuple of the relation COURSE-DETAIL (figure 4.24(b)) or MODULE (figure 4.25(b)) looking for 'status' = Basic and updating the associated 'unit-points'. All tuples have to be searched, because in the relational model the order of the tuples is of no significance. This increases the likelihood of inconsistencies and errors in the database. We have ordered them in the text only to make the normalisation process easier to follow.

Another reason concerns the possible inconsistency of the data. This does not occur in the TNF relations above, but in figure 4.24(b), the first and last tuples could have had module names 'Program 1' and 'Basic Programming' respectively as names for the same module (B741). This would cause confusion, but the normalisation process would detect the problem and the analyst will form the relation shown as figure 4.25(b).

Deleting information will also cause problems. If it is decided to drop the B74 course, we may still wish to keep details of the modules which make up the course. Information about modules might be used at another time when designing another course. The information would be lost if we deleted the course B74 from COURSE-DETAIL (figure 4.24(b)). The information about these modules will be retained in the module relation in TNF. The TNF relation COURSE will now consist only of one tuple relating to the 'crse' B74 and the TNF relation COURSE-MODULE will consist of the three tuples relating to the COURSE B94. However, the MODULE relation (figure 4.27 (c)) will remain the same.

We have previously regarded third normal form as the end of the normalisation process and this is usually satisfactory. However, much of the database literature discusses further levels or extensions of normalisation. Kent (1983) and Date (1995) describe these extensions.

Boyce-Codd Normal Form (BCNF) is one such extension. One criticism of third normal form is that by making reference to other normal forms, hidden dependencies may not be revealed. BCNF does not make reference to other normal forms.

In any relation there may be more than one combination of attributes which can be chosen as primary key, in other words, there are candidate keys. BCNF requires that all attribute values are fully dependent on each candidate key and not only the primary key. Put another way, it requires that each determinant (attribute or combination of attributes which determines the value of another attribute) must be a candidate key. As any primary key will be a candidate key, all relations in BCNF will satisfy the rules of third normal form, but relations in TNF may not be in BCNF.

STUDENT- MODULE

STUDENT	MODULE	LECTURER
Bell	B741	Dr Smith
Bell	B742	Dr Jones
Martin	B741	Dr Smith
Martin	B742	Prof. Harris

Fig. 4.28: Relation in TNF but not BCNF

It is best explained by an example. In fact, the third normal form relations in figure 4.27 are also in BCNF, so we will extend the example used so far. Assume that we have an additional relation which is also in

TNF giving details about the students taking modules and the lecturers teaching on those modules. Assume also that each module is taught by several lecturers; each lecturer teaches one module; each student takes several modules; and each student has only one lecturer for a given module. This complex set of rules could produce the relation shown as figure 4.28.

LECTURER

LECTURER	MODULE
Dr Smith	B741
Dr Jones	B742
Prof. Harris	B742

STUDENT-LECTURER

STUDENT	LECTURER
Bell	Dr Smith
Bell	Dr Jones
Martin	Dr Smith
Martin	Prof. Harris

Fig. 4.29: BCNF

Although this relation is in TNF because the 'lecturer' is dependent on all the key (both 'student' and 'module' determine the lecturer), it is not in BCNF because the attribute 'lecturer' is a determinant but is not a candidate key. There will be some update anomalies. For example, if we wish to delete the information that Martin is studying B742, it cannot be done without deleting the information that Prof. Harris teaches the module B742. As the attribute 'lecturer' is a determinant but not a candidate key, it is necessary to create a new table containing 'lecturer' and its directly dependent attribute 'module'. This results in two relations as shown in figure 4.29. These are in BCNF. Now, deleting the information that Prof. Harris teaches Martin (the second relation will now have three tuples) will not lose the information that Prof. Harris can teach on module B742.

Fourth normal form can be illustrated by looking at a relation which is in first normal form and which contains information about modules, lecturers and text books. Each tuple has module name and a repeating group of text book names (there could be a number of texts recommended for each module). Any module can be taught by a number of lecturers, but each will recommend the same set of texts.

MODULE-LECTURER

MODULE	LECTURER	TEXT
B741	Dr Smith	Database Fundamentals
B741	Dr Smith	Further Databases
B741	Dr Jones	Database Fundamentals
B741	Dr Jones	Further Databases
B742	Dr Smith	Database Fundamentals
B742	Dr Smith	Systems Analysis
B742	Dr Smith	Information Systems

Fig. 4.30: BCNF but not fourth normal form

The relation seen as figure 4.30 is in BCNF (and therefore TNF) and yet it contains considerable redundancy. If we wish to add the information that Prof. Harris can teach B742, three tuples need to be added to the relation. The problem comes about because all three attributes form the composite key: there are no functional determinants apart from this combination of all three attributes.

MODULE-LECTURER

MODULE	LECTURER
B741	Dr Smith
B741	Dr Smith
B741	Dr Jones
B741	Dr Jones
B742	Dr Smith
B742	Dr Smith
B742	Dr Smith

MODULE-TEXT

MODULE	TEXT
B741	Database Fundamentals
B741	Further Databases
B741	Database Fundamentals
B741	Further Databases
B742	Database Fundamentals
B742	Systems Analysis
B742	Information Systems

Fig. 4.31: Fourth normal form

The problem would be eased by forming from this relation the two all-key relations shown as figure 4.31. There is no loss of information, and there is not the evident redundancy found in figure 4.30.

The transition to fourth normal form has been made because of multivalued dependencies that may occur (Fagin, 1977). Although a module does not have one and only one lecturer, each module does have a pre-defined set of lecturers. Similarly, each module also has a pre-defined set of texts.

Although these examples are valid, in that they do show relations which contain redundancy and yet are in TNF and BCNF respectively, the examples are somewhat contrived. The reader will have seen that in both examples it was necessary to make a number of special assumptions. The implication is that such problems will not be found frequently by analysts when carrying out data analysis and therefore that TNF will normally be a reasonable stopping point for normalisation. However, many academics take normalisation even further. In order to provide an example of fifth normal form, Date has to bring in what he calls a 'bizarre constraint'. He goes on to suggest that 'such relations are pathological cases and likely to be rare in practice' (Date, 1986, page 390). We will stop at fourth normal form.

4.7 DATA FLOW DIAGRAMMING

The data flow diagram (DFD) is fundamental to structured systems methodologies and was developed as an integrated part of those methodologies. However the DFD has been adopted and adapted by a number of other methodologies, not all of the structured systems type, including Multiview and ISAC. In these methodologies the DFD or similar is not the major technique of the methodology but is used in conjunction with other techniques in the analysis of processes. Like entity modelling and normalisation, DFDs are an important technique in a variety of systems development methodologies.

The DFD provides the key means of achieving one of the most important requirements of structured systems, that is the notion of structure. The DFD enables a system to be partitioned (or structured) into independent units of a desirable size so that they, and thereby the system, can be more easily understood. In addition, information is graphical and concise. The graphical aspect means it can be used both as a static piece of documentation and as a communication tool, enabling communication at all levels: between analyst and user, analyst and designer and analyst and analyst. The fact that the DFD has proved amenable to users means that it is easier to validate for correctness and the probability of a successful information system resulting is increased. The graphical nature of the DFD also means a more concise document, as it is argued that a picture can more quickly convey meaning than more traditional methods, such as textual narrative. The DFD also provides the ability to abstract to the level of detail required. Thus it is possible to examine a system in overview and at a detailed level, whilst maintaining the links and interfaces between the different levels.

The DFD provides the analyst with the ability to specify a system at the logical level. This means that it describes what a system will do, rather than how it will be done. Considerations of a physical and implementation nature are not usually depicted using data flow diagrams and it is possible for the logical DFD to be mapped to a variety of different physical implementations. This provides further contrast, therefore, with the flowcharts used in the traditional systems development life cycle approach described in section 2.2 which depict the physical implementation. The benefit of this is that it separates the tasks of analysis (what is required) from design (how it is to be achieved). This separation means that the users can specify their requirements without any restrictions being imposed of a design nature, for example, the technology or the type of access method. There exists a logical and physical independence, the hardware can be changed or upgraded without changing the functions of the system. Alternatively, if as often happens a functional change is required, the relevant part of the logical specification is changed and a new mapping to the physical system is designed. The change is thus effected at the logical level, which is the correct place, and the implications of the change are agreed, and only then the necessary design changes made. This improves and speeds up the maintenance process which, as we have seen, is a major time and resource consuming activity.

The form of DFDs differs between the various proponents of structured systems analysis. The differences are relatively small and the basic concepts are the same. For example, the symbol used to represent a process differs. Gane and Sarson use a rectangle with rounded corners (a 'soft box') whereas many other authors use a circle. This means that superficially the DFDs look different but in practice the differences are relatively minor.

A logical DFD represents logical information, not the physical aspects. A data flow specifies what flows, for example, customer credit details. How it flows, for example, by carrier pigeon or via twisted copper wires, is immaterial and not represented in a logical DFD. A DFD is a graphical representation and is composed of four elements:

- *The data flow*: Data flow is represented by an arrow and depicts the fact that some data is flowing or moving from one process to another. A number of analogies are commonly used to illustrate this. Gane and Sarson suggest that we think of the arrow as a pipeline down which 'parcels' of data are sent, Page-Jones (1980) states that data flow is like a conveyor belt in a factory which takes data from one 'worker' to another. Each 'worker' then performs some process on that data which may result in another data flow on the conveyor belt. These processes are the second element of the DFD.

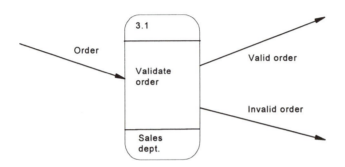

Fig. 4.32: Process box and data flows

- *The processes*: The processes or tasks performed on the data flows are represented in this example by a soft box (see figure 4.32). The process transforms the data flow by either changing the structure of the data (for example, by sorting it) or by generating new information from the data (for example, by merging the data with data obtained from another data source). A process might be 'validate order', which transforms the order data flow by adding new information to the order, that is, whether it is valid or not. It is likely that invalid orders flow out from the validation process in a different direction to valid orders. In this example the conventions used for the process symbol are as follows. The top compartment contains a reference number for the process, the middle compartment contains the description of the process, and the lower compartment indicates where the process occurs. Strictly speaking, this is not at the logical level and so would not always be used. A process must have at least one data flow coming into it and at least one leaving it. There is no concept of a process without data flows, a process cannot exist independently.

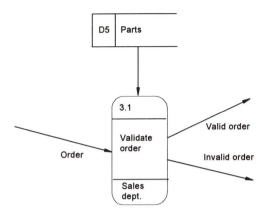

Fig. 4.33: Data store

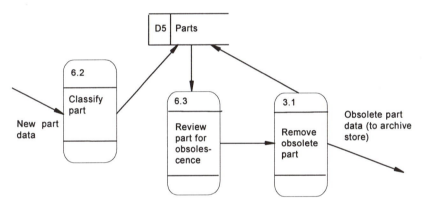

Fig. 4.34: Maintenance of parts data store

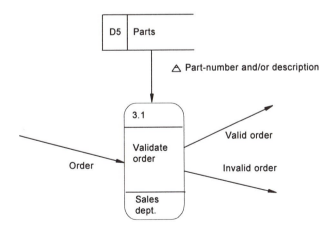

Fig. 4.35: Access to data store

- *The data store*: If a process cannot terminate a data flow because it must output something, then where do the data flows stop? There are two places. The first is the data store, which can be envisaged as a file, although it is not necessarily a computer file or even a manual record in a filing cabinet. It can be a very temporary repository of data, for example, a shopping list or a transaction record. A data store symbol is a pair of parallel lines with one end closed, and a compartment for a reference code and a compartment for the name of the data store. For example (see figure 4.33), the process of validating the order may need to make reference to the parts data store to see if the parts specified on the order are valid parts with the correct current price associated with it. The data flow in this example has the arrow pointing towards the process which indicates that the data store is only referenced by the process and not updated or changed in any way. In this example we would expect to find another process somewhere on the DFD which maintained the parts data store. For example, in figure 4.34, the new part data from process 6.2 is used to update the parts data store (the arrow points to the data store). The manner in which the access to the data store is made is usually regarded as irrelevant. However, it may be information which a designer needs, and therefore in cases where it is not obvious, this information may be added to the DFD. In the example (figure 4.33), it may be assumed that access to the parts file is via the part number. However, if a customer makes an order without specifying the part number, access via the part description may be required. This should then be specified (see figure 4.35).

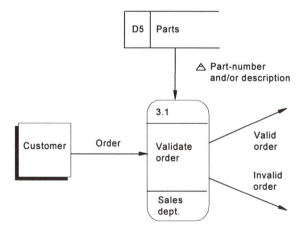

Fig. 4.36: External entity customer (source of order)

- *The source or sink* (external entity): The second way of terminating a data flow in a system is by directing the flow to a sink. The sink may, for example, be a customer to whom we send a delivery note. The customer is a sink in the sense that the data flow does not necessarily continue. The customer is a sink down which the delivery note may fall forever. The Department of Trade may be a sink to which a company may legally be required to provide information but never receive any in return. Sinks are usually entities that are external to the organisation in question, although they need not be, another department may be a sink. It depends on where the boundaries of the system under consideration are drawn. If a DFD for the sales department is being constructed, any data flows to the production department would be represented as a sink. However, if the whole organisation were being depicted, then the same data flows would go to a process within the production department. The original source of a data flow is the opposite to a sink, although it may be the same entity. For example, a customer is the source of an order and a sink for a despatch note. Sinks and sources, are represented by the same symbol which is a shadowed rectangle (see for example figure 4.36). Sources and sinks are often termed 'external entities'. Figure 4.38 is an example of a data flow diagram that illustrates the combination of the four elements discussed above.

Mason and Willcocks (1994) suggest the following rules for drawing data flow diagrams assuming the analyst has described the whole logical system in narrative form:

- Read the whole process a few times to get a clear picture of the system being described
- Identify the sources and sinks, identifying them by circling a key word for each of them
- Identify data stores (perhaps through the use of verbs such as 'store' or 'check' and underline them)
- Draw a source entity box for the external entity which seems to start the process off and name it (entities can only link with processes)
- To its right, draw the first process box
- From that draw an arrow representing the primary data flow and name it
- Name the process
- Link it with any data stores and name these as well as the data flow entering or exiting it to and from the process (data cannot go from store to store, there must be an intervening process)

- Link the next process and so on until all the external entities are drawn.

We would add that all these steps will be repeated as we iterate to improve our model following discussions of it with the users.

One of the most important features of the DFD is the ability to construct a variety of levels of DFD according to the level of abstraction required. This means that an overview diagram (frequently referred to as a level 0 diagram) can be consulted in order to obtain a high-level (overview) understanding of the system. When a particular area of interest has been identified, then this area can be examined at a more detailed level.

The different levels of diagram must be consistent with each other in that the data flows present on the higher levels should exist on the lower levels as well. In essence it is the processes which are expanded at a greater level of detail as we move down the levels of diagram. This 'levelling' process gives the DFD its top-down characteristic.

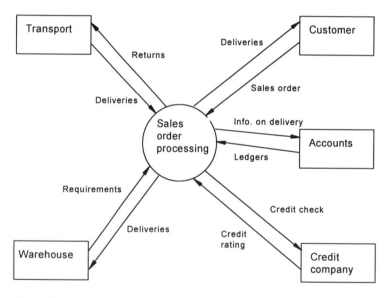

Fig. 4.37: Level 0 diagram

At the level 0, the process named in a circle is like a 'black box', it has inputs and outputs but we do not specify how the process converts the inputs to the outputs, only that it does. At lower levels (level 1 and below) these are analysed. The 0 level diagram is frequently referred to as the context diagram because it shows the context for the system. This is not a data flow diagram as such, but the diagram shows the system boundaries, external entities and inputs and outputs from the system (see figure 4.37). In effect, it has one very large process.

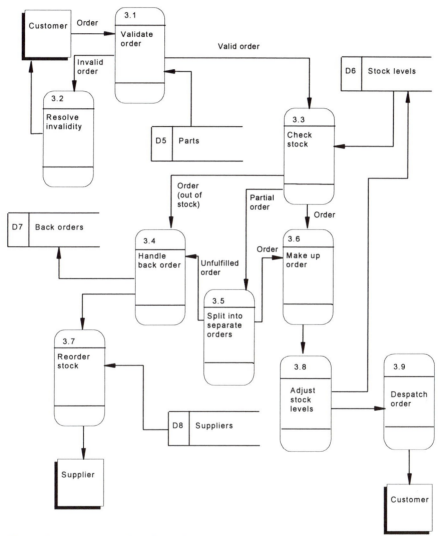

Fig. 4.38: Example data flow diagram

Figure 4.38 shows a level 1 data flow diagram, but we may well require more detail of any of the processes on the diagram. Figure 4.39 is the next level down (or an explosion) of the Validate Order task. The overall process is expanded into five tasks with various data flows between them.

However, all the data flows in and out of Validate Order (reference 3.1) in figure 4.38 can be found on figure 4.39. The new data flows are either flows that only exist within the Validate Order process, that is they are internal to it and are now shown because we have split this down into separate components (for example, Amended Order), or because they are concerned with errors and exceptions.

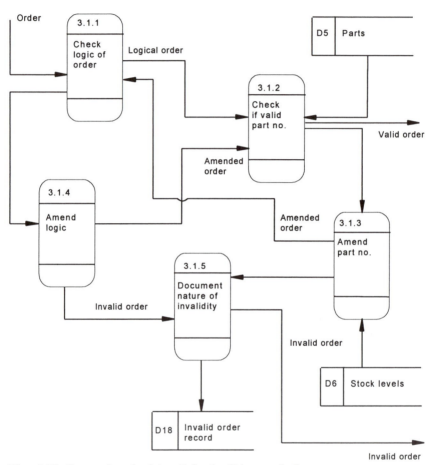

Fig. 4.39: Lower level of detail for 'validate order'

The details of errors and exceptions are not shown on high-level diagrams as it would confuse the picture with detail that is not required at an overview level. What is required at an overview level is 'normal' processing and data flows. To include errors and exceptions might double the size of the diagram and remove its overview characteristics. For example, figure 4.39 shows new processing concerned with amending an order and a new data store which did not appear on the higher level diagram. The problem is that it is sometimes difficult to decide what constitutes an error or an exception, and what is normal. Some common guidelines suggest that if an occurrence of a process or data flow is relatively rare, then it should be regarded as an exception. However, if it is financially significant, it should be taken as part of normal processing. Overall, it depends on the audience or use to be made of the DFD as to exactly what is included. At the lowest level, all the detail, including errors and exceptions, should be shown.

The question arises as to how many levels a DFD should be decomposed. The answer is that a DFD should be decomposed to the level that is meaningful for the purpose that the DFD is required. There comes a level, however, when each process is elementary and cannot be decomposed any further. No further internal data flows can be identified. At this point each elementary process is described using a form of process logic. The techniques of representing process logic are described in the next four sections.

Data flow diagrams are generally logical and not physical, but they can be used to represent a physical implementation. For example, the process 'Document nature of invalidity' (figure 4.29) might be 'type nature of invalidity on form B36 in sales office' with the B36 forms being output as part of the data flow and 'click on reason for invalidity using screen 5'. The physical DFD may then describe the present system or a proposed system.

Although it is a fundamental technique of some information systems development methodologies, DFDs have limitations even for addressing information about processes, such as how long data takes to get from one process to another and the detail about the data that passes between processes in terms of peaks and troughs. Also little detail is provided about decision aspects, that is, why data flows in one way and not another. However, other techniques can be used to provide this information although none of the process techniques are as well used.

Techniques and tools closely associated with DFDs include data dictionaries as well as process logic descriptions. The data dictionary is the central place where all the details of data flows, data stores and processes are stored in an ordered and logical fashion. The dictionary may be manual or computerised but the important thing is that it must be the centralised resource of the structured analysis project. It is not the same thing as the data dictionary of a database which is concerned with physical aspects of the data, although the two may very well be integrated.

The way in which process logic is described is by the use of decision trees, decision tables, structured English and action diagrams (sections 4.8, 4.9, 4.10 and 4.11). They are not complementary. A particular process is not described using all four techniques but by using whichever is the most appropriate, given the characteristics of the process concerned. Thus each technique has particular strengths and weaknesses but they all provide simple, clear and unambiguous ways of describing the logic of what happens in the elementary processes identified in the DFD.

4.8 DECISION TREES

Decision trees and decision tables are tools which aim to facilitate the documentation of process logic, particularly where there are many decision alternatives. A decision tree illustrates the actions to be taken at each decision point. Each condition will determine the particular branch to be followed. At the end of each branch there will either be the action to be taken or further decision points. Any number of decision points can be represented, though the greater the complexity, then the more difficult the set of rules will be to follow.

When constructing a decision tree, the problem must be stated in terms of conditions (possible alternative situations) and actions (things to do). It is often convenient to follow a stepwise refinement process when constructing the tree, breaking up the largest condition to basic conditions, until the complete tree is formulated.

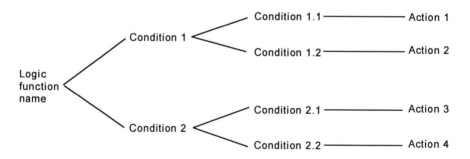

Fig. 4.40: General format of a decision tree

The general format is shown in figure 4.40. An example of a decision tree is given in figure 4.41. At the first decision point (or node), the customer is classified into one of two types, private or trade. If the customer is trade, then a second decision point is reached. Has the customer been trading with us for less than five years or five years or more? If the customer has been trading for five years or more, then the customer can obtain up to £5,000 credit, otherwise up to £1,000 credit can be given. If the customer was deemed private at the first decision point, then the action is to offer no credit at all.

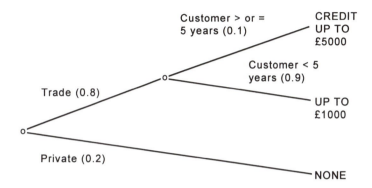

Fig. 4.41: Example of a decision tree

Decision trees are constructed by first identifying the conditions, actions, and unless/however/but structures of the situation being analysed gained from a written statement or during interviews with users. Each sentence may form a 'mini' decision tree and these could be joined together to form the version which will be verified by the users. Sometimes it is not possible to complete the decision tree because full information has not been given in the statement. For example, one branch of the decision tree may be identified, but no indication is given on the action to take on this branch. In such cases the analyst has to carry out a further investigation by interviewing staff or by using another method of systems investigation. The technique therefore helps validate the information as well as proving a useful documentation aid.

Decision trees prove to be a good method of showing the basics of a decision, that is, the possible actions that might be taken at a particular decision point and the set of values that leads to each of these actions. It is easy for the user to verify whether the analyst has understood the procedures.

Sometimes it is possible to associate probability scores for each branch once the decision tree is drawn. With this information it will be possible to compute expected values of the various outcomes and hence to evaluate alternative strategies. For example, in figure 4.41, research may have indicated that trade customers outnumber private customers by 4:1 and that 90 per cent of trade customers are less than five years standing. By extrapolating these figures with the total number of customers, it may be possible to evaluate the amount of total credit to be made available to all customers.

Where the number of decision nodes becomes greater than, say, ten, drawing the decision tree begins to be overly complex and a decision table is usually chosen as it can more readily represent processes with very complex decision structures. People are usually very familiar with the meaning of decision trees and these are more graphical, therefore decision trees are often preferred for communicating less complex logic.

4.9 DECISION TABLES

Decision tables are less graphical, when compared to decision trees, but are concise and have an in-built verification mechanism so that it is possible to check that all the conditions have been catered for. Again, conditions and actions are analysed from the procedural aspects of the problem situation expressed in narrative form.

When constructing a decision table, the narrative is analysed to identify the various conditions and the various actions. The various actions to be executed are then listed in the bottom left hand part of the table known as the action stub. In the top left hand part, the conditions that can arise are listed in the condition stub. Each condition is expressed as a question to which the answer will be 'yes' or 'no'. There can be no ambiguity. All the possible combinations of yes and no responses can be recorded in the upper right hand part of the table, known as the condition entry. Each possible combination of responses is known as a rule. In the corresponding parts of the lower right-hand quadrant, an X is placed for each action to be taken, depending on the rule associated with that column.

Figure 4.42 shows the decisions that have to be made by drivers in the UK at traffic lights. The condition stub (upper left section) has all the possible conditions which are 'red', 'amber' and 'green'. Condition entries (upper right quadrant) are either Y for yes (this condition is satisfied) or N for no (this condition is not satisfied). Having three conditions in the condition stub, there will be 2 to the power of 3 ($2 \times 2 \times 2 = 8$) columns in the condition entry (and action entry). The easiest way of proceeding is to have the first row in the condition entry as YYYYNNNN, the second row as YYNNYYNN and the final row as YNYNYNYN. If there were four conditions, we would start with eight Ys and eight Ns and so on, giving a total of 2^4 ($2 \times 2 \times 2 \times 2 = 16$) columns.

All the possible actions, are listed in a concise narrative form in the Action Stub (bottom left). An X placed on a row/column coincidence in the Action Entry means that the action in the condition stub should be taken. A blank will mean that the action should not be taken. Thus, if a driver is faced with Red (Y), Amber (Y) and Green (Y), the first column indicates that the driver should stop and call the police (a particular combination of conditions may lead to a number of actions to be taken). All combinations, even invalid ones, should be considered. The next column Red (Y), Amber (Y) and Green (N) informs the driver to stop. Only the Red (N), Amber (N) and Green (Y) combination permits the driver to go with caution.

CONDITION STUB   CONDITION ENTRY

Red	Y Y Y Y N N N N
Amber	Y Y N N Y Y N N
Green	Y N Y N Y N Y N
Go with caution	X
Stop	X X X X X X X
Call police	X X X X

ACTION STUB   ACTION ENTRY

Fig. 4.42: Decision table for UK traffic lights

Once the table is completed, rules which result in the same actions can be joined together and represented by dashes, that is, 'it does not matter'. The result of this is a consolidated decision table. Figure 4.43 illustrates an example decision table before and after the consolidation process. In the decision table represented in figure 4.43 (b), rules 3 and 4 of the decision table seen in figure 4.43 (a) have been merged into a consolidated rule 2, expressing a 'doesn't matter' condition. This is because the same processing needs to be executed whether or not condition 2 is 'yes' or 'no'.

In systems analysis, there are likely to be requirements to specify actions where there are a large number of conditions. A set of decision tables is appropriate here. The first will have actions such as 'go to decision table 2' or 'go to decision table 3'. Each of these may themselves be reduced to a further level of decision tables. The technique therefore lends itself to functional decomposition.

(a)

	1	2	3	4	5	6	7	8
Invoice > £300	Y	Y	Y	Y	N	N	N	N
Account overdue > 3 months	Y	Y	N	N	Y	Y	N	N
New customer	Y	N	Y	N	Y	N	Y	N
Put in hands of solicitor	X		X	X	X	X		X
Write first reminder letter		X	X	X		X		X
Write second reminder letter					X			
Cancel credit limit			X	X		X	X	

(b)

	1	2	3	4	5	6	7
Invoice > £300	Y	Y	Y	N	N	N	N
Account overdue > 3 months	Y	Y	N	Y	Y	N	N
New customer	Y	N	-	Y	N	Y	N
Inform solicitor	X		X	X	X		X
Write first reminder letter		X	X		X		X
Write second reminder letter				X			
Cancel credit limit			X		X	X	

Fig. 4.43: Consolidating decision tables

Sometimes the values of conditions are not restricted to 'yes' or 'no', as defined in the limited entry tables described. There can be more than two possible entries, and extended entry tables are appropriate in this case. For example, the credit allowable to a customer could vary according to whether the customer had been dealing with the firm for 'up to 5 years', 'over 5 and up to 10 years', 'over 10 and up to 15 years', and so on. The rule for obtaining the right number of combinations will need to be modified. If condition 1 has two possibilities and condition 2 five possibilities, then the number of columns will be 2 x 5, that is, ten columns.

Whereas decision trees are particularly appropriate where the number of actions is small (although it is possible to have large decision trees), decision tables are more appropriate where there is a large number of actions as they can be decomposed into sets conveniently, in other words decision tables can better handle complexity. However, decision trees are easy to construct and give an easily assimilated graphical account of the decision structure. Decision tables have good validation procedures and, further, can be used to generate computer programs which carry out the actions according to the rules. Here the processing is specified by the analyst in terms of decision tables. These are transferred to computer readable format, and the programs generated automatically. Decision tables do, however, suffer from the disadvantage that no indication is given regarding the sequence that the actions are to be followed (it cannot be assumed that they are to be followed in the sequence given in the decision table itself).

4.10 STRUCTURED ENGLISH

Structured English is very like a 'readable' computer program, it aims to produce unambiguous logic which is easy to understand and not open to misinterpretation. It is not a natural language like English, which is ambiguous and therefore unsuitable. Nor is it a programming language, though it can be readily converted to a computer program. It is a strict and logical form of English and the constructs reflect structured programming. Like a conventional programming language, the sequence of the commands expressed in structured English is important. It reflects the sequence in which the instructions should be followed. Although decision trees or decision tables are more suitable tools to document aspects where the system has many decision points, structured English proves to be a very useful technique to express logic in a system.

```
CREDIT RATING POLICY
IF the customer is a trade customer
        and IF the customer is customer for 5 or more years
                THEN credit is accepted up to £5000
        ELSE credit is accepted up to £1000
ELSE the customer is a private customer
        SO no credit given
ENDIF
```

Fig. 4.44: Structured English example

Structured English is a precise way of specifying a process, and is readily understandable by a trained systems designer as well as being readily converted to a computer program. An example is given in figure 4.44. Structured English uses only a limited subset of English and this vocabulary is exact. This ensures less ambiguity in the use of 'English' by the analyst. Further, by the use of text indentation, the structure of the process can be shown more clearly. As with all these techniques, however, although the logic can be formally expressed, there is no guarantee that the expression in syntactically correct structured English is semantically correct. That will depend on the systems investigation in the first place.

Structured English has a construct as follows:

> IF condition 1 (is true)
> > THEN action 1 (is to be carried out)
> ELSE (not condition 1)
> > SO action 2 (to be carried out)
> ENDIF

Functional decomposition can be supported in structured English by a construct using the IF ...THEN construct:

> IF condition
> > THEN do a named set of operations
> > (specified at a lower level)

Conditions can include equal, not equal, greater than, less than, and so on. The words in capitals are keywords in structured English and have an unambiguous meaning in this context.

The logic of a structured English construct is expressed as a combination of sequence, selection, case and repetition structures. Any logical specification can be written using these four basic structures (figure 4.45) :

1. *Sequencing* shows the order of processing of a group of instructions, but has no repetition or branching built into it.
2. *Selection* facilitates the choice of those conditions where a particular action or set of actions (or another decision and selection) are to be carried out.
3. *Cases* represent a special type of decision structure (a special kind of selection), where there are several possibilities, but they never occur in combination. In other words, they are mutually exclusive.
4. *Repetition* or loop instructions, facilitate the same action or set of actions to be carried out a number of times, depending on a conditional statement.

The actual keywords and their number will vary according to the particular conventions used, indeed, structured English is not a 'standard', but the basic structures of sequence, selection and repetition will be common to all.

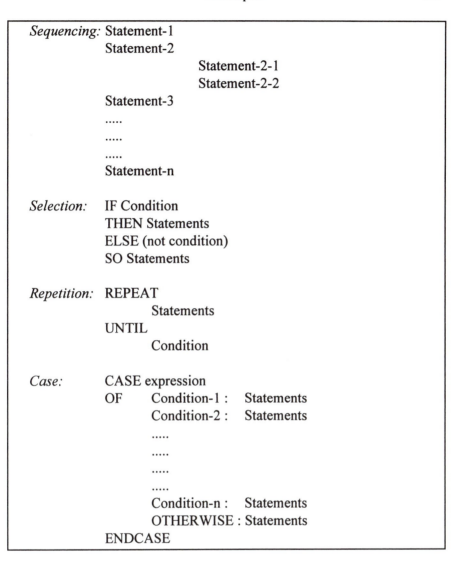

Sequencing: Statement-1
Statement-2
Statement-2-1
Statement-2-2
Statement-3
.....
.....
.....
Statement-n

Selection: IF Condition
THEN Statements
ELSE (not condition)
SO Statements

Repetition: REPEAT
Statements
UNTIL
Condition

Case: CASE expression
OF   Condition-1 :   Statements
Condition-2 :   Statements
.....
.....
.....
.....
Condition-n :   Statements
OTHERWISE : Statements
ENDCASE

Fig. 4.45: The basic structures of structured English

The layout of structured English is as follows:
- The use of capital letters indicates a structured English reserved word such as IF, THEN, ELSE and GET or the operators ADD, DIVIDE and so on, which have particular meanings in the context of structured English.
- Any data elements which are included in a data dictionary are normally underlined and these will include those items associated

with the particular application, such as <u>credit</u> or <u>customer</u> in the credit rating example (figure 4.44).

- Indentation is used to indicate blocks of sequential instructions to be created together, and hierarchical structures can be built by indenting these blocks

- Blocks of instructions can be named and this name quoted in capital letters to refer to the block of instructions elsewhere in the code and this will be particularly necessary where there are a number of places in the logic where this set of instructions needs to be performed.

When creating structured English statements, it is best to break down complex statements into a number of simple ones. Named blocks of these simple statements can be thought of as a way of effecting functional decomposition because the block can be performed in a REPEAT statement a number of times.

DeMarco (1979) discusses structured English at length and he argues that though there are many advantages of its use, in particular its ability to describe many aspects of analysis, its conciseness, precision and readability, and the speed with which it can be written, there are disadvantages as well. DeMarco highlights the time it takes to build up skills in its use and the fact that it is alien to many users (despite being English-like). Indeed, the terminology might be misleading to users because the structured English meanings are not exactly the same as their natural language counterparts.

There are alternative languages to structured English such as, 'pseudo code' and 'tight English'. These vary on their nearness to the computer or readability to users. For example, pseudo code has a DO-WHILE and END-DO loop structure which is similar to constructs of some conventional computer programming languages, and is obviously more programming-orientated than structured English. Tight English code seems nearer natural language than computer programming language, though it can also be interpreted by computer programs. When there are a number of decision points, in tight English it is usual to use to a decision table or decision tree.

4.11 ACTION DIAGRAMS

Action diagrams are also ways of representing the details of process logic, the business rules, and are not dissimilar to structured English in that a limited subset of the natural language is used to specify a sequence of actions. They are designed to represent both the detail and the overview levels. Action diagrams are used in a number of methodologies, most notably, Information Engineering (section 6.4). CASE tools can be used to check their internal consistency and also check the contents against the data dictionary as well as to make for easy construction. Tools are available which will generate computer code from action diagrams.

People have sometimes commented when seeing an action diagram for the first time that 'it doesn't look much like my idea of a diagram', and it has to be said that, compared to a data flow diagram or an entity model, it is rather lacking in diagrammatic features, such as boxes and circles. However, the basic construct of an action diagram is a bracket, which is diagrammatic. A bracket surrounds a group of actions. Actions are broadly defined, and can be parts of program code, a subroutine, a program itself, an operation, a procedure or, at the highest level, a function.

The bracket indicates control. The actions within the bracket are performed in linear sequence like structured English or a computer program, and the brackets may be nested to indicate hierarchical structure.

```
┌─── Admit a student
│
│        Review application
│        Interview student
│        Make decision
│        Communicate decision
└───────────
```

Fig. 4.46: Action diagram

Figure 4.46 represents the processing that is carried out in the function 'admit a student'.

```
┌─ IF AGE > 60
│     MOVE 'AGE ALLOWANCE' TO STATUS
│     MOVE 'AGEIND' TO CODE
├─ ELSE
│     MOVE 'STANDARD' TO STATUS
│     MOVE 'STANDIND' TO CODE
```

Fig. 4.47: Selection and case in action diagrams

Action diagrams support the structured programming constructs of condition, case, repetition, repeat....until. Figure 4.47 shows selection in an action diagram having an IF....ELSE construct. If there is more than one dash in the bracket this indicates that these parts are mutually exclusive, that is, it represents the CASE structure of structured English.

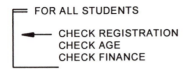

Fig. 4.48: Repetition in action diagrams

The execution of a loop is indicated by a double dash at the top of the bracket and a thicker than normal line for the bracket. Figure 4.48 illustrates some repeated actions, the arrow indicates a next iteration construct, that is, skip the remaining actions and go to the next student. The arrow can also be used to indicate an escape from this bracket completely, if it points not to the bracket itself but goes through the bracket to point to an earlier bracket in a nested set.

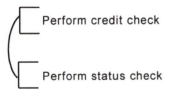

Fig. 4.49: Process action block

The do....while and do....until construct is shown in figure 4.49. The block of instructions to be performed is known as the process action block.

```
┌   Perform credit check
│   └──
│
│   ┌   Perform status check
└   └──
```

Fig. 4.50: Action diagram concurrency

Although brackets indicate sequences of actions, the technique can be used to show the notion of concurrency. Figure 4.50 illustrates concurrency, the link indicates that the actions in the two brackets can be performed in parallel.

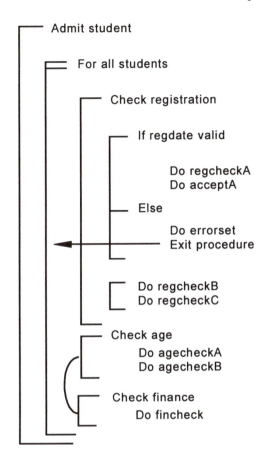

Fig. 4.51: Action diagram with a number of constructs

Figure 4.51 extends figure 4.48 and shows a number of the constructs of action diagramming in use.

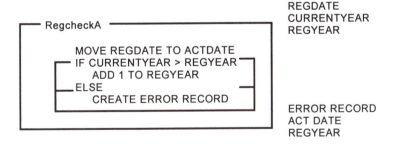

Fig. 4.52: Action diagram including inputs and outputs

In order to increase their applicability, action diagrams have been extended in two ways. First, so that the data required for each function or process can be identified, inputs and outputs are added to the diagrams. This requires the brackets to be extended to form a box with the required inputs for the action added to the top right-hand side and the outputs that the action produces added at the bottom right-hand side. Figure 4.52 provides a simple example.

The second extension to action diagrams is to accommodate the fact that actions often need to relate to database operations. The database operations of create, read, update and delete are added to the action diagram conventions and the name of the record that the operation refers to is enclosed in a box. These database operations relate to a single occurrence of the record, more complex operations relate to many occurrences and/or more than one record type, and these are represented by a double box. Searching or sorting, or, in relational environments, select and join, are examples of complex or compound database operations.

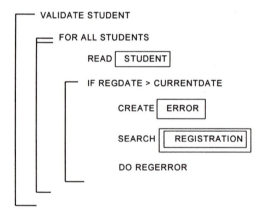

Fig. 4.53: Action diagram incorporating database operations

Figure 4.53 shows an example of an action diagram using the database conventions. Student records are read and processed a single occurrence at a time, 'read student' is only a single box. If we find an error, we wish to find the corresponding registration record and search the registration file. Thus 'registration' is in a double box.

Many argue that action diagrams are easy to construct and utilise, both by analysts and users. As well as being able to represent and communicate logic in the traditional systems development process, they are also advocated as being useful to end-users when developing their own systems using fourth generation languages and also by information centre staff when working with users. A particular benefit of using action diagrams is that it is possible to use the same technique for representing high level functions right through to low level process logic.

4.12 ENTITY LIFE CYCLE

This technique also varies slightly from methodology to methodology and is called by different names, but in essence what is being achieved is substantially the same. The entity life cycle is used at a variety of stages in a number of methodologies and is one of the few attempts to address changes that happen over time (most of the other techniques represent static views of a system).

The entity life cycle is not, despite its name, a technique of data analysis, but more a technique of process analysis. It does show the changes of state that an entity goes through, but the things that cause the state of the entity to change are processes and events, and it is these that are being analysed. The objective of entity life cycle analysis is to identify the various possible states that an entity can legitimately be in. The sub-objectives, or by-products, of entity life cycle analysis are to identify the processes in which the entity type is involved and to discover any processes that have not been identified elsewhere. It may also identify valid (and invalid) process sequences, not identified previously, and it can form the outline design for transaction processing systems. Thus, as well as being a useful analysis technique in its own right, it is a useful exercise to perform as a validation of other process analysis techniques.

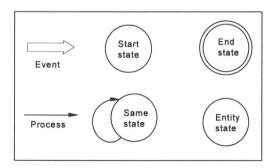

Fig. 4.54: Entity life cycle symbols

The documentation of the states of an entity in a diagram is one of the most powerful features of entity life cycle analysis. The diagram provides a pictorial way of communication that enables users to validate the accuracy or otherwise of the analysis easily.

The documentation conventions differ from methodology to methodology but the concepts are fairly consistent. In the following example the conventions used in D2S2, a precursor of Information Engineering, are used. Figure 4.54 shows the different symbols used and figure 4.55 shows a simplified example of an entity life cycle for the entity 'student' in the context of a university environment.

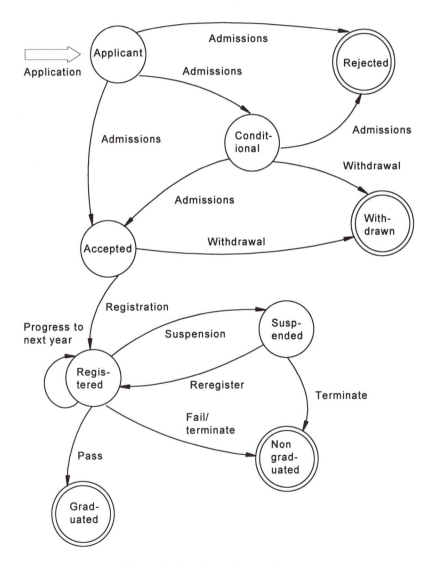

Fig. 4.55: Entity life cycle for the entity student

There is always a starting point, an event, which sets the entity into its initial state. There are three types of event. Events can be caused by external factors, such as a prospective student applies for admission into a degree programme (the starting point in the example); internal, such as a decision is made to accept the student for admission; or time-related, for example a candidate has not replied to our offer one month following our letter and is deemed to have withdrawn. There is also always (or should always be) at least one end or terminating point to finish the life cycle. In between, there may be many different states of the entity.

In the example, the initial state of the entity is as 'applicant'. This is triggered by an event, which is the receipt of an application for admission for one of the courses at the university. The entity changes state as a result of the admissions function which either causes the applicant to be rejected, conditionally accepted (which means that the applicant is accepted provided certain examination grades will be achieved) or unconditionally accepted. The resultant entity states are rejected, conditional or accepted. At any time, conditional or accepted applicants may withdraw their applications.

The accepted applicants start their courses and become registered. They may or may not graduate. Registered students may suspend their registration for a wide variety of reasons at any time, for example, ill-health, and may either return as registered or terminate as non-graduated. It should be noted that a function can be depicted that does not change the state of the entity. In this case the arrow points back to the same entity state. 'Progress to next year' is an example of this, it is a function that does not change the state of the entity, the student is still registered. An alternative perspective, that of three entity states named first year students, second year students and final year students, would have led to a change of state. Again, we see that systems analysis is not a science where all analysts following the same approach will result in exactly the same diagram.

In this example there are a number of terminated states, some conventions suggest that there should be only one. In this case we would add an extra state, called, for example, archived, and draw arrows from all our terminated states to this archived state.

It can be seen that the technique is useful in identifying the states of an entity, the processes that cause the states of an entity to change, and any sequences that are implied. Like many other of these process techniques, the key aspects of sequence (of business rules as the applicant proceeds through the university system to graduation); selection (between alternatives, such as, accept, conditionally accept or reject); and iteration (for example, registered students progressing from year to year) can be expressed clearly.

It is also important to identify the terminating states of the entity. Some information systems have not always done this and found that at a later date they have no way of getting rid of entity occurrences, which leads to obvious inefficiencies. One example found by the authors concerned the vehicle spare parts system of a large organisation. In this system vehicles require specific parts, but what is not known is which parts support specific vehicles. The result is that when vehicles become obsolete there is no way of withdrawing the parts that support that vehicle only. It is too dangerous to withdraw all the parts required by the obsolete vehicle, as many of these parts will be used by other vehicles. This results in a database which is continually growing as new vehicles and their parts are added. If there had been an entity life cycle analysis performed on the entity part, it would have been discovered that the entity occurrence did not terminate. The likelihood is that the organisation would then have designed a function to associate parts with vehicles and thus be able to terminate the entity and not have these ensuing problems.

The entity life cycle diagram is a good communication tool that enables users to validate the accuracy of the analysis. It can form an outline design for transaction processing systems. A by-product of the analysis process is that functions in which the entity type is involved are identified. The process is therefore useful as a validation of the other process analysis techniques. These charts should be drawn for all the entities.

Some approaches develop entity life cycle diagrams further. For example, they might use separate symbols for various states of entities, such as, set up, amended and deleted states. Others show functions on the diagram as well as entity states. The functions implied in figure 4.55 might include 'reject student', 'conditionally accept student' and 'student withdraws'. Their explicit inclusion, by labelling arrows in more detail, might help full understanding of the overall documentation set and also might identify functions which have been omitted. Events occurring in the problem situation which have not been modelled anywhere else in the documentation may also be identified.

In SSADM the entity life cycle is termed an entity life history. The diagram looks more like a hierarchical structure with the entity under consideration as the root or parent of the tree. The resulting diagrams are rather more complex than that shown in this section because they express more. These are described, with an example, in section 6.5. The entity life history diagram is also very similar to the structure diagram of Jackson Systems Development (JSD). In JSD this is the central modelling technique used in the methodology and there are also extensions to the entity life cycle technique and these are described in the methodologies section on JSD (section 6.7).

In some approaches, entity life cycles are developed to provide information related to events which are not possible to provide on standard entity life cycle diagrams. McDermid (1990) suggests three possibilities:

- If incorrect information has been added to an entity at what stages can it be corrected?
- What effect does premature termination have on the life of an entity?
- Can events happen out of sequence and, if so, which ones are permissible?

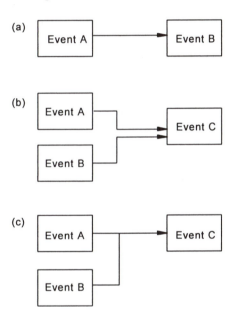

Fig. 4.56: Dependencies between events

Unless such questions are considered, the analysis work is not complete and this may lead to an information system unable to deal with all possible situations. The extra information can be captured on these state dependency diagrams which can be created alongside entity life cycle diagrams as they show the dependencies between events. In figure 4.56(a) event A must take place before event B; in figure 4.56(b) event A and event B must take place before event C; in figure 4.56(c) event A or event B must take place before event C. Of course, the inter-dependencies need to be thought out first and therefore the technique encourages proper analysis. Such analysis can be a vital input to the design of the information system in terms of the application logic following events.

4.13 OBJECT ORIENTATION

In section 3.9 we suggested a number of advantages claimed by the proponents of the object-oriented approach. In section 6.8 we look at the Coad and Yourdon methodology for information systems development. In this section we define the concepts and techniques used in the object-oriented approach.

An *object* is something to which actions are directed, it has an identity, a state, and exhibits behaviour. The identity enables us to distinguish it from other objects; the state is the current value of the dynamic properties of the object; and the behaviour is the actions that the object can itself undertake. Definitions of objects are not very satisfactory because basically an object can be anything, so we add that it should also be an abstraction of something in the problem domain, that is, it is of interest to us, given the context of our current objectives.

We are on more certain ground when we think of a group of objects that makes up a class of objects. All the objects in the class exhibit a common set of object attributes, such as, structure and behaviour. Classes are structured in a hierarchy based on inherited properties. Inheritance is a relationship among classes, such that one or more classes share the structure or behaviour of another class.

We will use an example to illustrate these concepts. Let us assume an object is an update transaction on a customer file and, further, that it is a particular transaction to update a specific record, in a specific way, made at a particular time. The object has a unique identity, for example, 16249, the number of the transaction on the transaction log, which is unique to this transaction. The object has a state, in this case the state is 'successfully completed', that is, the update has been properly made, other possible states of the object are 'unsuccessfully completed', 'in-process' or 'awaiting processing'.

In practice, the object may have many potential states combining a variety of attributes. The behaviour of the object is the actions (or operations) that it can undertake: it can trigger an error message, it can change the contents of the field, it can access the status of the file, it can update the log file and so on. The behaviour of the object is completely defined by its actions.

The class is the group of objects that share a common structure and behaviour. In this example the class might be that of 'customer file update', that is, the general class of which our object 16249 is an instance (the terms object and instance are in fact the same and are interchangeable). So the class is the general form of the object.

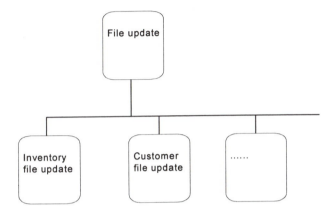

Fig. 4.57: Hierarchy of classes

If we compare the concepts used in the entity-relationship model, we can make an analogy in terms of the relationship between entity types and classes and entity occurrences and objects. The entity type is the general form and an entity occurrence is a specific instance of the entity type. For those with a programming background, class is analogous to type. Classes are structured in a hierarchy which share certain properties. Our class of 'customer file update' might be a subset of a class called 'file update', which might have more than one sub-class, for example, inventory file update, that is, there is a hierarchy of classes. This is shown in figure 4.57.

Hierarchy is not simply about classes and subsets or supersets of classes but also includes the object-oriented notion of inheritance to define the relationships in the hierarchy. Inheritance implies that the relationship is such that the hierarchy goes from classes of a general type down to classes of a more specific type, or from classes that exhibit commonality down to those with differences at the lower levels. The classes at the top are general and are then extended with more detail at the lower level in the hierarchy. In an object-oriented programming language, a lower class in the hierarchy can be produced from the higher class 'inheriting' all the higher class's structure and behaviour. The programmer can then add some more specifics and detail to the lower level class to make it perform more specifically than the higher class from which it was derived.

In our example, the class 'file update' would contain all the general features required to update a file. The class 'customer file update' would have inherited all these and then extended the class specifically to update the customer file. The benefits of inheritance include the fact that code can be re-used, instead of separately writing 'customer file update' and 'inventory file update' we just inherit most of it from the higher class of 'file update'. This saves time and has other efficiency benefits and also ensures that the processing (or rather behaviour) in both 'customer file update' and 'inventory file update' is the same, that is, there is a standard approach.

MS Windows-based programs, for example, use a hierarchy of windows classes to ensure that each window behaves in the same way no matter which application is running in order to achieve a consistent windowing environment. Inheritance is central to the object-oriented concept and provides much of its benefit. Indeed Booch (1991) suggests that 'if a language does not provide direct support for inheritance, then it is not object oriented'.

It needs to be pointed out that the object-oriented notion of code re-use is different from that in traditional programming, where code is often shared and borrowed. There are program libraries in organisations to enable code to be shared and re-used. However the code is copied or cloned and used as a basis for a new program, which is then amended and extended as necessary for the new application. This may indeed save time and effort but the difference is that there is no returning to the original code again, there are now two separate entities: the original program and the new program, and they develop along quite separate paths in different ways. If an error is found in the original code or a change is required in the original code, such change has to be made in both programs as the new program has changed and evolved so much it is usually impossible to go back to the original and start again.

With object orientation, and true inheritance, the higher-level class could be changed and the lower level classes would inherit that new behaviour by a simple re-compilation. The processing specifically required for the lower level, in this case the code to make it specific to customers, would not be effected, because the object-oriented inheritance has ensured that they are treated and organised separately. In the example, we would change the class 'file update' and then recompile the classes 'customer file update' and 'inventory file update' and this effectively means that they then inherit the changes. Thus the change is made only once and it is applied to all the sub-classes, and in this way really effective re-use is achieved.

It should be obvious that inheritance does not just happen, it needs careful design. We will examine this in more detail later, but at the very least the decision as to the classes and their hierarchical structure needs to be thought about very carefully such that the common aspects are included in the higher level objects and the special cases at the lower level. This is not always easy nor intuitive.

There is a further aspect of the definition of object-oriented programming that needs exploring, and that is the organisation of the program as '...a co-operative collection of objects'.

In the object-oriented world, objects simply sit around waiting to be activated, and that activation happens when the object receives a message from another object. Messages are the only way that objects communicate and we may think of objects being fired off by messages from other objects or by an initial event. So an object-oriented program is a series of objects organised to interrelate in particular ways to produce the functionality that is required of the program. This is referred to as cooperage. Communicating through messages makes objects independent, that is, they do not need any other object, or knowledge, to perform their job. They are triggered by a message from another object, they perform their job, which may itself involve the triggering of other objects, and when complete they usually return a message to that effect to the triggering object. This might be compared to a traditional subroutine.

For example, we might have an object, 'menu-selection', which displays a menu and obtains a selection of an option from the user. The object is fired or invoked by another object passing a message to 'menu-selection', part of the message might be the menu that should be used in 'menu-selection' itself. The object 'menu-selection' is basically the method for selection from a menu and all the associated error checking, and the message is the menu text to be used. On completion, 'menu-selection' returns the option that has been selected as a message or it may invoke another object. In this way the object is independent of other objects and exists as an independent entity, the exact way that it performs its task and the data used by the object is unknown to other objects, indeed it may be written in a different programming language. This also means that the internal workings or implementation of the object can be changed without it causing any problems to the other objects.

What is important is the external interface of the object, that is, its messages. Furthermore, data and procedures are not allowed to exist externally or independently of an object, unlike in most other programming approaches. The fact that the internal processing and the details of the data are hidden (or private) is known as encapsulation and this is described by Booch (1991) as one of the fundamental elements of the object model. Daniels and Cook (1992) use the analogy of an egg to illustrate the concept, see figure 4.58. An object is an egg, the yolk is the data surrounded by the white which is the processing or operations that act on the data. The shell of the egg surrounds the whole thing and keeps it all together. It effectively hides the data and processes from the outside world, the shell is the interface and the only thing that is seen. The data and processes are said to be encapsulated in the object. The analogy breaks down a little because the shell of the egg, although nicely encapsulating the contents and hiding them from the outside world, does not interface very well with other eggs, nor does it send messages, or have an identity. Encapsulation is radically different to that of most other approaches and methodologies, for example, SSADM, Merise and Information Engineering, where the data and processes exist, but are analysed, separately.

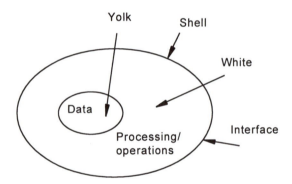

Fig. 4.58: Objects represented by an an egg (modified from Daniels and Cook (1992))

An object-oriented program is simply a collection of interrelated objects where the connections are unidirectional paths along which messages are sent. The program begins with an initiation from outside, often an event, which triggers an object, from then on that object initiates others and so on. The activations form a network of objects that together make up the program, see Figure 4.59. Programming an application consists of defining the network or the co-operation between objects and in an ideal world most of the objects will already exist. The path that the program takes is called the thread, or the thread of control, and it can be difficult to identify once the program is invoked, because of course the objects are interacting in ways dependent on events and user responses.

Also, in theory, the objects may be operating concurrently, and even on different processors.

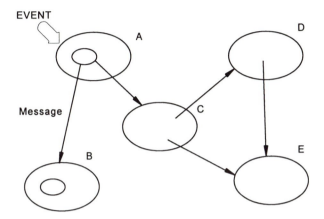

Fig. 4.59: A network of objects make up a program

In the discussion above the concept of an object is that of an instance or particular occurrence of a class, which means that objects are unique and yet it is only the data, the identity and state, that are different. The general behaviour of objects in the same class is the same. Clearly it would be foolish to write the code for each object separately, so what is done in practice is that the code is written once and shared by objects of the same class. The object is really just a data structure holding the relevant data for the object and when the object is asked to perform an operation the run time system ensures that the data is connected with the relevant code. This means that in practice the implementation of object-oriented programs differs somewhat from the theory.

Having discussed the theory, we will now summarise the benefits that the proponents of the approach argue it generates. As discussed above, the concept of inheritance leads to the realistic re-use of code. In theory, the organisation will develop a library of object classes that deal with all the basic activities that the organisation undertakes. Programming becomes the selection and connection of existing classes into relevant applications, and because those classes are well tried and tested as independent classes, when they are connected they provide immediate industrial-strength applications that run correctly the first time. Only completely new classes will need to be developed or perhaps purchased. Proponents of object orientation believe that, eventually, there will exist international libraries of object classes that developers will be able to browse to find the classes they require and then simply buy them. The classes in these libraries will be guaranteed to perform as specified, and so new applications are easily developed. Further, existing (object-oriented) applications are modified and extended in functionality just as easily. Software development is not only quicker and cheaper but the resulting applications are robust and error free. This attacks the problems that have bedevilled the software industry for so long, such as, projects being delivered late, over budget, and full of errors. The implication is that the information systems developed using object-oriented techniques will be equally robust and error-free and quicker and cheaper to develop.

4.14 STRUCTURE DIAGRAMS

Like other techniques described in this chapter, structure diagrams are used by a number of methodologies. The structure chart is a series of boxes (representing processes or parts of computer programs, usually referred to as modules) and connecting lines (representing links to subordinate processes which show the way that data and control can be passed between processes) which are arranged in a hierarchy. Each module should be small and manageable. Structure charts therefore exemplify the functional decomposition aspect of many of these process techniques. The basic diagram is shown as figure 4.60. This structure chart shows that:

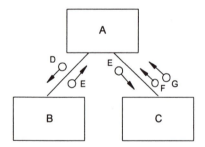

Fig. 4.60: Structure diagram

- Module A can call module B and also module C. This is shown by lines joining the boxes, which represent modules. No sequencing for these calls nor whether they actually occur is implied by the diagramming notation. When the subordinate process terminates, control goes back to the calling process.
- When A calls B, it sends data of type D to B. When B terminates, it returns data of type E to A. Similarly, A communicates with C using data of types E and G.

- When C terminates, it sends a flag of type F to A. A flag is used as a flow of control data. The difference in the symbol between data types and flags lie in the circle being filled in or not.

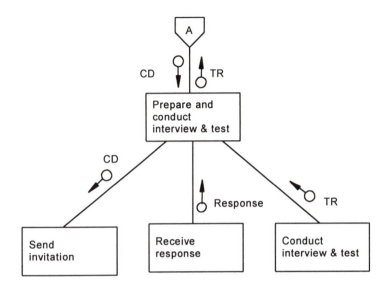

Fig. 4.61: Structure diagram for interviewing and testing applications

Figure 4.61 shows the structure chart for part of the processing of applications to undergraduate courses. It concerns those students that we propose to interview. CD refers to the candidate document and TR to the test result. Other charts in the set will be constructed for the processing of examination results and final assessment amongst others. The top-level diagram (A in figure 4.61) will show each of these processes as a box and refer to the lower-level structure charts.

The structure of a program should be designed to ensure that it minimises the interdependence between modules (known as coupling), it encourages module re-use (known as cohesion) and eases the programming task and later maintenance. By minimising the connections between modules and therefore their independence, coupling reduces the risk of errors or changes in one module affecting another. But cohesion, that is, the way the activities within a module are related should be maximised as it helps ease of understanding the module and encourages re-use because the elements of the module concern similar activities which may be used in other applications. The designer should be aware of the possibilities of code re-use, both in terms of designing modules that may be re-used and using modules already written.

We have so far suggested that the structure diagram relates to computer program design. But of course the basic technique applies to the overall design of information systems, of which the design of computer programs play a part. This sequence is not a coincidence, for many of these process techniques were used first in the structured programming movement which influenced many systems analysts. Structure diagrams are used in SSADM, Multiview and elsewhere, but they are described fully in section 6.7 when discussing the entity structure step, a phase in Jackson Systems Development (JSD). Jackson has been a notable contributor to both the software world (through Jackson Systems Programming) and the technical aspects of information systems.

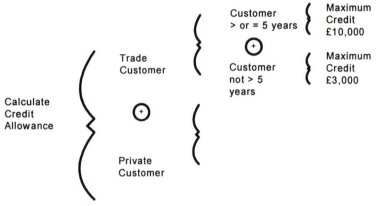

Fig. 4.62: Warnier-Orr diagram

A further variation on the same structure chart theme, but looking somewhat different, is the Warnier-Orr diagram which has been well used in the United States. As shown in figure 4.62, the diagrams use brackets to show the hierarchical structure and they are read from left to right rather than from top to bottom. Sequential processes are therefore shown within one bracket. There is a selection construct, where a choice is made between mutually exclusive possibilities, according to a condition. This is denoted by the symbol which has a plus sign within a circle. The example provided reflects a low level decision-making process, but Warnier-Orr diagrams can be used and are used to reflect the higher level processing of the structure charts of the above examples.

4.15 MATRICES

One of the most common of techniques in all walks of life is the matrix, a simple tabular expression of the relationship, usually between two things (three things would require a three-dimensional set of matrices, feasible on computers). We find them in many aspects of many methodologies. In this section we give a few examples of their use.

Function Name / Event name	Process enquiry	Selection	Enrolment	Accounts
Student enquiry	X	X		
Student accepted		X	X	
Student registers				X

Fig. 4.63: Function/event matrix

A common matrix is that showing the relationship between functions and events. In figure 4.63, which is a part-formed matrix relating to the acceptance of students on courses, the Xs at the intersections between an event row and a function column shows that an event triggers a particular function. Thus the event that a student makes an enquiry triggered the 'process enquiry' and 'selection' functions. Every event should trigger at least one function and every function should have a triggering event.

Entity Name / Function name	Staff member	Group session	Location	Programme type	Patient
Group session attendances	R	U	R		C/U
Programmes	R			R	R
Contacts traced	R				U
Assessments	R				R/U/D
Programme costing	R			U	R

Fig. 4.64: Entity/function matrix

A second matrix used by many methodologies is that associating functions with entities, that is, what entities are used by each function to enable that function to be carried out? Figure 4.64 shows the entities used by functions carried out in a hospital. Frequently in methodologies, the entries reveal more than this, that is, the manner of access. Thus C (create), R (read), U (update) and D (delete) entries will give more information about the relationship between the two than a mere X. This is often referred to as a CRUD matrix. There should be represented in the full set of matrices all the entities contained in the entity model and all the functions contained in the data flow diagrams. The CRUD matrix can also be used for the relationship between entities and events. Some cells may contain more than one entity, for example, C and R. This will show that a particular event leads to an entity occurrence either being created or simply accessed.

The systems designer may well also create a key attribute and entity matrix. This will ensure that all entities have a key. Where there are two or more Xs, this will show that an entity has a composite key. Other matrices commonly found are those associating entities and attributes and that showing the relationship between user roles and functions.

Document Department	Order	Note	Invoice
Sales office	1		2
Warehouse	2/4	1/3	
Production	3	2	
Accounts	5		1
Post room			3

Fig. 4.65: Document/department matrix

A matrix showing the document flows through the system, such as that relating the data elements in the data dictionary system with the various documents that record their existence is also commonly in use. Figure 4.65 shows the sequence by which documents are processed in various departments.

FURTHER READING

Avison, D. E. and Wood-Harper, A. T. (1990) *Multiview: An Exploration in Information Systems Development*. McGraw-Hill, Maidenhead.

Checkland, P. and Scholes, J. (1990) *Soft Systems Methodology in Action*. John Wiley, Chichester.

The description of rich pictures, root definitions and conceptual models given in this chapter are their interpretation in the Multiview methodology (Avison and Wood-Harper) and this text discusses the professional association case and five others in detail. Examples of the use of the three techniques in various projects as designed for soft systems methodology are given in the Checkland and Scholes book.

McDermid, D. C. (1990) *Software Engineering for Information Systems*. McGraw-Hill, Maidenhead.

Mason, D and Willcocks, L. (1994) *Systems Analysis, Systems Design*. McGraw-Hill, Maidenhead.

Both these companion texts in the series give a good description of many of the techniques described in the chapter, in particular, data flow diagrams, entity life cycles, entity modelling, normalisation, state dependency diagrams and structure charts in McDermid and data flow diagrams, decision tables, structured English and matrices in Mason and Willcocks.

Chapter 5
Tools

5.1 INTRODUCTION

We saw in section 3.10 various attempts at automation including the ambitious ISDOS project. Since these early attempts at automation, technology has continued to develop, becoming more powerful, cheaper and more widely available. Improved graphics facilities and windowing environments have also had an impact. Indeed, some people have suggested that the latest generation of automated tools are, potentially, a panacea for the problems of systems development. Brooks (1987) describes automation as one of many 'silver bullets' that are being suggested as the key to improving systems quality and eliminating the development and maintenance backlog of information system applications. We think that there are no panaceas in information systems, but that the quality of the software has improved over time so that automated tools are potentially beneficial.

There are a growing number of automated tools in the market place that support some aspects of the systems development process. It has been estimated that in 1985 there were about 80 such products, by the beginning of 1990 this had grown to about 700, and that there were over 1,000 by 1993. Today there is such a plethora of tools that the field is characterised by confusion. In this chapter we attempt to classify (and thereby define) tools into some significant groupings.

We first look at project management tools. Project planning and control is a key factor in the success or failure of the information systems development project, increasing the likelihood that it is completed on time, within budget and with the required functionality and quality.

We then consider database management systems. Data is a key resource of the organisation and database software can enable the sharing of data, ensure its quality and improve standards. By having a database of the whole organisation, or at least significant large groupings within the organisation, we have the potential for providing management information and decision support and therefore improving management decision making. We also look at ways in which various types of user can access the database with the help of the database management system. These 'routes to the database' include extensions to conventional programming languages, query languages or soft copy forms and menus.

An important piece of software which runs with the database is the data dictionary system. A data dictionary provides information about the database, such as, what data is held, which users have rights of access, what are the permitted values and so on. It is therefore a software tool which helps manage the data resource. Although data dictionary systems have proved useful, it became obvious that there is a parallel need to store information about processes and programs, and data dictionaries became extended to hold information about both data and processes. These extended data dictionary systems developed to systems repositories which, as we shall see, became an important integral tool of the CASE toolset. Systems repositories contain the information necessary to integrate the different stages of information systems development. In theory at least, the high end systems repositories will contain all the information required to support the creation and maintenance of information systems in the organisation.

Before looking at CASE tools in detail, consideration is also given to drawing tools (which may, in any case, be part of the CASE 'toolset'). Drawing tools support the drawing of one or more commonly used diagrams, such as data flow diagrams, entity-relationship models and the like. They help drawing the diagrams initially and also their revision which is otherwise a tedious task and sometimes not carried out. Additionally, they help ensure diagrams are accurate, consistent and conform to any particular standard.

The main focus of this chapter is on CASE tools. This presents a software environment, a complete toolset, to support the information systems development process. Amongst the facilities of a CASE tool may be all the tools discussed in previous sections of this chapter (project management, database management, systems repository and drawing tools) along with many others. But CASE tools differ greatly in their functionality. Some provide toolsets which address the early stages of systems development (strategy, planning and analysis); others address physical design, programming and implementation stages; yet others integrate the two into a single, fully integrated development and support facility.

As well as consider the facilities of CASE tools in general, we look in more detail at Information Engineering Facility, which is one of the best known examples of a high end, integrated, CASE tool. This description reveals the complexities of such a tool, both in its design and use. Again, because of the importance of CASE tools, we look in detail at their potential benefits and provide an overall evaluation of their potential contribution. Such a detailed commentary provides a possible checklist of areas in which to evaluate other tools. An opportunity is taken to discuss the human, social and organisational aspects. The likelihood of successful adoption of CASE or other software tools (or, indeed, information systems) is as much to do with people and organisational factors than the qualities of the software tool itself. An opportunity is taken to discuss the factors which increase the likelihood of success of using a CASE-based approach to information systems development in general.

Finally, we look at the possible ways that CASE tools may develop, and an opportunity is taken to contemplate possible short-term and long-term facilities that may be provided in most CASE tools of the future. Readers should not expect the latter to be available in CASE tools for some time. Further, there are many disadvantages (as well as advantages) of being associated with the leading edge of software and applications development.

We have therefore taken the opportunity in the chapter to look at a number of wider considerations in adopting software. Although we use CASE tools as our example, many of the ideas discussed are relevant to information systems development in general as is the importance at looking at these developments in their social and organisational context. Some of these debates are developed further in Chapter 7, which follows the descriptions of a number of information systems development methodologies in Chapter 6.

5.2 PROJECT MANAGEMENT TOOLS

One of the most important tasks of a manager of an information systems project is project planning and control. The aim of this activity is to ensure that a project is completed on time, within budget and with the required functionality and quality. Projects which take longer than scheduled, for example, cause a loss of money as well as embarrassment, particularly if they are unexpected or cannot be explained easily. By using project management techniques it should be possible to ensure that there are a series of checkpoints that can be readily identified so that systems development staff can work to these and hence control the project. These checkpoints can be used to provide the interface to a project control package. This can help to ensure that projects are scheduled at the earliest possible date, with the least drain on resources and that there is a good chance that this date will be met. If there are delays, then at least the managers of the project and their clients have information about the delays and could, for example, increase resources to make up time (assuming resources are available).

In this section we will make reference to PRINCE (PRojects IN Controlled Environments), a project management method which is the standard used in UK government computing departments, to illustrate aspects of project planning and control. PRINCE is often used along with SSADM (section 6.5), which is the standard UK government information systems development methodology. An information systems development methodology can work well with a project planning methodology by identifying the standard phases, tasks, techniques and deliverables, and ensure that they occur in the standard sequence.

Project control techniques start with an attempt to break down the large and complex project into tasks, normally called activities. This process is called the work breakdown structure. Once the activities have been identified, a time and resource requirement is assigned to each of these. This is not an easy task, and although there are algorithmic methods available, particularly in estimating resources for software development, the main method is to use experience of similar activities in past projects. The inter-relationships between activities is also established. In other words, those tasks which are dependent on the completion of other tasks are identified. These activities, and information about them, can be entered into a project control tool which may be stand alone or part of the toolkit included in a CASE tool.

Project control, even with the help of a good computer package, requires careful and detailed work in:
- Identifying tasks
- Establishing the inter-relationships between tasks
- Allocating their resource needs.

This is an analytical process which is time consuming.

Using this information, the computer package can draw up a network. In a network, the activities are represented by arrows which join the nodes (circles). The latter represent events, that is, the completion of activities. Figure 5.1 shows a network. The arrows represent the activities, though the length of the arrow does not indicate the time taken for each task. Arrows drawn from the same node indicate tasks that can be carried out simultaneously. Arrows following others indicate tasks that are dependent on the completion of these other tasks.

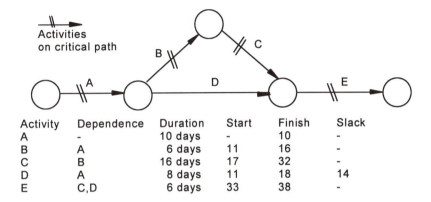

Activity	Dependence	Duration	Start	Finish	Slack
A	-	10 days	-	10	-
B	A	6 days	11	16	-
C	B	16 days	17	32	-
D	A	8 days	11	18	14
E	C,D	6 days	33	38	-

Fig. 5.1: Project control - the network and critical path

The manual development of networks is lengthy, and project control software can make the task much easier. It can draw the network and highlight critical activities on which any slippage of time will cause the whole project time scale to suffer. The path of the critical activities joined together forms the critical path, and it is useful for the package to highlight these activities. In figure 5.1, the activities A-B-C-E are on the critical path. Each activity on the critical path has been marked with short parallel lines. If it is possible to reduce the time of these activities, possibly by moving resources allocated from other activities to them, then the overall project time should decrease. Activity D is not on the critical path, and there is a slack of 14 days on this activity. In other words, there can be a delay of up to 14 days on D without delaying the overall project. If feasible, it might be expedient, therefore, to move resources from activity D to an activity on the critical path. This change can be entered into the package and the results recalculated.

Many computer tools will aggregate the various resources, such as the number of people working on the activity, automatically and attempt to level the use of these resources throughout the project. These outputs can be particularly useful as management reviews and when approving plans. The philosophy embedded in PRINCE is that these management products (and associated quality products) are just as important as the technical products of the project.

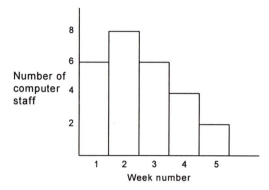

Fig. 5.2: Project control - resource schedule

It is usually better to use resources as smoothly as possible in the lifetime of the project, otherwise staff will be used efficiently for only part of the project. However, unless the resources are taken out of non-critical activities, this process is likely to lead to an increase in the overall project time. Once smoothing has been done, a bar chart showing the resource allocation over time can be displayed and printed (see figure 5.2).

Many project control packages will report on inconsistencies within the network, such as, the same resource being used at the same time. A typical package will also convert days into calendar dates and to allow for weekends, bank holidays and other holidays.

We have made reference to the quality of the products, that is, quality procedures, product reviews and associated documentation. PRINCE suggests that quality is about fitness for purpose. In terms of quality software, it may refer to mean time between failure, testing standards or some other technical norm. Frequently it is less tangible, but quality review procedures, including the setting of quality norms and their verification through structured walkthroughs and audit trails, need to be established.

Normally there is a trade-off between time and cost (assuming the same quality), in other words, the more resources allocated (and the more costly the project), the quicker it can be finished. The user may like to input resource availability in terms of:

- Minimum
- Most likely
- Maximum.

figures, according to estimates based on past experience.

This will give three different results for time/cost comparisons. Such an exercise would be very laborious if drawn by hand. Although the plan

should allow for minor deviations, project planning packages may permit the user to ask 'what if?' questions so as to see the consequences of more major deviations, for example:
- Re-allocation of staff
- Unexpected staff leave
- Machine breakdown.

The user can also use the package to highlight the results of following different work patterns or changing other aspects of the business.

Useful reports from the package might include a list of activities presented in order of:
- Latest starting date
- Earliest starting date
- By department
- By resource
- By responsibility.

Information relevant only to a particular department or sales area can be created so that people are not given unnecessary information. Furthermore, a computer package may simulate the effects of the following possibilities to the total project time:
- Prolonging an activity
- Reducing resources applied to it
- Adding new activities.

Similarly, it may be used to show the effects of changing these parameters on project costs. The manager may be faced with two alternatives: a resource-limited schedule, where the project end date is put back to reflect resource constraints, or a time-limited schedule, where a fixed project end date leads to an increase in other resources used, such as human and equipment resources.

Once the project has started, there will be progress reporting. This can be used to:
- Compare the time schedule with progress made
- Compare the cost schedule with progress made
- Maintain the involvement of the users and clients
- Detect problem areas
- Replan and reschedule as a result
- Provide a historical record which can be used for future project planning.

Another way of displaying the information is as a Gantt chart (figure 5.3). In this chart, the estimated duration for each activity is shown in clear boxes and the actual duration in patterned boxes. The Gantt chart is particularly good at showing progress graphically. In this case, a delay in

activity A by three days is recorded and this has set the start of activity B
and C back by three days. Unlike the losses caused by delaying activity A,
the gains made in reducing activity D by three days shown in the chart will
not make any difference to the overall project time. However, the analyst
is informed of any delays and can replan resource use or inform the users
about the delay in good time. Indeed, with PRINCE, users are seen as part
of the organisation component of the project where the roles of the users
are 'built in' to the project organisation and will be part of the decision-
making process. Delays in a project need to be communicated elsewhere in
an organisation with many projects as the delay of one project may lead to
a delay in another if, for example, it uses the same resources.

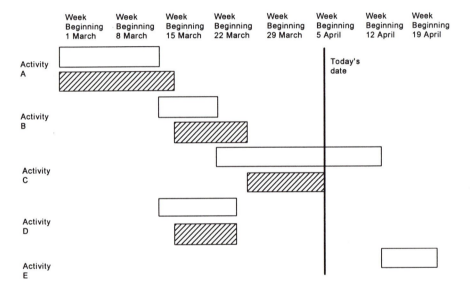

Fig. 5.3: Gantt chart

The computer system may also give information about how to act to put
right any deviation from the schedule. This may be achieved by, for
example, increasing the resources on some activities and re-scheduling
others. This goal-seeking type of analysis, almost impossible manually, is
a feature of a number of computer packages.

Any change in a manual system will require the chart to be re-drawn.
This is an inconvenience at best, more often a task left undone, which
makes it likely that any previous planning is ineffective.

The development of an information system is likely to be a large scale
project, particularly where the overall plan has a number of inter-related

subsystems that need to be completed separately but integrated later. 'Large' here refers to time, cost, people and equipment resources, as well as to the complexity of the inter-relationships between activities in the project. The use of a project planning tool makes an up-to-date and informative plan far more likely and such large scale integrated information systems projects feasible. It helps the analyst to identify the risks associated with project planning and act to reduce them. To set against these gains, however, allowance has to be made for the costs of maintaining and updating the project control systems.

Successful project management aims to ensure projects are completed within the allocated time period, within the budgeted costs, of the required quality and with all the specified functionality. Further, it aims to do so without disturbing the other work of the organisation. However, although we have stressed here the need for project planning tools, we must also admit that the difficulty of meeting deadlines on computer systems development projects has always been a major problem. Although project planning tools help, particularly by enforcing a planning stage and thereby enabling the discovery of important aspects of the project that might otherwise go unnoticed (such as conflicting objectives), their effect sometimes only seems to have been in measuring our lack of success at meeting deadlines. However, the use of project planning procedures and tools may help us in the process of organisational learning and lead to improved project planning and control in the future.

5.3 DATABASE MANAGEMENT SYSTEMS

A large collection of books owned by the local council is not a public library. It only becomes one when, amongst other things, the books have been catalogued and cross-referenced so that they can be found easily and used for many purposes and by many readers. Similarly, a database is more than a collection of data. It has to be organised and integrated. It is also expected to be used by a number of users in a number of ways. In some companies the whole organisation is modelled on a database, so that, in theory at least, users can find out any information about any aspect of the organisation by making enquiries of the database.

In order to make this feasible, there needs to be a large piece of software which will handle the many accesses to the database. This software is the database management system (DBMS). The DBMS will store the data and the data relationships on the backing storage devices. It must also provide an effective means of retrieval of that data when the applications require it, so that this important resource of the business, the data resource, is used effectively. Efficient data retrieval may be accomplished by computer programs written in conventional programming languages such as Cobol, C and Fortran accessing the database. It can also be accomplished through the use of a query language, such as SQL, or in other ways which are more suitable to people who are not computer experts. In this section we discuss both the potential facilities of a DBMS and the data retrieval methods to access the data on the database using the DBMS.

The ANSI/X3/SPARC committee, established in 1972 with reports in 1975 and 1978 (ANSI/X3/SPARC, 1975, and Tsichritzis & Klug, 1978) suggested an architecture for database systems which has proved very influential in the database field. As seen in Figure 5.4, which is a modified version of this basic architecture, it has three views of the database. These are the external, conceptual and internal views.

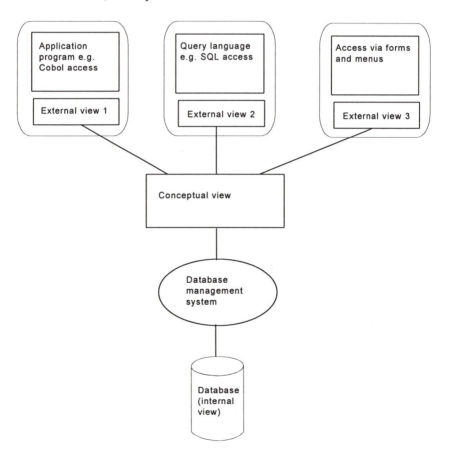

Fig. 5.4: Database architecture

The external view is the view of the data as 'seen' by application programs and users. It will be a subset of the conceptual (entity) model which is a global or organisational view of the data. There may be a number of different user views and different users and programs may share views. In the diagram, two users share the same view, but their methods of accessing the data are different. This arrangement enables aspects of the database to which a user may not be interested or does not have rights of access to be hidden from that user. This reduces the complexity from the users' points of view. There is a series of mappings or transformations between the external views and the conceptual view which is handled by the software.

The conceptual view here is different to that referred to in section 4.4 'conceptual models'. Readers have been warned about the non-standard terminology in information systems! Indeed, the conceptual view expressed in this model is that of the entity models described in section 4.5. In those sections we described techniques to provide a global data model. The conceptual model discussed in this section is the data model of the organisation. It describes the whole database for the organisation (or community of users) in terms of entities, attributes and relationships.

The internal view of the data describes how and where the data is stored. It will describe the access paths to the data storage and provide details of the data storage. There is again a mapping between the conceptual and internal views. The latter will not be of direct interest to the users, but will be of indirect interest in terms of speed of access and the general efficiency of database use. It will be of direct interest to the database administrator who is normally responsible, amongst other things, for the efficient organisation and effective use of the database.

There is a series of mappings from the conceptual view (the overall data model) to the various external views (the subsets of the data which applications or users have access) and internal view (the way in which the data is organised on computer storage devices). These mappings will be carried out by the database management system software.

The separation of different views of the data enables data independence, which is so crucial to the database approach. This means that it is possible to change the conceptual view without changing the external views (and the application programs that use them). It is also possible to change the internal view without changing the conceptual and external views. In other words, data independence provides flexibility.

Data independence represents only one of the hoped-for advantages of the database approach which are to:

- *Increase data shareability:* Large organisations, such as insurance companies, banks, local councils and manufacturing companies, have for some time been putting large amounts of data onto their computer systems. Frequently the same data was being collected, validated, stored and accessed separately for a number of purposes. For example, there could be a file of customer details for sales order processing and another for sales ledger. This 'data redundancy' is costly and can be avoided by the use of a database management system. In fact some data duplication is reasonable in a database environment, but it should be known, controlled and be there for a purpose, such as the efficient response to some regular database queries. However, the underlying data should be collected only once, and verified only once, so that there is little chance of inconsistency. With conventional files, the data is often collected at different times and validated by different validation routines, and therefore the output produced by different applications could well be inconsistent. In such situations the data resource is not easily managed and this leads to a number of problems. With reduced redundancy, data can be managed and shared, but it is essential that good integrity and security features operate in such systems. In other words, there needs to be control of the data resource. Furthermore, each application should run 'unaware' of the existence of others using the database. Good shareability implies ready availability of the data to all users. The computer system must therefore be powerful enough so that performance is good even when a large number of users are accessing the database concurrently.
- *Increase data integrity:* In a shared environment, it is crucial for the success of the database system to control the creation, deletion and update of data and to ensure its correctness and its 'up-to-dateness'; in general, ensure the quality of the data. Furthermore, with so many users accessing the database, there must be some control to prevent failed transactions leaving the database in an inconsistent state. However, this should be easier to effect in a database environment, because of the possibilities of central management of the data resource, than one where each application sets up its own files. Standards need only be agreed and set up once for all users.
- *Increase speed of implementing applications:* Applications ought to be implemented in less time, since systems development staff can largely concentrate on the processes involved in the application itself rather than on the collection, validation, sorting and storage of data. Much of the data required for a new application may already be held

on the database, put there for another purpose. Accessing the data will also be easier because this will be handled by the data manipulation features of the database management system.

- *Ease data access by programmers and users:* Early database management systems used well-known programming languages, such as Cobol and Fortran to access the database. Cobol, for example, was extended to include new instructions which were used when it was necessary to access data on the database. These 'host language' extensions were not difficult for experienced computer programmers to learn and to use. Later came query languages and other methods to access the data which eased the process of applications development and data access in a database environment. Once the database had been set up, applications development time should be greatly reduced.

- *Increase data independence:* There are many aspects to data independence. It is the ability to change the format of the data, the medium on which the data is held or the data structures, without having to change the programs which use the data. Conversely, it also means that it is possible to change the logic of the programs without having to change the file definitions, so that programmer productivity is increased. It also means that there can be different user views of the data even though it is stored only once. This separation of the issues concerning processes from the issues concerning data is a key reason for opting for the database solution. It provides far greater flexibility.

- *Reduce program maintenance:* Stored data will need to be changed frequently as the real-world that it represents changes. New data types may be added, formats changed or new access methods introduced. Whereas in a conventional file environment all application programs which use the data will need to be modified, the data independence of a database environment, discussed above, circumvents the necessity of changing each program. It is necessary only to change the database and the data dictionary (or systems repository) which will contain information about the data in the database (amongst other things). We discuss data dictionaries and systems repositories in sections 5.4 and 5.5.

- *Provide a management view:* With conventional systems, management does not get the benefits from the expensive computing resource that it has sanctioned. However, managers have become aware of the need for a corporate view of their organisation. Such a view requires data from a number of sections, departments, divisions and sometimes companies in a larger organisation. This corporate view cannot be gained if files are established on an application-by-application basis and not integrated as in a database. With decision-support systems using the database, it becomes possible for problems previously considered solvable only by intuition and judgement to be solved with an added ingredient, that of information, which is timely, accurate and presented at the required level of detail. Some of this information could be provided on a regular basis whilst some will be of a 'one-off' nature. Database systems should respond to both requirements.

* *Improve standards:* In traditional systems development, applications are implemented by different project teams of systems analysts and programmers and it is difficult to apply standards and conventions for all applications. Computer people are reputed to dislike following the general norms of the organisation, and it is difficult to impose standards where applications are developed piecemeal. With a central database and database management systems, it is possible to impose standards for file creation, access and update, and to impose good controls, enabling unauthorised access to be restricted and providing adequate back-up and security features.

Much of the success or failure of a DBMS lies in the role of the database administrator (DBA) who will be responsible for ensuring that the required levels of privacy, security and integrity of the database are maintained. The DBA could be said to be the manager of the database and, because the design of the database involves trade-offs, will have to balance conflicting requirements and make decisions on behalf of the whole organisation, rather than on behalf of any particular user or according to a particular departmental objective. The role is multi-varied and is usually carried out by a database administration team. In some organisations, the information resource is regarded as one of the key elements of success (which it surely must be) and there is a high-level data administration function which includes the lower-level database administration team, responsible for the computer data.

DBMSs are software packages which manage complex file structures. They make databases available to a variety of users and the sharing of data can reduce the average cost of data access as well as avoiding duplicate and therefore possibly inconsistent or irreconcilable data. Databases hold large amounts of data and operations required to use the database are complex. Correspondingly, DBMSs are large and complex pieces of software. Users of databases do not directly access the database. Instead they access the DBMS which interprets the data requirements into accesses to the database, makes the accesses required, and returns the results to the user in the form that the user requires.

There are many database management systems available. These include DB2, Ingres, Oracle, Access, Paradox and dBase. Some of these run on mainframe computers, others on PCs. Most DBMSs are relational, in that the data model which the user sees is the relational one, that is, sets of tables. Ontos is an example of an object-oriented database management system. As would be expected, in Ontos the data management and programming language aspects are integrated: all data is represented as objects and the programming language manipulates objects. The Ontos DBMS supports complex objects, object identity and inheritance, features which we identified as important for object-oriented systems.

Of particular interest to users developing systems in a database environment is the way by which access can be made to the database. We will now turn our attention to this aspect of DBMSs. Data access may be made using soft copy forms, menus, conventional computer languages and query languages.

Many DBMSs provide alternative ways of access. This can be a useful facility as there are many types of user. Users can be untrained and intermittent in their use of the system. These casual users should be encouraged by its ease of use. Regular users may make frequent, perhaps daily, use of the database and are usually willing to learn a simple syntax. Other users will be professional users who are data processing people and will apply their long experience as computer and database users, and be concerned about efficiency of their work.

As reference to figure 5.4 makes clear, the external view of the database is that which each user 'sees'. It is derived from the conceptual view. The external view is the subset of the database which is relevant to the particular user, and although it may be a summarised and a very restricted subset, the user may think that it represents the whole view, because it is the whole view as far as that user is concerned. The presentation and sequence of the data will also suit the context in which it is presented. The format will depend on the particular host language, query language, report writer or other software used. Indeed, there may well be several external views, perhaps as much as one per application or user that accesses the database. But whatever the description given to the user, the underlying data will be the same.

External views may be presented to the user through the use of host language programs or a query language but they may also be obtained through a dialogue, which approaches a natural language dialogue, or via a menu. In figure 5.5, the users pass through a number of menus to indicate which system and which part of the system they require to use. The options are provided in the menu. The user has only to select the option required by pressing the appropriate key (following the question mark) or pointing the cursor to the appropriate option and clicking the mouse.

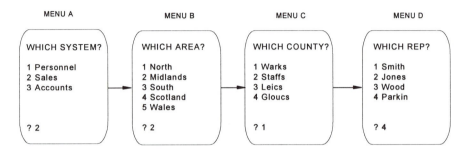

Fig. 5.5: Sequential menus

Alternative approaches which are also easy to use include a soft copy form presented to the users who 'fill in the forms' to state their requirements (figure 5.6) or icon/mouse systems, which use graphical symbols (icons) representing such requirements as filing (file cabinet), deleting (wastepaper basket) and so on. The user specifies the option required by pointing the mouse to the icon and clicking a button on the mouse to activate that process.

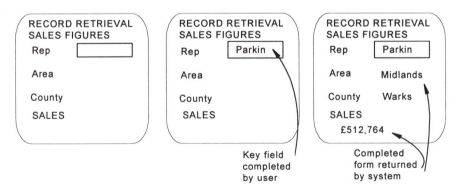

Fig. 5.6: Form processing

As we have already noted, the user programs 'see' only subsets of the logical schema, called external views, as appropriate to the particular users. Casual users may find menu-driven and form-driven systems appropriate whereas more experienced users may access the database using query languages or user programs written in a conventional programming language which act as host languages. These host languages are procedural and the computer programmer has to know the set of logical procedures required to fulfil a particular request. This requires an in-depth knowledge of the computer programming language, but if the knowledge of the language has been acquired, the requests for database access need not be complicated nor difficult to write.

Query languages are usually somewhat less procedural and may be suitable for regular users (who may find menu-driven or form-driven applications slow and inflexible). A useful distinction between programming languages and query languages is to consider the former as being concerned with the *how* of an operation, and the latter as being concerned with the *what*. One area of possible confusion is that 'query' languages offer ways of updating (changing) the data as well as accessing it in the form of a query. Many query languages are more difficult to use and more procedural than the 'ideal'. The most well known, SQL, is somewhat disappointing in these respects though it has become a standard.

Perhaps the most 'user friendly' are those which use a near natural language interface. However there are lots of 'false friends', that is, words in the query language having different meanings to their natural language meanings. Natural language is ambiguous, and computers require unambiguous input. Natural language computer systems usually require long and tedious clarification dialogues.

Most systems support a set of standard queries which are stored for regular use, and easily included in particular applications. Results may be displayed in tabular form on a screen or as a number of windows on a screen. *Ad-hoc* queries are also supported. Dialogue can be of the question-and-answer type, such as:

System   WHICH CUSTOMER RECORD DO YOU WANT TO SEE?
User   PARKES HOLDINGS

This dialogue may be computer initiated (as above) or user initiated. However, true natural language interfaces are not available as yet. the problem is so complex.

Query-by-Example (QBE) is designed for users with little or no programming experience. The intention is that operations in the QBE language are analogous to the way people would naturally use relations (tables). The QBE query is formulated by filling in templates of relations. The user enters the name of the table and the system supplies the attribute names. The system then details the attributes in which the user has indicated an interest. Many systems are menu-based and it is not necessary to recall the relation names. This means that minimal training is necessary.

However there are severe limitations to retain flexibility unless the database is not very complex.

A session could begin with the user requesting a template of the EMPLOYEE relation (figure 5.7 (a)). The user marks with a P (for print) any entries for that employee that are required (figure 5.7 (b)). The system then supplies the information required (Figure 5.7 (c)). The user can specify retrievals to be made on a range of values, for example, all tuples where the salary is greater than a fixed amount and also use ANDs, ORs and so on. In the example, a conditional statement has been added: that only tuples where the attribute 'pay to date' is more than 1500 are printed. This 'pay to date' figure is not included in the report (there is no 'P'). Further, the tuples are listed in ascending order of employee number ('AO'). Aggregation functions, such as sum, average, minimum and maximum values can also be requested using QBE. If data is required from a number of relations, then more than one relation is requested and completed in this manner.

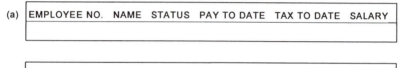

(a)

EMPLOYEE NO.	NAME	STATUS	PAY TO DATE	TAX TO DATE	SALARY

(b)

EMPLOYEE NO.	NAME	STATUS	PAY TO DATE	TAX TO DATE	SALARY
P:AO	P	P	>1500		P

(c)

EMPLOYEE NO.	NAME	STATUS	SALARY
756	SMITH	FULL	23,500
787	JOHN	FULL	18,500

Fig. 5.7: QBE example

Updates can be performed on the tables. Thus to effect a salary increase of 20 per cent 'across-the-board', the user keys in the statement '_x 1.2' under the column headed 'salary'. In this case the user has to denote this by 'U' for update (as against 'P' for print). Queries that are frequently required can be stored by name and recalled as required.

QBE proves much easier to use than SQL. Commands are not necessary. It is 'natural' to use because the user interface is in the form of tables. The user specifies requests in a tabular form and receives responses in that form. Again, compared to SQL, there are far fewer key strokes necessary. But the requests need to be relatively simple for QBE to be effective.

Query languages such as SQL prove to be able to handle the more complex queries and we will now look briefly at SQL (Structured Query Language) which is used on many DBMSs, both on small and large computers, including DB2, Oracle and Unify.

There are three basic commands for data definition: CREATE TABLE, ALTER TABLE and DROP TABLE (SQL uses the more inexact but more popular language of table, row and column, rather than relation, tuple and attribute). The CREATE TABLE command specifies a new relation; the ALTER TABLE command is used to add attributes to a relation: and DROP TABLE is used for deleting a relation. For example, the command to set up a relation called STUDENT with a field of up to 40 characters called STUDENT_NAME with a field signifying the year of the course for the student would be:

CREATE TABLE STUDENT
    (STUDENT_NAME   VARCHAR(40) NOT NULL
    YEAR_OF_COURSE   INTEGER);

The SELECT command is the only command for querying the database, but it has many variants. The command has three clauses as follows:

SELECT attribute list
FROM table list
WHERE condition.

The attribute list contains the names of those attributes whose values are to be retrieved; the table list is a list of the relation names required to process the query; and condition is a Boolean expression which identifies the tuples to be retrieved.

Thus, in order to find the name of the students who take the third year of the course, the SELECT statement will be:

SELECT STUDENT_NAME
FROM STUDENT
WHERE YEAR_OF_COURSE ='3'

There are a large number of variants to this statement, for example, those requiring complex Boolean expressions in the WHERE statement consisting of combinations of ANDs, ORs, =, + and - and even BETWEEN and LIKE in some implementations where BETWEEN gives a range of values and LIKE helps the searcher where the exact value is not known; or those requiring some computation on the SELECT expression, so that fields are added using SUM or the minimum, maximum or average values of an attribute which are obtained using MAX, MIN or AVG; and those requiring large numbers of relations to obtain all the information. Some queries require whole series of conditional statements in order to derive the required result. Hence, although SQL is powerful, it requires expertise to use satisfactorily.

5.4 DATA DICTIONARY SYSTEMS

A data dictionary system (DDS) is a software tool for managing the data resource. It enables the recording and processing of 'data about the data' (meta data) that an organisation uses. DDSs were originally designed simply as documentation tools, ensuring standard terminology for data items (and sometimes programs) and providing a cross-reference capability. Over the years they developed in their capabilities and often became fully integrated with particular database management systems. The basic concepts of DDSs have subsequently been developed into system repositories for CASE tools which are discussed below.

A DDS is a central catalogue of the definitions and usage of the data within an organisation. Information systems development is helped by the clear definition of the data already in the database. The DDS can be accessed by each new information system as it uses the database and therefore the DDS eases the sharing of data. If used alongside a DBMS it could be said to be a directory of the database, 'a database of the database', although this might be better expressed as an 'information source about the database'. Thus DDSs are documentation aids providing reference to data in the database. A DDS holds definitions of all data items, which may be any objects of interest, their characteristics, as well as how and where the data is used, and how it is stored. The kind of information stored in a DDS for each data item is typically as follows:

- The various names associated with that data item. There are often different names used in different parts of the organisation
- A description of the data item in natural language
- Details of ownership (normally the department which creates the data)
- Details of the users that refer to the data item
- Details of the systems and programs which refer to or update the data item
- Details of any privacy constraints that should be associated with the item
- Details about the data item in the database, such as the length in characters, whether it is numeric, alphabetic or another data type, and what logical files or tables include the data item

- The security level attached to the data item in order to restrict access to it
- The storage requirement
- The validation rules (for example the range of acceptable values)
- Details of the relationship of the data items to others.

Some form of query language (section 5.3) is usually provided to enable questions to be easily asked of the DDS. Where the DDS was associated with a particular DBMS, then this would be the query language of that DBMS, and the data dictionary would itself be held on the database, just like any other database file.

From the point of view of the organisation, management is now becoming more aware that the data of the enterprise is a valuable and important resource which must be properly managed. There must therefore be a knowledge of what data exists and how it is used. There must be control over modifications to the database and the processes that use it. There must also be control over plans for new uses of data and over the acquisition of data. The DBMS may well achieve some of these objectives itself, but in order to gain full control over the data resource, it is necessary to collect and store information about the data.

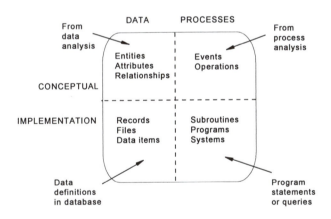

Fig. 5.8: The four quadrants of a DDS

Although such DDSs prove very useful, it did not take long for people to realise that this was only half the story and that a similar dictionary was required for documenting the processes and programs in the organisation. Further, given the fact that cross referencing between the data and the processes was required, it obviously makes sense to integrate the two. The DDS was therefore also used as a storage base for process and program information including code (sub-programs) which enabled the reuse of sub-programs in a number of applications. Many DDSs therefore became more than mere points of reference about data, but also held information about processes. As a result, some people thought that the term DDS was no longer appropriate (as it was no longer just about data) and so the term system dictionary, and later system encyclopaedia or system repository emerged.

This wider view of a DDS is reflected in figure 5.8. The horizontal divide is between logical information and physical (or implementation) information. The vertical divide is between information concerning data and that between processes.

The logical view identifies elements from the analysis and design stages, such as entity models, functional hierarchies, process and event models. The physical view provides details of the applications in computing terms. The processes are defined in terms of systems, programs and sub-programs (modules), and the data in terms of files (or relations), records and fields. Some systems also include an operations view as part of the implementation level. This described information relating to the operation of the system, such as the schedule for running the computer information system and its hardware requirements.

5.5 SYSTEMS REPOSITORIES

In section 5.4, we saw that the original idea of the data dictionary to support databases has been expanded to include information to support wider aspects of information systems development, in particular, the inclusion of process information. With these changes, the term data dictionary was no longer appropriate and the term systems repository or systems encyclopaedia evolved. It has been suggested that there is a difference of meaning between the two terms: that a repository is an empty shell which becomes an encyclopaedia when it is filled with information. This distinction is not particularly useful, nor universal, so we treat the two terms as interchangeable. Arbitrarily, we shall use the term repository in the rest of this chapter.

The systems repository is a development from the data dictionary discussed in the previous section, even those holding information about processes as well as data. In section 5.7 we discuss CASE tools, and these will have a systems repository as part of their integrated 'toolbox'. The systems repository is a much more sophisticated tool than the DDS, for example, it contains the information necessary to integrate the different stages of information systems development. To give an example of this potential, it may hold the logical design information and enable this to be used (usually in conjunction with a CASE tool) as the input to physical design and then code generation. Using the systems repository in the CASE tool, it may even have the ability to then go back and change the logical design and automatically regenerate code to accommodate the changes.

The repositories for Lower CASE tools (which address the later parts of the life cycle) would contain this type of data relating to the physical and operational elements of data and process, for example, physical data items, processes, modules, code and test data. The repository for an Upper CASE tool (which address the earlier parts of the life cycle) would contain data referring more to the logical and functional levels, for example, data and process models and diagrams. Integrated CASE tools contain a combination of the two. The different types of CASE tools are discussed in section 5.7.

More sophisticated repositories (sometimes termed active repositories) also contain information that enables the rules of a technique or even an information systems development methodology to be applied. These may permit cross checking, analysis, validation, consistency and completeness checking. There may be a separate element of the repository (sometimes known as the repository manager), in which knowledge or rules are embedded. In some repositories these 'rules' (which could be the rules of the methodology) are locked (or hard coded) into the repository. In other repositories they are more flexible and easily changed or defined in an expert systems language. In theory at least, therefore, a repository contains everything needed to support the creation and maintenance of information systems in organisations.

An example of a repository is Application Development Workbench (ADW). This is designed to support the Information Engineering methodology (section 6.4). The repository contains information about all the objects in the development project, be these entities, processes, relationships, attributes or whatever. The repository is structured in two parts, one dealing with the organisation's information and data, and the other with the organisation's functions or activities. Information is also kept about the relationships between one part to the other, for example, the data used by the various functions. In addition, there is the Knowledge Co-ordinator which sits between the diagramming support tools (section 5.6) and the repository. This is a kind of expert system containing the rules of the Information Engineering methodology and it enforces the correct standards for completeness, consistency, and accuracy of the diagram content.

A further interesting feature of ADW is the approach adopted to the storage of the various diagrams. It does not store them as diagrams, for example, as data flow diagrams, but as a series of definitions about the objects in the diagrams. This means that objects that appear in more than one diagram are only stored once and that diagrams are generated from the current information in the repository, as and when they are needed. When changes are made to one diagram, the effects of that change are automatically reflected when other diagrams in which the object appears are generated. It is the Knowledge Co-ordinator that translates the repository information into diagrams and displays, and vice versa. This enables the basic repository information to be displayed in a number of ways according to needs. For example, an object may feature in an entity model, a data flow diagram and an action diagram.

Some repositories contain information about the wider framework in which these systems sit as well as information that enables software to be created and maintained. This means that the repository should contain information beyond that which is needed to create software systems and include models of the organisation and environment, that is, the framework of the systems. This is a view of the organisation in terms of business areas, functions, hierarchies, departments, locations, strategic relationships, critical success factors, objectives, plans and so on.

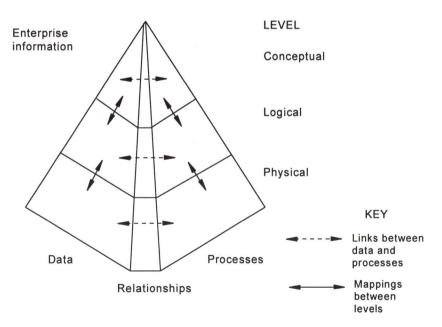

Fig. 5.9: Repository with enterprise information

Some commentators have termed this the 'enterprise model'. McGaughey and Gibson (1993) use an analogy with architecture. An individual house does not stand alone but is part of a wider environment of shared plumbing, sewers, countryside and so on, which are all of critical importance. The same is true in information systems, where the individual system is a house and the wider environment is the organisation and its environment. If the wider blueprint does not exist, then we have no idea if the system is out of synchronisation with the rest of the environment. This blueprint is the enterprise or organisational model, which has successive layers each providing greater detail and eventually enabling the developers to build a system. Figure 5.9 illustrates the addition of the enterprise information at the conceptual level to a traditional repository containing information about data and processes, and also emphasises the relationships between them.

So far we have discussed repositories as conceptually one large database containing all the information required to model and support the enterprise and the development of software. Some are indeed like this and are large centralised databases (usually relational) residing on a mainframe computer. Others, however, are distributed, and elements of the repository can reside in different places on different servers or machines in a distributed or client/server environment. It is more difficult to control the repository in a distributed environment.

With a centralised repository there is usually only one copy of the data and there are few problems concerning updates, as the only copy is the latest. In some situations, carefully controlled versions of part of the repository can be downloaded to a PC or workstation so that an individual or group can work on a limited subset of the repository and, when the work is finished, upload and update the main repository. A fully distributed environment presents more updating and consistency problems, but this is increasingly the type of environment that developers require. Other important issues of decentralisation concern the handling of security, integrity and recovery.

As indicated, most repositories are themselves databases of the relational type, which reflects the dominant technology when they were developed. Yourdon (1992) suggests that this is likely to prove inadequate for the future, and that a repository based on an object-oriented database management system is likely to be better for dealing with the wide variety of data types in the future, for example, the unstructured data such as image, voice and graphics of multi-media systems. The implication is that repositories should themselves be developed using object-oriented techniques.

A CASE tool usually comes with its own integral repository provided by the tool manufacturer, but there have been attempts to make repositories more open, such that they can be used with a variety of different tools from different vendors. The most notable example of this type has been IBM's product Repository Manager (RM) which supports IBM's life cycle development environment called AD/Cycle. This is designed to allow a wide range of different development approaches for different requirements and different platforms. A number of third party tools can be used with RM.

The concept of the open repository is attractive as it allows an organisation to choose the best tools available without sacrificing integration between the results. However, there is a degree of scepticism about the ability of vendors to achieve such objectives.

5.6 DRAWING TOOLS

Drawing tools help developers to create and maintain the various diagrams required in systems development. Most are designed to support the drawing of one or more of the common diagramming techniques and do not usually support any particular methodology. However, sometimes they offer the same technique drawn under the conventions of different methodologies. For example, they may give the options of drawing data flow diagrams following the Gane and Sarson, Yourdon or other conventions.

Drawing tools are important, first, as a useful automated aid in systems development, and second, as a precursor to more sophisticated tools, in particular, CASE tools. Drawing tools help in the drawing of many of the diagramming techniques described in Chapter 4, such as data flow diagrams, entity models, process logic techniques, entity life cycles and so on. They are sometimes described as documentation support tools, being designed to take the drudgery out of drawing and revising documents, and thus making the implementation of changes easier.

Additionally, they contribute to the accuracy and consistency of diagrams. For example, in a data flow diagram, a drawing tool can check that levels of the diagram are consistent with each other. This will include ensuring that all data flows on a higher level diagram appear on the lower level ones and that the descriptions are consistent across a set of diagrams. Drawing tools can also be used to ensure that certain documentation standards and conventions are adhered to. For example, processes must have inputs and outputs and data stores must be specified in terms of contents and flows.

Most drawing tools will prevent users from doing things which are not permitted in that diagrammatic convention. With an on-line reference manual and context sensitive 'help' facilities available in some tools, users can be guided through the technique.

Probably the greatest benefit of their use is that analysts and designers are not reluctant to change diagrams, because the change process is simple. Frequent manual re-drawing is not satisfactory, because of the effort involved and the potential of introducing errors when re-drawing. Without such drawing tools many a small change required by a user would not be incorporated into the documentation of the system.

Such drawing tools usually support only a single technique or a few basic techniques (most commonly, data flow diagrams and entity-relationship diagrams). Although useful, these are somewhat limited in the sense that, for example, much of the information required for a data flow diagram is also required, in a slightly different form, for process logic representation and elsewhere. It makes more sense to have a central repository of all the information required for the development project. This will ensure consistency between techniques as well as within a diagramming technique. This provides the potential for integrated, co-ordinated and consistent support throughout all aspects of the life cycle, ensuring a smooth transition from one phase to the next. With these ambitions the simple drawing tools began to evolve into more sophisticated products known as CASE tools.

5.7 CASE TOOLS

CASE is the term used to describe software support tools that help in the applications software and information systems development process as a whole, not just in the drawing of some individual diagrams. We use CASE as a generic term applying to systems development in general (computer aided *systems* engineering) although it is more usual to define it more narrowly (computer aided *software* engineering) which was the original wording of the term (section 3.12).

The definition of CASE, as we find with much of information systems terminology, does not prove to be easy. Many users of the term fail to define it at all and those that do are somewhat inconsistent. Others define it so broadly that almost any software could be argued to be a CASE tool. We attempt a definition of CASE, based on that provided by Thompson (1990), but significantly modified as:

> any computer software and/or system which is specifically
> designed to support any sub-stage, stage or stages of the

information systems development process of a computer-based information system or any aspect of the management of these tasks and processes.

This is a fairly wide definition but does not include generic software, such as word processors or spreadsheets because although they might conceivably be used in the process and management of software development, our definition excludes them as they are not 'specifically designed' to support the software development process. Individual drawing tools ought not to be included, as these support a technique and not a sub-stage, stage or stages of the information systems development process. Compilers, interpreters, editors, and other software for the design and programming of software would also be excluded because they are not integrated with other parts of the development process, they are not 'support' for the process but are the process themselves.

Our use of the term CASE includes a wide variety of tools and support environments that are also known by a variety of other names, such as workbenches, builder tools, project support environments, integrated project support environments (IPSEs) and system factories.

A distinction is often made between Upper and Lower CASE tools (or as they are sometimes known, front end and back end tools). The purpose of this distinction is to indicate which stages of the life cycle they address. Upper CASE includes tools which are used at some point in the strategy/planning, analysis or logical design stages, whereas Lower CASE tools are concerned with aspects of physical design, programming, and implementation, including automatic code generation. A third category of tools, known as Integrated CASE (sometimes abbreviated to I-CASE) are tools that integrate both Upper CASE and Lower CASE elements into a single, fully integrated development and support facility.

This suggests more agreement about terminology than exists. For example, Iivari (1993) describes a tool as integrated when three or more phases of development are supported. Another definition of an integrated tool is that it supports all phases of a particular information systems development methodology.

Integration in CASE is particularly important and we shall examine this further. Kurbel and Schnieder (1993) suggest there are a number of elements of integration. The first is horizontal integration, which is the integration of different tools at a particular stage of the development cycle. At the analysis stage, for instance, there are a number of different techniques which are supported by different tools, for example, data flow diagrams, entity models, function decompositions and so on. These are regarded as horizontally integrated if information is shared between the tools and if changes made in one diagram (using the diagram support tool) are reflected in the other diagrams, where appropriate.

Vertical integration, on the other hand, is the integration of tools between different stages of the life cycle, and this means that the results from one stage should be available in an automated form to the other stages. The results should be capable of being passed forward to subsequent stages of the life cycle (known as forward integration). Further, for highly integrated tools, the results or changes made at later stages of development should be capable of being passed backwards, and be reflected in, earlier stages. This is known as reverse integration.

Although a number of tools exhibit forward vertical integration (of varying degrees of comprehensiveness) the same cannot be said of reverse integration which hardly appears to exist in any form. This presents problems if changes are made to the physical design or code in the later stages of the life cycle that impact on (or render inaccurate) the models of earlier stages, because the documentation becomes inaccurate. Further, when maintenance or systems enhancements have to be performed, the developers cannot simply go back to the analysis models, make the necessary amendments and re-generate, because the changes to the design or code would also need to be made to the re-generated version. Integration is about achieving consistency across all aspects of information systems development and ensuring that changes to one diagram, for example, will be reflected in all other related diagrams.

Kurbel and Schnieder identify another important aspect of integration which they term interpersonal integration. System development, in organisations of any size, is a matter of co-ordinating the work of many people and perhaps many development teams, frequently in different locations. Additionally, the work may be performed in parallel, at least within development stages, and possibly in parallel across stages. CASE tools, it is argued, should ensure that the work is co-ordinated and consistent. An important aspect of this is version control, that is, the organising and handling of the large numbers of different versions of systems that exist. Again, this is not a feature found in most CASE tools available.

Figure 5.10 presents an idealised integrated CASE tool with all features. In order to make our description of CASE tools more tangible, we will next look at Information Engineering Facility, and then look at potential benefits and costs and attempt to provide an evaluation of the CASE approach. Finally, we look at the directions in which CASE tools are developing.

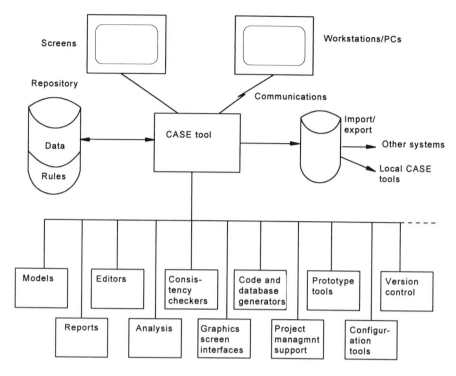

Fig. 5.10: A schematic for an integrated CASE tool (generic)

There are well over 1,000 CASE products on the market with about 500 vendors world-wide, and this figure is still increasing. Existing CASE products are continually changing and evolving. New products, especially object-oriented tools, seem to be appearing monthly. Some other tools are being re-positioned and renamed as CASE tools. Other CASE vendors and products have changed ownership as a result of company acquisitions and disposal policies. In this environment, examples of CASE tools are inevitably dated almost before they are written. Therefore, in this book, we have decided not to describe in any detail the wide variety of products available in the marketplace. Rather we will list a few of the common products (mainly I-CASE and Upper CASE) and then describe in more detail Information Engineering Facility (IEF). This supports the Information Engineering information systems development methodology and although it has been available for many years, it has developed into one of the most sophisticated tools available.

Products available at the time of writing include:

• Information Engineering Facility (IEF) from TI Information Engineering

- Oracle CASE from Oracle Corp.
- Systems Engineer from LBMS (Learmonth and Burchett Systems)
- Pro-IV from MDIS
- Excelerator from Intersolv
- Application Development Workbench (ADW) from KnowledgeWare (this is the successor of IEW which was described in the first edition of this text)
- Software Through Pictures from Interactive Development Environments
- Foundation from Andersen Consulting
- SDW from Cap Volmac/Hoskyns
- Maestro II from Softlab
- Method Manager from Manager Software Products
- System Architect from Popkin Software and Systems
- Modeler from Casewise
- Virtual Systems Factory (VSF) from Systematica
- Select from Select Software Tools
- MetaEdit from MetaCase Consulting.

Information Engineering Facility (IEF) is an integrated CASE tool developed by Texas Instruments to support the Information Engineering (IE) methodology (section 6.4). IEF was originally conceived in 1984 with the objective of 'simplifying application development, improving systems quality and enhancing productivity'. IEF implements the main diagrams and standards of IE, transfers information between diagrams, performs consistency checks, evolves objects from one stage to another, and can generate a database and source code. IEF is widely recognised as being one of the best known examples of a high-end, integrated, CASE tool. IEF comprises five toolsets as follows:

- The planning toolset
- The analysis toolset
- The design toolset
- The construction toolset
- The implementation toolset.

These are all integrated together via the encyclopaedia (the IEF term for the repository). The main encyclopaedia is designed to be held on an IBM mainframe with a variety of local encyclopaedias that can be held on workstations and PCs running various operating systems, including MS Windows and OS/2. The toolsets can also be run on workstations and are mouse-driven with multiple windowing capabilities. This conforms to the IBM defined SAA common user access standards. IEF can generate applications for a variety of target hardware and software platforms. Along with the IBM mainframe, these include Digital, HP, Unix and OS/2.

IEF is described by Texas Instruments as being 'integrated'. In their terms this means that across the tool there is a common look and feel, the use of an encyclopaedia to enable consistency, and the automatic generation of objects and code from higher-level objects. We will consider each toolset in turn.

Planning toolset: The planning toolset is designed to support a strategic approach to systems development following the identification of high-level business requirements. The planning toolset helps in the production of three architectures: the information architecture, the business system architecture and the technical architecture. First, planners build a subject area diagram which is a high-level entity model of the subject areas of interest to the business. The data modelling tool in the planning toolset enables the users (planners) to build the subject model and view the whole model or individual subject areas which can then be exploded to reveal the component entity types of the subject area. The tool ensures consistency across the diagrams and shows any relationship aggregation lines. A relationship aggregation line represents a relationship between entities in different subject areas.

Planners also construct high-level function hierarchy diagrams using the activity hierarchy diagramming tool which enables the functions and their position in the hierarchy to be captured and displayed in various ways. The activity dependency diagramming tool is then used to document the sequence in which the functions must be performed, with arrows indicating the direction of the dependency showing what function must occur before another. There is also an organisational hierarchy diagramming tool which enables the construction and manipulation of an organisation chart. Finally in the planning toolset is the matrix processor which enables various interactions to be identified and depicted.

The matrix processor can be used to show the interactions between any two sets of things, but particular support is provided for 40 standard matrices, which include objectives, strategies, goals, critical success factors (CSFs), entities, subjects and functions. The tool is basically a two dimensional matrix (section 4.15) with, for example, functions as rows and objectives as columns, and an entry is made in the relevant cell when a particular function supports a particular objective. The matrix may reveal that some objectives are not supported by any functions, and therefore that objectives have been lost, or that only one or two functions support the majority of the objectives and therefore these are the key functions.

Planners use such matrices in a variety of ways, and the tool makes the comparison and analysis that much easier. One important use of the matrix processor is to plot the interactions between functions and entities, that is which functions affect which entities. This may be create, delete, read or update (CRUD) as shown in figure 4.64. The matrix processor is also able to cluster, and this might lead to the automatic production of a new matrix where all entities affected by a particular function are clustered together.

The diagramming tools in the planning toolset are integrated. Once, for example, the entities are entered in the data modelling tool, they will appear automatically in the matrices concerned with entities. This means that elements are only entered once and when they appear in a matrix none will be forgotten.

Analysis toolset: Analysis refines a particular area of the business identified in the planning stage, and any information captured in the planning toolset is automatically available to the analysis toolset using the encyclopaedia. In this stage (business area analysis in IE), analysts enter further details, often using the same toolsets as in the planning stage. For example, the subject area diagram is expanded into an entity-relationship diagram using the data modelling tool. Relationships between entities are now defined, including their cardinality and optionality. This is done by the use of dialogue boxes within the tool, as is the definition of the entity attributes and properties. The user can specify entity subtypes and the diagram can be viewed with these expanded or contracted. Functions are further refined into a series of lower-level processes using the activity hierarchy diagramming tool. For each process, the tool enables details to be entered. These details will include name, description, type, and any entities that the process uses. Similarly, the function dependency diagrams from the planning stage are expanded, where necessary, into a series of process dependency diagrams using the activity dependency diagramming tool and details of events that trigger processes and relationships with external objects are added. Some further matrices are also produced using the matrix processor.

The elementary processes are then defined as action diagrams (section 4.11) using the process action diagramming tool. The action diagram defines the steps required for each elementary process and the way that they interact with the entities. The tool automatically begins constructing the action diagram using information derived from other diagrams, such as the entity model, the process hierarchy and dependency diagrams, and the analyst can insert extra actions and manipulate the action diagram as required. The process action diagramming tool applies action diagramming rules and only allows entries that are semantically and syntactically correct. The information for checking semantic correctness is derived from the earlier diagrams in the encyclopaedia. In addition, the toolset allows processes to be 'synthesised'. This may be achieved by the tool asking certain relevant questions to clear up areas that have not been completed or, in certain circumstances, by automatically generating process logic based on the entity model and information from the matrices. For example, if an entity REGISTRATION has a compulsory one-to-one relationship to entity STUDENT, then certain action logic is implied by this that can be generated by the tool. All this provides the detail of the process logic upon which any subsequent code generation will be based. In certain situations, IEF can be used in analysis without the preceding planning stage having been undertaken, in which case all information is entered from scratch in the analysis toolset.

Design toolset: In IE terms, this includes support for both business systems design and technical design and the design toolset enables the designer to take the results of analysis and transform them into designs. The first stage is for each process to be transformed into a set of procedures by the dialogue design tool. Dialogue flow diagrams are produced and the tool allows the designers to specify control and sequences of screens. Next, the screens themselves are designed using the screen design tool. This will automatically produce an initial attempt at the screen design, based on the information in the encyclopaedia, which can then be modified. The tool also attempts to provide previously defined screen elements to the designer and so encourage consistency in design. Indeed, designers can create templates that enforce standards across applications. The screens are then prototyped, showing the layout and flow of screens, and input validation can also be performed using the tool.

The process action diagrams from analysis can also be converted into procedure action diagrams using the action diagramming tool. Designers can specify the detailed logic statements and associate processes with commands from the keyboard, specify runtime error routines and so on. Again, as in the process action diagrams, the tool identifies potential errors and prevents them occurring.

Up to this stage, the design has been at the logical level and not dependent on any target hardware or software environment. Technical design now requires the business systems design to be taken to the next stage which is physical design where the target environment is specified. Using the toolset, the designers can specify the target environment and the constraints that this implies. Common specifications are already stored in the encyclopaedia. Physical design is then initiated, and the data model is transformed into a physical database design of records, fields, linkages, entry points and so on, depending on the actual database specified in the physical environment. These would be tables, rows, columns and foreign keys in a relational database environment. The physical design can be modified and tuned by the designers as necessary, usually for performance reasons. Such changes do not modify the conceptual designs and only changes as business requirements change the logical models.

Construction toolset: The next stage is construction, and the automatic production of complete application systems by the generation of code. The tools support the generation of either COBOL or C source code and embedded SQL calls to the database. The tool also produces screen definitions, graphical user interfaces, database definitions, referential integrity triggers to control database deletions and a transaction control program. The code generated is based on the logic specified in the action diagrams and the entity models.

Developers can test and modify the code without tampering with the source code. Changes are made to the code by changing the action diagrams (or further back in the analysis stage) and regenerating the code. The toolset 'remembers' any changes made for tuning purposes and then reapplies them to the regenerated code. Further, for small changes, not all the code needs to be regenerated and, using various dialogues, the developers can specify which components require regeneration.

Implementation toolset: Finally, the implementation toolset (which resides on the target hardware) enables the installation of code and database on the computer. This includes the compilation, linking and binding of the application and the allocation and building of the database, plus a facility that enables the running and testing of the application on the target computer.

The encyclopaedia: The encyclopaedia is the key to IEF and enables the storing of models, the concurrent use of these models, the progression from one stage to the next, and the transformation of the models ultimately into complete developed applications. The main encyclopaedia, known as the host encyclopaedia, resides on a mainframe and is implemented as a DB2 relational database. Subsets of the models on the main encyclopaedia can be downloaded to workstations and PCs to enable individual developers to work on them in a workstation environment for convenience, and to enable teams to work on developments concurrently. Any changes to models are then uploaded back to the main encyclopaedia in a controlled way from the local encyclopaedias to ensure consistency. The encyclopaedia also provides version control, which means that multiple copies of models for different purposes can be stored and used without confusion.

The encyclopaedia stores definitions of models rather than graphical representations of the model, and therefore the toolsets use these definitions to produce the diagrams that are required. Many of the diagrams are based on common information, and this enables a high degree of consistency between diagrams as they are constructed as and when necessary from the stored definitions. This also enables any changes or updates made by developers to one diagram to automatically be reflected in other diagrams. For example, in an entity diagram, a change to a relationship between two entities from optional to mandatory will automatically be reflected in all other diagrams that use this information when they are next displayed, because they will access the one definition of that relationship in the encyclopaedia. The storing of definitions in the encyclopaedia rather than the storing of graphical representations not only enables models to share information but allows the information to be easily passed forward to subsequent stages.

The encyclopaedia also enables consistency checking. This can be initiated at any stage, and on the whole development or subsets, such as an entity model, functions, processes, action diagrams and so on. Some checks are enforced when information and definitions are entered. For example, an attempt to use the same name for two functions or entities would be highlighted as they were entered. Other checks will only be made upon request or before proceeding to a subsequent stage. An entity model, for example, would be checked for completeness, in that it has attributes and relationships, and that these are consistent with the functions and activities defined, before proceeding from analysis to design. This concludes our examination of IEF, and we will now proceed to examine some of the benefits derived from CASE tools.

The common benefits cited for CASE are outlined below. These are generic benefits of integrated automation and do not necessarily apply to all CASE tools. We have divided the benefits into major groupings as follows:

Improvements in management and control: Applications development, particularly for large projects, is inherently difficult to manage and control. The process must therefore be tightly managed, and the IT profession has not historically been very good at this, particularly in the areas of estimation and keeping to budgets and schedules. CASE tools can help in this process by providing a central repository of information concerning the project, including rules and standards to be followed, and experience from other projects, such as the length of time certain activities actually take in the organisation. They can also help with estimation, risk analysis, project planning and the monitoring of project progress. Particular support for techniques may be included, such as function point analysis, COMOCO, PERT and critical path analysis. Some of these were discussed in section 5.2. CASE tools can also structure the work of developers, for example, by handling the devolvement of tasks to developers so that work can be completed in parallel and the subsequent recombination of the work put into a coherent whole. They can also support the change control process and ensure that new versions and releases are well organised and managed.

Improvements in system quality: The problems associated with specifications have already been discussed. It is argued that CASE tools can help overcome these problems by providing better and more complete specifications through the use of diagrammatic representations that are easily modifiable by developers and users. These should also represent the real requirements better, partly because there is less resistance to changing diagrams because the tools make it easy.

Improved designs, better reflecting the specifications: Another problem of systems development is that the designs produced do not always accurately reflect the requirements specification. This can be the result of incomplete or conflicting information, and designers often make guesses or opt for whatever is the easy solution. CASE tools can help by providing the necessary information from the specification, having it available to designers from the repository and also by automating some of the process. This also helps to produce consistent designs, including that across applications.

Automated checking for consistency according to the rule base: CASE tools can automatically check the consistency of information input at the analysis and design stages, including information input using models and diagrams. They can highlight information or areas that have been missed or interfaces that do not match as well as incomplete information. This kind of automated consistency checking can be based on a set of rules concerning the methodology as a whole or those of the various constituent diagramming techniques. These rules would be included in the systems repository. Such checking should improve the consistency and quality of deliverables and thus the final information system, leading to less re-working and change at later stages.

Greater focus on analysis rather than implementation: It is often argued that the use of CASE enables and encourages the focus of development to be changed from the later stages of the life cycle, such as design, coding and implementation, to the earlier stages of analysis and requirements determination. Such a change of emphasis is likely to lead to better quality systems, as problems are detected and corrected at an earlier stage than with purely manual methods. The earlier in the development process that problems or errors are detected, the cheaper they are to correct. In one study (Stone, 1993), it is argued that the traditional development planning and analysis stages took 25 per cent of the total time, whereas with CASE it accounted for 55 per cent of the time; but, at the other end, construction took 50 per cent of total time in traditional development but only 20 per cent with CASE.

Enforcement of standards and consistency: CASE tools can also help with the definition and enforcement of various standards in development. The tool itself often embodies certain conventions and standards which can help to ensure consistency in the development of individual projects and across different projects in an organisation. The tool may enforce discipline by not allowing developers the freedom to ignore or contravene certain rules, procedures and standards. It can also ensure consistency in the use of techniques, definitions, and terminology in the organisation. As well as enforcing standards, some CASE tools may ensure that the rules and objectives of a particular information systems development methodology are followed. Again, this may be achieved by the tool not allowing the developers the freedom to diverge from the requirements of the methodology and by the tool itself adhering to the methodology in what it does.

Improvements in productivity: Perhaps the most cited benefit of CASE tools is that of improved productivity due to a number of factors. First, it is argued that information systems are developed more quickly than with conventional methods. This is obviously a very attractive benefit, given the enduring problems of systems development in this respect. Faster systems development is achieved by a combination of a number of the characteristics of CASE tools, some of which have already been discussed above. These include improved management and control, the use of graphics and other tools that make it quicker to create and change diagrams and specifications and the automation and elimination of various manual stages, including the automatic generation of some aspects of design and the automatic generation of code. This latter benefit can potentially make a significant improvement to productivity, as the writing of code has always been a very labour-intensive part of the development process.

The second element is the ability to develop systems with fewer people. The automated support for much of the process and the automation of some tasks, it is argued, means that fewer developers are required. With fewer people, there is the added benefit that the number of interfaces and the communication required between developers is also reduced, which is also likely to enhance the speed and quality of development. The law of diminishing returns applies to systems development. This suggests that after a certain figure is reached, the more developers that are added to a project, the longer it will take to finish. This indicates that a kind of inverted economy of scale may apply to systems development, because the more interfaces there are between people, the slower things happen.

The third element contributing to improved productivity is the reduced costs of development. This is essentially an effect of the first two factors, that is, faster development and fewer people.

A fourth element of improved productivity is the ability to re-use existing development objects. The information captured by a CASE tool over a number of projects may eventually provide a repository of models or objects of various kinds that can be used again. These may include analysis and design models of all types and libraries of common code that can be utilised in future developments. Depending on circumstances, these models may be used in their entirety or are amended according to the requirement of the new project. This can save a significant amount of development effort and also help achieve consistency between applications, as the standards in the original models will be incorporated into the new developments.

Reductions in maintenance: It is argued that CASE tools help reduce the large degree of effort required for both maintenance and enhancement of existing systems. First, CASE tools can produce good and consistent documentation which can lead to easier maintenance. Second, the better quality specification and analysis provided by CASE tools means there will be less change and thus less maintenance. Third, the improved design and implementation, including some automation, results in fewer errors at the programming, testing and implementation stages, thus requiring less maintenance.

Accurate and effective testing is also an important element in reducing the maintenance load. Traditionally, this has been carried out by a separate group of people because the developers themselves were not trusted to test a system that they had developed effectively. Programmers were thought to be the worst people to test their own systems, as they assumed the system would work because they had written it. The consequence of having a different set of people performing the testing is that they have no knowledge of how the system was developed. Further, they could not take advantage of any verification and validation potential that the development methodology might have to offer. CASE tools can provide this enhanced testing by helping to administer and control the activity and by generating test data, applying (or even simulating) it, and analysing and comparing the results. The type of test data generated can be derived from, and reflect, the requirements of the analysis and design stages. For example, if there is a requirement that when a particular process is invoked then a subsequent process must also be performed, then this can be captured as knowledge in the repository and a particular set of tests and relevant data automatically produced. An example may be that a debit from one account must be accompanied by a credit to some other account. Such testing is more likely to test the requirement effectively than random testing, and the benefits apply both to the initial systems development and any subsequent maintenance or enhancement to the system.

Further, with CASE, the traditional form of maintenance is made obsolete because changes are not made by directly re-working the code in response to errors and changing requirements, but by going right back to the analysis and design stages and amending the original diagrams and specifications and re-generating the code automatically. This helps eliminate the frequently-encountered problem of introducing new errors as a result of correcting existing ones.

Re-engineering (or reverse engineering) of existing systems: The problem in many organisations is not so much that of developing new systems, but the maintenance and enhancement of their old systems, some of which are based on 1960s designs, third generation languages, and dated file and access methods. These systems are commonly termed 'legacy systems'. An indication of the problem is provided by the claim that there is currently still a total of over 77 billion lines of COBOL code in applications around the world (Niessen, 1990). Re-engineering is 'the application of tools, techniques and methods to cost-effectively extend the useful life of application systems' (IBM GUIDE Committee). Re-engineering changes the underlying technology of a system without affecting the functioning of that system. Thus, for example, the hardware platform and environment, including the programming language of the applications to reduce maintenance costs, may be changed. The rapid developments in technology have rendered many existing systems, even some that are relatively recent, obsolete, not in the functional sense, but in the programming language used and the hardware on which the system runs. Such systems can carry a high maintenance workload and be difficult to enhance. Further, many manufacturers refuse to maintain old hardware, and it is difficult to integrate legacy systems with more recently developed systems. For many organisations, the cost and resource implications of scrapping these old systems and re-developing them from scratch on new hardware platforms is prohibitive. The normal use of CASE tools provides support for 'forward development', that is, the top-down, linear approach to the development of new systems. Some CASE tools are designed to support reverse engineering as well, providing the ability to capture the primary elements from current systems, such as their process logic and the data they use, including entities, attributes, names, locations, sources, edit criteria and relationships. From this captured information, the tool can help to clean up the data definitions, produce entity models, re-structure the process logic and build process hierarchies, and construct the repository for the old system. The CASE tool can then be used in the normal forward development mode to produce new systems. The degree of automation of the reverse engineering process varies and there still needs to be manual input at most of the stages. Nevertheless, CASE tools can be very helpful in this context. Niessen (1990) cites an organisation that re-engineered 1.5 million lines of code in a month which resulted in the maintenance support required being reduced by half, from 28 to 14 people.

Strategic contribution: The potential for information systems to contribute to the achievement of the strategic objectives of the organisation has been discussed in section 3.3, and the use of CASE tools can help by improving the quality of systems and the speed at which those systems are developed and enhanced. Additionally, the planning elements of a CASE tool may help to identify and prioritise those systems which are most likely to contribute to the business strategy.

Improved responsiveness: This is really a function of improved maintenance and enhancement of systems that have already been discussed, but the particular element emphasised here is that systems developed using a CASE tool are likely to be more easily and quickly enhanced leading to improved responsiveness to changing and evolving business needs.

Portability: Some argue that the use of CASE makes it easier to move systems from one hardware platform or environment to another. This is really a function of the ability of some CASE tools to generate code for a variety of different languages on different hardware platforms. For a particular application, the code can be re-generated for a different environment without affecting the functionality of the application. Whilst this makes it easier, it does not in practice mean that there is no manual intervention required.

Keeping up with the state of the art: Some people argue that CASE tools enhance the credibility of the information systems group. It indicates to the rest of the organisation, and perhaps the world at large, that they are at the cutting edge of the latest technological developments. This is perhaps not a totally justifiable benefit, as it seems to be an argument for technology for technology's sake. A better justification is that it helps to attract and retain good information systems staff and increases satisfaction among developers.

Having identified the potential benefits of CASE tools, we will discuss the other side of the equation. The most obvious cost is that for the CASE software itself. Buyer guides indicate a very wide range of costs from about £5,000 to £250,000, and the authors know of at least one example where the cost was £500,000. These differences arise partly because at the lower end they might refer to a single PC user and at the upper end to a mainframe-based system for multi-project developments in very large organisations.

However, the cost of the software is really only the beginning, and other costs frequently get ignored. There is the cost of the hardware upon which the software will run and this can be very expensive. Even if the hardware already exists in the organisation, there is the opportunity cost of using it for the CASE tool. Frequently, the CASE tool takes more of the hardware resources 'than was originally envisaged and may eventually require a dedicated machine.

There are then a series of softer costs associated with the adoption of CASE tools. The first of these is the not insignificant staff education and training costs. These are costs that apply not only for the professional developers, but for users and user management as well. These costs are not just one-off, as is sometimes assumed, but are a long-term requirement, because of staff turnover and new versions of the tool. Consultancy and training is required, sometimes from the tool vendor which can be particularly expensive. In some cases, the training may only be available from the vendor, and vendors may make most of their profits from consultancy and training. Consultancy related to CASE may absorb most of the IT consultancy budget. A further problem is that staff turnover tends to increase as experienced CASE tool developers are currently much sought after.

Parkinson (1991) identifies further soft costs:

- *Development of appropriate conditions:* The setting of standards, working practices and the resolution of conflicts all need to be sorted out and an appropriate environment and culture for the use of the CASE tool developed. Again this takes time and effort and is an initial cost of adopting a CASE tool.
- *Integration of the new tool:* The CASE tool needs to be integrated into the existing development environment so as not to cause conflict. It is very unlikely that any organisation will be able to change to a CASE tool development environment except in well thought out and managed stages. All this may require organisational change and will certainly take management time and effort.
- *Customisation of the tool:* There can be quite a major effort required to tailor the tool for use in the particular organisation. This will take time and other resources, and often expensive systems support and consultancy from the tool vendor.

We would add the following costs to Parkinson's list:

- *People's time:* Often the time of people using CASE tools is not properly costed, as time is often assumed to be free and to have no opportunity cost. For example, if someone goes on a course, the cost of the course is usually included but not the time lost by that person. The time put in by users is also frequently ignored.
- *The cost of recruiting experienced staff:* Probably an organisation will not train everyone from scratch but will seek some developers from outside who are experienced in using and managing the CASE tool. Recruiting costs can be particularly expensive as such staff are well sought after.
- *Other hardware costs:* It is frequently the case that the hardware needs upgrading and more workstations are required as more projects are developed. Again, these costs may not have been included in the initial estimates.

Far too many organisations have ignored the softer costs in their cost-benefit analyses and concentrated solely on the direct hardware and software costs associated with CASE tools. An indication of the importance of these other costs is an estimate by Ernst and Young (1993) that they 'can easily amount to $2-$5 for every one dollar spent on the CASE tools themselves'. Based on experience from about 200 implementations, Parkinson (1991) suggests that true CASE tool costs average around £60,000 per workstation.

The evaluation of costs versus potential benefits is difficult and the whole area of CASE is surrounded, as are many IT developments, by a degree of 'hype', much of it emanating from vendors trying to market their products, but also, perhaps, from over-enthusiastic developers seeking quick solutions. The IT community is characterised by the greeting of new approaches and products with great enthusiasm and a belief that, contrary to previous experience, this latest innovation is going to solve all known problems. Inevitably, there is then a backlash against this overly optimistic view and a certain pessimism sets in whereby people begin to condemn the innovation as either worse than useless or nothing new and simply 'old wine in new bottles'.

Organisations that jump on the bandwagon, expecting a panacea, experience difficulties and problems. They then turn against that particular innovation and vehemently condemn it before rushing on to the next one. The truth with any innovation is usually somewhere in the middle of the two extremes. With CASE, it is not yet determined exactly where that boundary lies, but we have already seen evidence of a backlash. This is perhaps not altogether surprising considering some of the hype, for example, one vendor suggests that productivity of 25 times that of traditional development can be achieved with their product.

There are further concerns related to the particular context of this book. The first is that the technology of CASE tools might be distracting people from the real issues of information systems development, that is, a concentration on the tool rather than the development approach that lies behind the tool, and as a result the tool being used indiscriminately and inappropriately. Second, the tool may force people to use some methods that are not relevant or well enough defined. It has been suggested that CASE tools are sometimes purchased and used without enough thought being given to the processes that they enforce. In other words, the tool enforces a particular approach to systems development and it is this approach that needs to be carefully considered rather than the look and feel of the software. Some companies have found themselves implementing a particular development methodology without quite realising it, due to their use of a particular CASE tool.

There are others (Yourdon, 1992 and Alavi, 1993), however, that argue that one of the main problems of CASE tools are that they do not sufficiently enforce adherence to, or support, a particular methodology. Indeed Yourdon suggests that a 'CASE tool without methodology support is nothing more than a glorified drawing tool'. We have consistently argued that the correct approach is to choose the development methodology first and only then the CASE tool that supports that methodology. A CASE tool without an underlying methodology upon which it is based will not be able to progress from stage to stage because the rules and methods for doing so will not be defined, it is unlikely to involve the whole systems development life cycle, it will probably not be able to automate some of the activities (in particular moving from analysis to design), and it will not lead to code or application generation because the information required is likely to be incomplete.

A further issue is the degree to which the methodology is supported and the way the support is implemented. Vessey *et al.* (1992) suggest that the rules of the methodology can be supported in either a restrictive, guided, or flexible way. Jankowski (1994) expands on this, and defines three important aspects of their implementation: the timing of the rule, the invocation of the rule and the enforcement of the rule. The timing relates to when the rule violation is presented to the user; invocation is the mechanism by which the violation is communicated to the user; and enforcement is concerned with the options available to the user after a violation has been communicated. A rule is considered restrictive if the user 'is automatically presented with the rule violation while using an operator or while terminating use of an operator, and is forced to correct the violation before proceeding' (Jankowski 1994).

A guided CASE tool is one which provides the user with either active or guided choice. Active choice is the provision of unsolicited information and suggestive advice concerning the rule violation which the user can ignore and proceed without correcting the violation. Passive guidance is where guidance is solicited by the user at various points, that is, the user asks for various checks to be made, and the tool presents advice and suggestions, as with active guidance. On the other hand, some tools do not provide any checking of particular methodology rules or even the rules of a particular technique. Some tools allow any processes to be defined and these are only checked automatically at a later stage, for example, before proceeding to design.

The evaluation of a CASE tool is thus not a simple process, and the statement that a CASE tool supports a particular methodology is only the beginning of the story. The way in which the rules are enforced is also critical. A further problem is that the CASE tool vendor may interpret the rules of the methodology somewhat differently to the author of the methodology or the organisation adopting the tool.

Yet another area of concern in relation to CASE tools is the degree to which the claims made are really achievable in practice. There are many problems associated with the implementation of CASE and the productivity benefits are not of the magnitude suggested. For example, a study by Xephon, a market research company, led to the conclusion that CASE users have an average backlog of thirteen months compared with the fourteen months of non-CASE users (*Computer Weekly*, April 26th 1990).

Such statistics are by no means conclusive, indeed there are serious worries about statistics in this area in general, because very few organisations appear to have in place the necessary mechanisms for effectively measuring information systems development productivity (or any other benefits) either before or after the introduction of CASE. Fischer *et al.*(1993) quote a study where nine out of eleven companies reported improvements in user satisfaction, although none of them had any user satisfaction measurement systems in place. Thus many of the findings are based on perceptions rather than actual benefits. This is not to say that perceptions are not important and relevant, but it is important to know whose perceptions they are. It could, for example, represent those of the person who champions an innovation and whose career prospects depends on its success.

Whatever the degree of improved productivity claimed, there appear to be a growing number of indications that achieving them is more difficult, and takes longer, than might be thought. I/S Analyzer (1993) quotes a US study which indicated that 70 per cent of the CASE tools purchased were no longer being used one year after implementation. This seems rather high, but might be explained by a number of factors. First, the long learning curve period necessary, second, the fact that many CASE tools are initially sold, not for serious use, but for learning and experimentation purposes, and third, that some organisations purchase a number of tools for evaluation and subsequently use only one of them.

The learning period for CASE tools is long. Both developers and users need time to learn, assimilate and become effective in using the tools. In terms of productivity, it has been suggested that the learning curve (productivity benefits plotted against time) may actually dip below that of non-CASE tools, in the early stages. The length of this early stage, before improved performance is reached, has been estimated to be between six months and two years. It perhaps makes more sense to measure the length of the learning curve, not in time, but in terms of numbers of projects, in which case it has been suggested that it is not until the third or fourth project when productivity benefits begin to accrue.

Clearly CASE is not an instant panacea and some organisations may not be prepared for the kind of long-term investment that is required, particularly as productivity may actually decline initially. As was shown earlier, the time spent on analysis tends to increase with the use of CASE tools, and whilst this was suggested as a positive benefit, there have been instances where CASE projects have been abandoned before completion of the analysis phase. Many people expect all phases of the systems development life cycle to be speeded up, but when they find that the earlier stages are taking longer, they are not impressed. This implies that the introduction of CASE must be handled with care and that peoples' expectations relating to benefits, problems and timescales need to be realistically managed.

Our second assertion, that many of the CASE tools are not seriously used is supported by a US study (Howard and Rai, 1993) that differentiated between depth of CASE injection and breadth of CASE diffusion. Injection entails a period of product acquisition, experimentation and evaluation, and diffusion is the widespread, everyday, routine use of the product. Whilst they identify that CASE products have sold well, with an overwhelming proportion of information systems departments actively experimenting with CASE (injection), few departments had broadly implemented CASE as their predominant standard systems approach (diffusion). They also discovered that the most used element of CASE were the diagramming tools and the least used were re-engineering facilities.

Up to now we have discussed CASE tools in terms of the software development project as the unit of analysis. However, there are other dimensions, one of which is concerned with the context of the organisation and its characteristics. Here the focus is much broader. Fischer *et al.* (1993) suggest that CASE is most successful when it is seen as part of a process of changing the culture of an IT department, that is, when it is seen as a process of organisational development.

It may also be seen in the context of changing the organisation as a whole. For example, CASE may be seen as part of a process of empowering users or it may result in the centralisation of power in the hands of an elite in an organisation. It may be analysed in terms of its effect on the hierarchical structure of an organisation or as an element of organisation learning. Even if people do not wish to consider the effects of CASE on the organisation, they cannot ignore the effects of the organisational characteristics on CASE. The tools are not implemented in an organisational vacuum and there are many indications that success or otherwise is heavily influenced by a range of organisational and human factors, for example:

- The management approach
- Power structures in the organisation
- The degree of organisational creativity
- The organisational culture
- Work patterns
- Teamwork
- The incentive and reward systems
- Perceptions of job security
- Job satisfaction levels
- The role of champions and sponsors
- Change agents
- The history of innovation and experimentation in the organisation.

In essence, these are characteristics of the organisational fit of the tool, or, as it is sometimes termed, the compatibility of the innovation with its context. There is a wide literature in organisation theory, and also in the diffusion literature, concerning compatibility and fit (see, for example, Iivari, 1992, Kwon and Zmud, 1987 and Rogers, 1983).

Another important aspect of the organisational dimension of analysis is the maturity of the IT department and the software development process in the organisation. There are a number of models of the 'stages of growth' of IT in organisations and there are indications that certain stages of maturity need to have been reached to allow the effective introduction of CASE. Further, the type of tool, and its objectives and justification, might be different depending on the stage of growth at which it is introduced.

In terms of the stages of growth framework of Galliers and Sutherland (1991), which built on the work of Nolan, 1979, an IT department at either of the first two stages (ad-hocracy and foundation building) might be argued to be organisationally incapable of successfully implementing a CASE tool. An organisation at stage 3 (centralised dictatorship) might benefit from a restrictive CASE tool and justify it as a means of exercising tight control over the development process and achieving credibility in the organisation. However, at stage 4 (democratic dialectic) the relevant tool might be of the guided type and the emphasis and justification might be on the communication and co-operative work aspects. At Stage 5 (entrepreneurial), the tool required might be flexible, with the focus on the strategic and planning aspects.

A further organisational dimension concerns the way that innovations are adopted and diffused in organisations. Wynekoop *et al.* (1992) identify three CASE introduction strategies. The first is *'laissez-faire'* in which the tool is expected to be adopted without any deliberate organisational encouragement. The second is 'cautious', which is a slow but deliberate approach. The third is 'active', which is fast and requires a high degree of organisational and managerial push.

They also identify three types of innovation in terms of the degree and nature of the change that occurs. The first is 'compatible' innovation if the tool fits in with and does not change current working practices, such as the methodology. The second type is 'incremental', where the tool involves only small changes to current working practices, and the third is 'radical', if it requires major change and differs significantly from current experiences.

A further dimension beyond that of the project is that of the individual affected by the CASE tool. These individuals, or stakeholders, are first, the developers (or the primary CASE tool users), and their perceptions and feelings are important aspects in determining success. These perceptions can be analysed in relation to the degree of change to work practices, job satisfaction, reward, communication, teamwork and so on. It has been argued that CASE tools sometimes require the primary users to unlearn old practices and learn new ones, and that this may result in a perceived loss of status (Fischer *et al.* 1993). Further, there are indications that such changes may be more difficult for older, more experienced developers to make (Norman *et al.* 1989). However, it is not always made clear whether these difficulties are the result of the introduction of CASE or whether it is due to the introduction of an associated methodology. We suggest that the introduction of a CASE tool together with a new methodology into an organisation is a more difficult innovation than the introduction of a CASE tool to support an existing and well-established methodology, simply because of the greater degree of learning (and unlearning) involved.

The reaction of individual developers is not always negative, as some perceive the use of CASE as enriching their work. Others perceive it as deskilling, reducing their creativity, and increasing the ability of management to exert control, in much the same way that supermarket checkout systems monitor and control their operators. The reaction of individuals appears to be difficult to predict, but it is likely to be a key element in the ultimate success or failure of the introduction of CASE in an organisation. The planning and management of CASE introduction needs to focus on addressing these personal perception and motivation issues. Some organisations have recruited new graduates and trained them from scratch in the use of the CASE tool, thereby attempting to circumvent some of the problems relating to unlearning and resistance.

A second set of individuals, who are potentially as important as the primary users, are the secondary users, that is the people who use (or manage) the systems that are developed with the CASE tool. They may be involved in the development process as well as being the users of the information system produced. Their perceptions of the tool, its impact and effects are also important, although frequently forgotten. If the secondary users perceive that the CASE tool results in better quality systems, or faster production of systems, or improved identification of their requirements, or enabling their greater participation, or whatever advantage, then this is likely to result in the organisation as a whole regarding CASE in a favourable light. Of course the reverse is also true. Unfortunately when CASE tools are introduced there seems relatively little emphasis on involving the secondary users or recognising them as an important component of success.

CASE tools are continually changing and evolving and this is likely to continue for some time yet. Thus we need to examine some of the trends and developments that might affect CASE tools for the future and identify a number of likely directions and developments. The first development concerns the way in which the diagramming tools can handle greater model complexity which leads on to meta CASE tools which are a most sophisticated development.

One of the benefits of CASE tools is to help handle the large volume and wide range of interrelated and intertwined information that is required in the development of an information system. The way in which the tool helps to reduce this complexity for the developer is a key issue. One way is by much of the information, at least in the upper or earlier stages of the development cycle, being collected and represented diagrammatically. This itself helps to reduce complexity on the basis that 'a diagram is worth a thousand words'. Nevertheless, diagrams can become large and complex with a high degree of interrelationships.

Brinkkemper *et al.* (1993) distinguish two types of diagram complexity. The first concerns the complexity within diagrams. The second concerns the complexity of the interrelationships between different types of diagram. The complexity within diagrams can be reduced in a number of ways by CASE tools:

- *By the provision of 'views':* This reduces complexity by the separation of functional information in a diagram. For example, a complex diagram might be reduced into a set of different functional views which are easier to understand individually. Brinkkemper *et al.* illustrate this by using a data flow diagram which also contains control information. The tool should enable the diagram to be broken down into a view of the data flow and a separate view of the control information.

*   *By supporting the showing and hiding of information:* This makes the diagrams easier to read and understand, an example being an entity-relationship diagram in which the tool enables attributes to be shown or hidden on the diagram.

*   *By supporting the decomposition of diagrams:* A data flow diagram may be decomposed into a number of levels and the tool can help ensure consistency between the levels.

*   *By colouring (or other means of identification):* The complexity of diagrams is reduced by the use of colour to enhance particular information. For example, all objects of the same type might all have the same colour.

The second category is concerned with reducing complexity in the interrelationships between different types of diagram. Brinkkemper *et al.* term this 'modelling transparency', which is an indication of the number of user operations required to transfer from one diagram type to another. A tool of degree zero modelling transparency is one without the functions of a repository, where there are no connections made between different types of diagrams. These will include stand-alone CASE tools or simple diagramming tools.

Level 1 occurs where there is a repository that enables any dependencies between diagrams to be stored, but connections can only be made and seen by closing one diagram tool and then loading another. The connections that can exist between diagrams of different types (or between objects within diagrams of different types) are defined by the CASE tool, which may itself be a function of any underlying methodology.

Level 2 is as Level 1 except that direct switching from one diagram to another is provided, even allowing both diagrams to be displayed concurrently, and changes made to one diagram are reflected, where appropriate, in the other. Brinkkemper *et al.* show that in ADW the user can be viewing a data flow diagram and then click on a data store to view the entity-relationship diagram immediately. This is model transparency, in this direction, of level 2. However, the reverse switch from the entity-relationship diagram to data flow diagram is not provided, so this is only Level 1. Thus it is not always easy to say whether a particular tool is Level 1 or 2. It is necessary to analyse the percentage of potential links at each level.

Level 3 allows the tool user to define connections and links between diagrams themselves, rather than, or in addition to, those defined and built into the tool. This is discussed below in relation to meta CASE tools.

Meta CASE tools are those which are not fixed in their functionality but have the ability to be changed according to the needs of the developers and the development environment. These are in Brinkkemper *et al*'s terms, tools at level 3. The tools Foundation and Excelerator provided a high degree of customisability. Martiin *et al.* (1992) termed tools with customising abilities CASE shells. These are tools 'to define a method or chain of methods', and they suggest that VSF, Paradigm+, RAMATIC, MetaView and MetaEdit go some way towards meta modelling.

Meta modelling is the specification of the data stored, represented and manipulated in the repository, and the process to specify the content and functionality of that repository. The tool itself has facilities that can be used to define the way in which the tool will function. For example, Excelerator can be used with a facility called customiser to define the information model, that is, the type of entities and relationships storable in the repository and constraints. It also has the presentation model, that is, the graphical objects, symbols and dialogues for displaying information. The third model is the behavioural model, that is, the actions and trigger points that determine the execution of the tool. Finally the verification model concerns the rules about verification and validation of the data in the repository.

The customiser enables Excelerator to be specified differently for each user or group of users or for each project or organisation. Currently there are limits to the degree of customisation that can be achieved, but in the future it may be possible to develop the meta modelling characteristics and power to the extent that developers will be able to easily specify a tool to reflect their own preferences and, in effect, their own methodologies. This flexibility will perhaps enable a more contingent approach to the use of CASE tools, that is, to customise the tool according to the characteristics and environment of the project being developed. It may also enable an improved ability to learn what works well in an organisation and what does not, by experimenting with different approaches and configurations of the tools. It may mean that developers do not have to adhere to one particular methodology, but may be able to mix and match elements from a variety of approaches and be able to specify how these should work.

Before these, perhaps more long term, possibilities, the development of meta CASE tools is being driven by the need for some vendors, particularly those not tied to a particular methodology, to be able to offer support for multiple methodologies. Rather than develop separate tools, they argue that it is more effective to develop a meta tool which can reflect different approaches or methodologies. MetaEdit, for example, already claims to support 'all the well known object-oriented methodologies', but more than that it also claims to provide 'the possibility of familiarising yourself with the lesser known methods, that could be even more suitable for your needs'. Other vendors perform the meta modelling themselves to produce different versions for different methodologies, sometimes under different names, to market as separate products. For example, SEtec is a separate CASE tool marketed by Syslab created with their meta CASE product, Maestro II.

The integration of CASE tools with expert systems concepts is likely to continue in the future. Such tools are referred to as Intelligent CASE. Some tools already have knowledge bases in their repositories that contain the rules of the methodology and use expert system techniques to perform a number of checks, to identify violations, to analyse the impact of changes and to transform from analysis models to design specifications. The potential, however, is perhaps even greater than this. One possibility is for the expertise of systems development to be incorporated. It may be possible to capture and use it to provide knowledgeable assistance to developers when building systems. This already happens to some extent, but may well be enhanced with interactive dialogue sessions between the developer and the tool. For example, the expert system element of the tool might ask whether the developer has realised that some critical design decisions hinge on the specification of a particular relationship between two entities, and the expert system might suggest that the developer check with the users that this is actually what was required. This type of expertise might be described as a kind of sensitivity analysis that the expert system can apply.

Obviously the range of expert support that the tool might be able to apply is dependent on the knowledge concerning information systems development that can be elicited from experts, and up to now there has been little agreement amongst experts. However, in the future, an expert system CASE tool itself may be able to capture some of this knowledge on the basis of experience of a number of projects in an organisation. Indeed, knowledge might be able to be captured from a number of organisations using the same tool.

In a similar way, knowledge concerning a particular application and a particular organisation might also be able to be captured and utilised by an expert system. For example, the tool might 'know' from past projects the characteristics of a particular department and be able to identify that some analysis information is already available in the system or that something in this system conflicts with information previously specified in an earlier project and therefore should be carefully re-checked. With expertise of information systems development along with domain information concerning the application and the organisation, the system may even be able to recommend an appropriate information systems development methodology.

Another aspect of intelligent CASE tools might be the education and training of new developers. Apart from guiding developers based on the knowledge in the repository, this may be able to tutor or teach by using simulations of systems based on actual data in the knowledge base of the repository.

Developments in multi-media systems and technology, together with reducing prices, have made an impact in many application areas. Methodology, tool and technical manuals could well be more effectively stored and retrieved on multi-media systems in the same way as conventional encyclopaedias, such as, Microsoft's Encarta. The multi-media facilities of graphics and imaging are obviously highly relevant to CASE where many of the models and objects are expressed graphically. Improved quality and presentation would be highly relevant. Speech, such as recordings of interviews with users or reviews with managers, could be digitised and stored with the potential for speedy analysis and retrieval. Video recordings of meetings and conferences or recordings of a user performing operations in practice relevant to an application, could be stored in the repository and analysed using ethnographic techniques. Further, in the future perhaps, virtual reality systems could be produced that might enable users to 'experience' the system or critical parts of the system, before it was physically implemented. This would provide benefits similar to simulations or prototyping, that is, the ability to 'see and feel' the system and request changes when they are still relatively easy and cheap to make.

The repository would need to be extended to cope with data in a variety of forms, including speech and video. These forms need to be integrated or at least related to the traditional repository information. Because of the large volume of information required for audio and video, even with compression techniques, recordable CD storage might be required.

As mentioned above, CASE tools have not in the past been particularly good at supporting interpersonal communication and co-ordination amongst the various groups involved in information systems development. Yet in other contexts there has been some advances in just these human interaction support capabilities, for example, in Group Decision Support Systems (GDSS) and Computer Supported Co-operative Work (CSCW).

These developments could be integrated with CASE developments in the future. Facilities like electronic mail, video conferencing, meeting support, and the co-ordination and communication of documents (and other objects such as graphs and spreadsheets) between developers is certainly relevant to CASE tools. Such groupware support can help working groups agree what has to be done, help in the allocation of tasks and perhaps most importantly support group activities, such as meetings with users, decision making concerning requirements and group design sessions.

Additionally, it can support the co-ordination of remote groups working in different locations, even in different countries. One organisation is attempting to utilise the world's different time zones to speed up the development of a project by having two groups of developers in different zones. After a day's work when the first group goes home the work is taken up by a new team of developers in a different time zone of the world. At the end of their day they then hand it back to the first group and so on. This obviously requires not only a great deal of work group co-ordination but also a shared repository of some kind. So far they have applied this experimentally to the detailed design and coding aspects of development.

With the advent of collaborative CASE tools and the support and co-ordination for remote users, there is no reason why this kind of working could not be effective in some situations. Researchers at the University of Beijing, China and Florida University, USA have developed a CASE environment called JadeBird/III which supports a geographically dispersed collaborative development environment where teams can 'co-specify, co-analyse, co-design, co-code, and co-debug information systems projects over a distributed, heterogeneous network of workstations and desktops' (Yang *et al*. 1994).

The systems architecture is shown in figure 5.11 and the key feature is the collaboration manager which supports synchronous multi-user tools and the multi-media repository. It controls concurrent access to applications, and establishes and terminates multi-user dialogues and all the information flows. It handles what is termed shared workspace support, which allows the dynamic sharing of tools. For example, a diagram can be viewed and updated on different remote workstations in a single session. Algorithms pass control between users ensuring input from only one source at a time. Collaborative sessions are supported and remote users may use a 'shared remote pointer' to supplement an audio dialogue. The repository is a standard storage system (except that it stores audio and video as well) and is unaware of any collaboration features, which are handled by the Repository Manager.

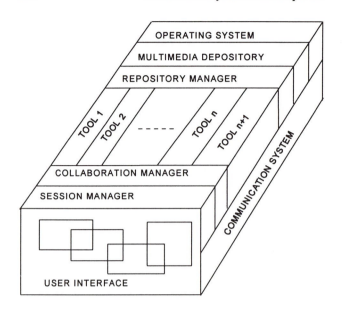

Fig. 5.11: Systems architecture of JadeBird/III (after Yang *et al.* (1994))

CASE tools for object-oriented approaches are already available and with the continuing development of the object-oriented paradigm, the demand for improved object-oriented CASE tools will undoubtedly grow. Indeed Yourdon (1994) sees CASE as the only possible way of successfully implementing object-oriented methodologies.

Object-oriented CASE tools need to support the new diagrams and their conventions, for example, Class-&-Objects. This might require a totally new tool but could possibly be supported by an existing tool that allowed flexibility in its graphical notation. Alternatively, a meta CASE tool which enables the diagrams and rules to be defined may be appropriate.

Yourdon also suggests that 'hiding and revealing' features are of great importance for object-oriented approaches, because the models 'can become incredibly complex' and it is necessary to check for errors and consistency between diagrams and objects. This is similar in principle to the requirements for non object-oriented development, although the rules may be somewhat different.

Because the object-oriented approach focuses so heavily on re-use for its benefits, any tool needs to provide the ability to search or browse for objects that might be reusable. Yourdon argues that these must be especially powerful and that it is not just a question of keyword searching. It must be available at any time irrespective of what the developer is currently doing and it must allow browsing up and down through the levels of a class hierarchy. It is clear that effective browsing is one of the keys to reuse because it must not be easier to create a new object or piece of code than finding a fully validated object that is already in the library. It should also be possible for the CASE tool to support browsing of libraries outside of those developed in the organisation. If object libraries are developed as a general resource by commercial companies in the way that some foresee, then CASE tools will need easy and effective access and consult such industry-wide class libraries. This probably means that they will need external communication links to other repositories or CD ROM libraries, rather than relying on internal repositories. Such browsing facilities may be add-on software to the main CASE tool. Nevertheless they must integrate and be available when needed.

It does seem, therefore, that there may be some extra facilities required of object-oriented CASE tools and for this reason a number of vendors have produced specific tools for object-oriented development. On the other hand, other vendors have produced extensions to their existing tools to handle object-oriented requirements. There are also meta tools that are beginning to enable the construction of tailored tools. Once again, the position in the market place is uncertain and subject to continuing change with no clear indication of the correct route for new object-oriented developers to go in their quest for CASE. As a result a piece of advice that Yourdon gives is to choose something 'small, simple, and cheap' in the first instance.

FURTHER READING

Bradley, K. (1993) PRINCE: A Practical Handbook. Butterworth-Heinemann, Oxford.
    Ken Bradley's book provides a good practical overview of the project management approach.

Avison, D. E. (1992) *Information Systems Development: A Database Approach*. McGraw-Hill, Maidenhead.
Elmasri, R. and Navathe, S. B. (1989) *Fundamentals of Database Systems*. Benjamin/Cummings, Redwood City, Ca.
McFadden, F. R. and Hoffer, J. A. (1991) *Database Management*. 3rd ed., Benjamin/Cummings, Redwood City, California.
    There are a number of texts which include a detailed description of various DBMSs. Avison's book looks at databases from an information systems perspective and the other two books are excellent technical books on databases.

Date, C. J. and White, C. (1988) *A Guide to DB2.* 2nd ed., Addison-Wesley, Reading, Mass.
This text is devoted to DB2 and SQL. General database texts also cover SQL to an extent.

Yourdon, E. (1992) *Decline & fall of the American programmer.* Yourdon Press, Englewood Cliffs, New Jersey.
McClure, C. (1989) *CASE is Software Automation.* Prentice Hall, Englewood Cliffs, New Jersey.
Any text book on tools is likely to be out of date. However, these two texts contain a thorough description of the principles of toolsets that have the potential to improve information systems development.

Chapter 6
Methodologies

6.1 INTRODUCTION

In this chapter, we look at a number of information systems development methodologies which are well used, respected, or which typify the approaches described in Chapter 3.

The first methodology described reflects the process modelling theme (section 3.7) and was proposed by Chris Gane and Trish Sarson. The main techniques used are the process-oriented ones of functional decomposition, data flow diagrams, decision trees, decision tables and structured English. Functional decomposition gives structure to the processes reflected in particular by the most important technique of data flow diagrams. This emphasis on structure gives the name of the methodology: Structured Analysis and Design of Information Systems (STRADIS).

Yourdon Systems Method (YSM) was originally very similar to STRADIS, indeed, Gane and Sarson were at one time colleagues of Ed Yourdon. However, more recent versions of YSM suggest that a 'middle-up' approach to analysing processes called event partitioning is more appropriate than the top-down approach (functional decomposition). Although, emphasis is placed on the analysis of processes, when compared to STRADIS, there is greater emphasis on the analysis of data.

Whereas STRADIS and YSM emphasise processes, Information Engineering (IE), based on the work of James Martin and Clive Finkelstein, has more emphasis on data (section 3.8). Similarly, whereas the fundamental techniques of the process-oriented approaches are functional decomposition and data flow diagrams, the basic approach in IE is the data-oriented entity-relationship approach. However, like the development of many methodologies described in this chapter, IE has been extended and has, for example, a planning phase, which is the first phase of the methodology, reflecting some of the discussion in section 3.5.

Structured Systems Analysis and Design Method (SSADM), a methodology originally developed by Learmonth and Burchett, and now a standard in most UK government applications, can be said to be the modern version of the traditional information systems development life cycle approach discussed in Chapter 3. For example, it includes the techniques of data flow diagramming and entity life histories, and recommends the use of tools, such as CASE tools and workbenches.

Merise is a widely-used methodology for developing information systems in France and elsewhere, and, like SSADM, may become very influential in any future European standard. Unlike other methods described above which emphasise either process or data aspects of information systems analysis and design, Merise has been designed so that both are considered equally important and these aspects are analysed and designed in parallel.

Michael Jackson's program design methodology, Jackson Structured Programming (JSP), has had a profound effect on the teaching and practice of commercial computer programming. Jackson Systems Development (JSD), is a development from JSP into systems development as a whole. An information system is seen, in effect, as a very large program. The approach is somewhat different from the methodologies described before as it concentrates on the design of efficient and well-tested software which reflects the specifications. It has links with formal methods (section 3.12) and is particularly applicable to applications where efficiency is paramount, for example, in process control applications.

Coad and Yourdon's Object-oriented Analysis is also significantly different from the approaches that have been discussed so far. It is an approach which reflects the view that in defining objects and their component parts (attributes) we capture the essential building blocks of information systems (section 3.9). It is also a unifying approach, as analysis and design can be undertaken following this approach, and applications developed using object programming languages and CASE tools. The object-oriented theme leads to consistency throughout.

Information Systems Work and Analysis of Change (ISAC), a methodology developed in Scandinavia by Mats Lundeberg and colleagues, seeks to identify the fundamental causes of users' problems and suggests ways in which the problems may be overcome (not necessarily through the use of computer information systems) by the analysis of activities and the initiation of change processes. It is therefore a people-oriented approach with emphasis on the analysis of change and the change process.

Effective Technical and Human Implementation of Computer-based Systems (ETHICS) is a methodology proposed by Enid Mumford. It is a people-oriented approach (based on participation, section 3.13) and, as the name implies, attempts to embody a sound ethical position. It encompasses the socio-technical view that, in order to be effective, the technology must fit closely with the social and organisational factors in the application domain.

Soft Systems Methodology (SSM), a methodology proposed by Peter Checkland, is influenced by the systems approach (section 3.2). Whereas many of the earlier approaches stress scientific analysis, breaking up a complex system into its constituent parts to enable analysis, systems thinking might suggest that properties of the whole are not entirely explicable in terms of the properties of the constituent elements. This is normally expressed as 'the whole is greater than the sum of the parts'. SSM addresses the 'fuzzy', ill-structured or soft problem situations which are the true domain of information systems development methodologies, not simple, technological problems.

Multiview is a hybrid methodology which brings in aspects of other methodologies and adopts techniques and tools as appropriate. In other words, Multiview is a contingency approach: techniques and tools being used as the problem situation demands. It has been influenced particularly by aspects of SSM and ETHICS, but also by the proponents of process modelling and data modelling approaches. However, readers will see aspects of a number of approaches described earlier in the chapter.

Davenport's process innovation does most to tie business process re-engineering (section 3.4) with information technology and information systems, IT being seen as the primary enabler of process innovation as it gives an opportunity to change processes completely.

The need to develop information systems more quickly has been driven by rapidly changing business requirements. Rapid Application Development (RAD) is a response to this need. It is based on the evolutionary, prototyping approach discussed in section 3.11, and is usually enabled by CASE tools and systems repositories which were discussed in Chapter 5. User requirements are often determined through joint applications development (JAD) which was introduced in section 3.13.

The expert systems approach KADS is the outcome of a European Union ESPRIT research project to develop a comprehensive, commercially viable methodology for knowledge-based systems construction. It therefore develops the them of expert systems discussed in section 3.14.

The last section discusses Euromethod which also results from a European initiative. It is more of a framework for the planning, procurement and management of services for the investigation, development or amendment of information systems than a methodology, though SSADM, Merise and Information Engineering, amongst others, have influenced its design. This framework and associated standards will, it is hoped, help overcome the problems posed by the current diversity of approaches, methods and techniques in information systems and help users and service providers to come to common understandings concerning requirements and solutions in information systems projects.

We have not described similar methodologies, even if both are well used, but reference this similarity where appropriate. The methodologies are described largely uncritically so that readers can follow their principles and practice, although we have commented on aspects of the methodologies where they reveal important features. However, the descriptions of the methodologies represent interpretations of the methodologies by the authors of this text, and these views may not correspond to those of the methodology suppliers. We return to the problem of interpretation in section 7.4.

6.2 STRUCTURED ANALYSIS, DESIGN AND IMPLEMENTATION OF INFORMATION SYSTEMS (STRADIS)

The major statement of Gane and Sarson's methodology of systems development called STRADIS comes in their book entitled *Structured*

Systems Analysis (Gane and Sarson, 1979). The development of this structured systems approach to analysis came as a result of the earlier development of a structured approach to design. The structured design concepts were first propounded in 1974 by Stevens, Myers and Constantine (1974) and these ideas were later developed and refined by Yourdon and Constantine (1978), and Myers (1975 and 1978). The work of Jackson (1975) was also influential.

Structured design is concerned with the selection and organisation of program modules and interfaces that would solve a pre-defined problem. However, it makes no contribution to the defining of that problem. This proves to be a practical limitation as the development of an information system requires both analysis and design aspects to be addressed, and whilst structured design was acknowledged to provide significant benefits, these benefits were wasted if the definition of the original problem was not well stated or inaccurate.

A number of people have therefore attempted to take the concepts of structured design and apply them to systems analysis, in order to develop a method of specifying requirements and to provide an interface to structured design. In this way the techniques of structured analysis were developed. Apart from Gane and Sarson's work, DeMarco (1979), Weinberg (1978) and more recently Yourdon (1989) are all texts on structured analysis covering some of the same ground and utilising very similar techniques within each approach.

Gane and Sarson only relatively briefly outline a methodology of systems development in their book. The majority of the book is devoted to descriptions of the techniques which the methodology utilises. This is in direct contrast to some other methodologies. SSADM (section 6.5) and ISAC (section 6.9), for example, lay out the steps of the methodology in great detail. Therefore the most important aspect of the Gane and Sarson methodology is the bringing together of many of the techniques which were described separately in Chapter 4 in this book. Nevertheless, we will continue to use the term 'methodology' in the context of STRADIS. These techniques are utilised, in some form or another, by many different methodologies, and therefore Gane and Sarson's methodology is not unique but, along with Yourdon Systems Method (section 6.3), may be regarded as epitomising those methodologies based on functional decomposition (section 3.7) and the use of the data flow diagram (DFD), described in section 4.7.

Gane and Sarson's methodology is conceived as being applicable to the development of any information system, irrespective of size and whether or not it is going to be automated. In practice, however, it has mainly been used and refined in environments where at least part of the information system is automated. The methodology is envisaged to be relevant to a situation in which there is a backlog of systems waiting to be developed and insufficient resources to devote to all the potential new systems.

1 Initial study

The starting point of the methodology is an attempt to ensure that the systems chosen to be developed are those that most warrant development in a competing environment. The most important criterion in this selection process is argued to be the monetary costs and benefits of each proposal. Systems are viewed as contributing towards increasing revenues, avoiding costs or improving services. The initial study to discover this information is conducted by systems analysts gathering data from managers and users in the relevant areas. The analyst is to review existing documentation and assess the proposal in the light of any strategic plans relating to systems development that may exist within the organisation. The initial study usually involves the construction of an overview data flow diagram of the existing system and its interfaces, and an estimate of the times and costs of proceeding to a detailed investigation. In addition, some broad range of final system development costs might be estimated. The initial study normally takes between two days and four weeks, depending on the size and importance of the application.

On completing the initial study, a report is reviewed by the relevant management and they decide on whether to proceed to a more detailed study or not. If they approve of the proposal, they are committing themselves to the costs of the detailed study but not necessarily to implementing the proposed system.

Gane and Sarson's initial study might be thought to be quite close to the traditional notion of a feasibility study outlined in section 2.2. However, there are some important differences. Gane and Sarson's methodology does not include a review of alternative approaches to the proposal and it is not, perhaps, as major or as resource intensive a task as a traditional feasibility study. Furthermore, a traditional feasibility study, if approved by management, is usually in practice a commitment to the implementation of the complete proposal. Gane and Sarson do address all these aspects, but at later stages within their methodology.

2 Detailed study

This takes the work of the initial study further. In particular, the existing system is examined in detail. As part of this investigation, the potential users of the system are identified. These users will exist at three levels:

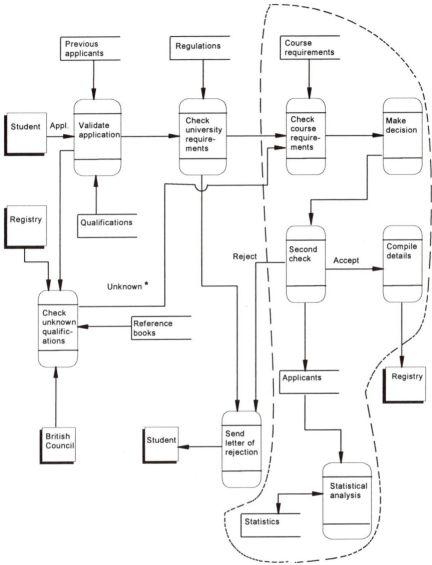

Fig. 6.1: Example of a system boundary

1. The senior managers with profit responsibilities, whom Gane and Sarson call the 'commissioners', whose areas will be affected. They initially commissioned the system proposal.
2. The middle managers of the departments affected.
3. The end users, that is, the people who will actually work directly with the system.

Having identified these three sets of users, the analysts ascertain their interests and requirements by interviewing them. Next, the analyst prepares a draft logical DFD of the current system. This will usually involve constructing a DFD that extends well beyond the current system under consideration, in order to be clear exactly what and where the boundaries are in relation to other systems, and to identify the interfaces between various systems.

Figure 6.1 depicts a data flow of part of a university admissions procedure. The system under consideration is that enclosed by the dotted line, but in order to appreciate the context, a larger system is depicted which enables the interfaces to be clearly identified. Any data flow that crosses the dotted line must be addressed by both the external system and the system under consideration. In this case the diagram has highlighted the fact that those applications where the qualifications are not known, require a decision to be made (see the data flow marked with the asterisk). This is a non-obvious interface which might otherwise have been neglected.

The boundary may be drawn in any place and could be moved. It may, for example, be more logical to include other processes within the boundary in order to minimise the number of interfaces to the external system. This is particularly important when the automation boundary is being chosen.

Gane and Sarson describe in detail the drafting of DFDs at various levels, showing how each level is exploded into lower levels through to the level where the logic of each process box in the low-level DFD should be specified using the appropriate process logic representation, for example, decision trees, decision tables or structured English (sections 4.8, 4.9 and 4.10). They suggest that DFDs and other outputs should be reviewed or 'walked through' with a number of users, so as to check their validity and alterations made where necessary.

The detail of the DFDs and the process logic is entered into the data dictionary. The data dictionary can be either manual or computerised. On the DFD, data flows and data stores are defined using a single name which is meaningful. All the details that the name represents must be collected and stored in the data dictionary (section 5.4).

The extent of detail that the analyst goes to at this stage in the methodology is not made clear, but it appears that not all low-level processes are specified in process logic and that not all data flows and data stores are specified in the data dictionary. It is usual to specify in detail only the most significant at this stage.

The initial study estimated the costs and benefits of the proposed system in outline. These estimates are further refined within the detailed study. The analysts need to investigate the assumptions on which the estimates were based, and ensure that all aspects have been considered. They also need to consider the effects and costs of the proposed system from the point of view of organisational impact. In other words, they need to have a better estimate on which a final decision can legitimately be made.

In summary, the detailed study contains:

- A definition of the user community for a new system, that is, the names and responsibilities of senior executives, the functions of affected departments, the relationships among affected departments, the descriptions of clerical jobs that will be affected, and the number of people in each clerical job, hiring rates, and natural attrition rates.
- A logical model of the current system, that is, an overall data flow diagram, the interfacing systems (if relevant), a detailed data flow diagram for each important process, the logic specification for each basic process at an appropriate level of detail, and the data definitions at an appropriate level of detail.
- A statement of increased revenue/avoidable cost/improved service that could be provided by an improved system, including the assumptions, the present and projected volumes of transactions and quantities of stored data and the financial estimates of benefits where possible.
- An account of competitive/statutory pressures (if any) including the system cost and a firm cost/time budget for the next phase (defining a menu of possible alternatives).

The results of the detailed study are presented to management and a decision will be made either to stop at this stage or proceed to the next phase.

3 Defining and designing alternative solutions

The next phase defines alternative solutions to the problems of the existing system. First, the organisational objectives, as defined in the initial study, are converted into a set of system objectives. An organisational objective is a relatively high-level objective having an effect on the organisation. This could include increased revenue, lower cost or improved service. A system objective is at a lower level, and relates to what the system should do to help management achieve the organisational objectives.

The system objectives should be strongly stated. This means that they should be specific and measurable, rather than be general. For example, 'improving the timeliness of information' would be a weakly-stated objective and it would be preferable to state this objective more strongly, for example, 'to produce the monthly sales analysis report by the fourth working day of the following month'.

The analyst uses these objectives to produce a logical DFD of the new or desired system. The existing system DFD would normally be used as the basis for this, and the desired system may involve the introduction of new or changed data flows, data stores and processes. The new DFD should be constructed to a level of detail which shows that the most important system objectives are being met.

The methodology then enters a design phase. At this time, analysts and designers work together to produce various alternative implementation designs which meet a variable selection of the identified system objectives. The alternatives should cover three different categories of designs. First, a low-budget, fairly quick implementation which may not initially meet all the objectives; second, a mid-budget, medium-term version, which achieves a majority of the objectives; and third, a higher budget, more ambitious, version achieving all the objectives. Each alternative should have rough estimates of costs and benefits, time-scales, hardware, and software.

The report of this phase of the project should be presented to the relevant decision makers and a commitment made to one of the alternatives. The report should contain the following:

- A DFD of the current system
- The limitations of the current system, including the cost and benefit estimates
- The logical DFD of the new system.

For each of the identified alternatives, the design will include statements covering:

- The parts of the DFD that would be implemented
- The user interface (terminals, reports, query facilities and so on)
- The estimated costs and benefits
- The outline implementation schedule
- The risks involved.

4 Physical design

The design team then refines the chosen alternative into a specific physical design which involves a number of parallel activities:

1.   All the detail of the DFD must be produced, including all the error and exception handling, which has not been specified earlier, and all the process logic. The content of the data dictionary is completed

and report and screen formats designed. This detail should be validated and agreed with the users.

2.   The physical files or database will be designed. They will be based on the data store contents previously specified at the logical level. Data stores are defined in the DFD as the temporary storage of data needed for the process under consideration. This has the effect of introducing many data stores scattered all over the DFD. Many of these will be very similar in content and have a significant degree of overlap.

3.   The data stores need rationalising, and the technique of normalisation (described in section 4.6) is utilised to consolidate and simplify the data stores into logical groupings. The actual process of mapping and the design of the physical files (or database) are not defined by Gane and Sarson.

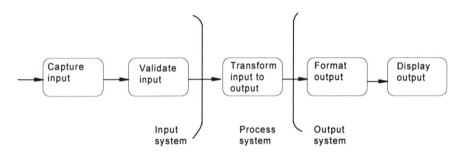

Fig. 6.2: Transform-centred system

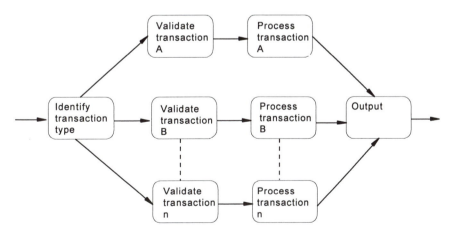

Fig. 6.3: Transaction-centred system

4.   Derive a modular hierarchy of functions from the DFD. The designer seeks to identify either of two structures that any commercial data processing system is thought to exhibit. The first structure is the simplest. Here all transactions follow very similar processing paths (figure 6.2). Such a system is termed a 'transform-centred' system. The second structure is one in which the transactions require very different processing. This is termed a 'transaction-centred' system and is illustrated in figure 6.3.

The first step therefore is to identify which type of system is being described. It is recommended that the raw input data flow is traced through the DFD until a point is reached at which it can no longer be said to be input, but has been transformed into some other data flow. The output is traced backwards in a similar fashion until it can no longer be considered to be output. Anything in between is termed the 'transform'. The transform is then analysed to see if it is a single transform or a number of different transforms on different transaction types. Once one or other of these high-level functional hierarchy types have been identified from the DFD, the detail of the modules in the hierarchy and the communication between them is constructed.

The final task in this phase is the definition of any clerical tasks that the new system will require. The clerical tasks are identified according to where the automated system boundary is on the DFD and according to what physical choice of input and output media has been made.

The above activities are pursued to a level of detail at which it is possible to give a firm estimate of the cost of developing and operating the new system. The major components of these costs are identified as:

- The professional time and computer test time required to develop the identified modules
- The computer system required
- The peripherals and data communication costs
- The professional time required to develop user documentation and train users
- The time of the users who interact with the system
- The professional time required to maintain and enhance the system during its lifetime.

Subsequent phases of the methodology are not clearly defined by Gane and Sarson as the methodology is effectively concerned mainly with analysis, to a lesser extent design and hardly at all with implementation. However, the following list indicates the remaining tasks that Gane and Sarson envisage as being needed to complete the development of the system:

- Draw up an implementation plan, including plans for testing and acceptance of the system
- Develop concurrently the application programs and the database/data communications functions (where relevant)
- Convert and load the database(s)
- Test and ensure acceptance of each part of the system
- Ensure that the system meets the performance criteria defined in the system objectives, under realistic loads, in terms of response time and throughput
- Commit the system to live operation and tune it to deal with any bottlenecks
- Compare the overall system facilities and performance to original objectives, and amend to resolve any differences, where possible
- Analyse any requests for enhancement, prioritising these enhancements, and placing the system in 'maintenance' state.

6.3 YOURDON SYSTEMS METHOD (YSM)

YSM was originally very similar to STRADIS, indeed, Gane and Sarson were at one time colleagues of Yourdon. Functional decomposition or top-down design, in which a problem is successively decomposed into manageable units, was the basis of the approach. However, although based on the structured approach, particularly its modelling techniques, the more recent versions of YSM (the most recent version of YSM (3.0) is described in Yourdon (1993)) use an approach known as event partitioning. This approach is neither pure top-down nor bottom-up, but is described as 'middle-out'. The analyst begins by drawing a top-level context diagram which indicates the system boundaries and thus the sources and sinks. Then, following interviews with the users, a textual list of the events in the environment to which the system must respond is constructed. Following this, most of the techniques described in Chapter 4 are used to further document the system.

YSM covers both the activities of the organisation (although this could be at department level as well as at the level of the organisation as a whole) and the system itself. Enterprise requirements need to be modelled as well as system requirements. For example, analysts may create an entity-relationship diagram and other information about data for the department, but only some of this will be appropriate for the system. Modelling at a department rather than system level will ensure consistency as well as avoiding the duplication of time and effort. Emphasis is therefore placed on modelling both the organisation and the system. Many of these modelling methods are appropriate to the use of support tools, particularly CASE tools (section 5.7).

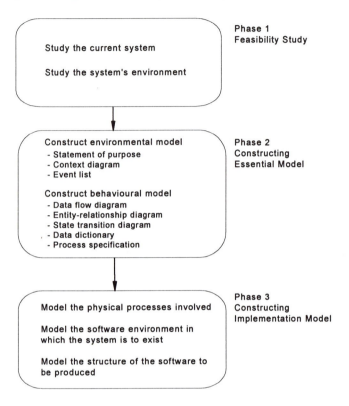

Fig. 6.4: The Yourdon Systems Method (YSM)

There are three major phases in the YSM approach. as shown in figure 6.4. The feasibility study looks at the present system and its environment. Phase 2, essential modelling, aims to describe the essence of a software system in terms of how the required system must behave and what data must be stored to enable this to happen. It assumes that there are no limitations affecting implementation, that is, it assumes unlimited resources, unlimited power of technology and so on. It is the major phase of the approach. The final phase, implementation modelling, aims to incorporate those features found in the customer's statement of requirements using the essential model and will be dependent on the appropriate use of available technology. We will look at these three phases in more detail.

Although the only enterprise model described in detail in Yourdon (1993) is the enterprise essential model, the creation of an organisation as well as system level model might suggest an enterprise or strategic planning phase for information systems development. Indeed, some followers of YSM include a strategic planning phase before the feasibility study of each proposed system in that plan. However, an enterprise implementation model is suggested and this will point to proposals for the hardware and software decisions for the organisation as a whole, not a decision which should be dominated by the needs of one particular application.

1 Feasibility study

The feasibility study looks at the present system, its environment and the problems associated with it. The objective here is to get a general understanding and an overview of the existing system. It is to understand what the existing system does (not how it works). The analyst will tend to draw an overview data flow diagram for the current system and its interfaces and the analyst may also start to put together an entity-relationship diagram. The information required will normally be obtained from interviewing the users. This phase is much the shortest of the three, normally taking only a few weeks to complete.

2 Essential modelling

This stage gains the most emphasis in YSM. There is both an enterprise and a system essential model. We will emphasise the latter in our description as it is essentially the sum total of the systems models. The same considerations and models are reflected at the organisational level where, unusually, the 'organisation' to be considered and the 'system' are the same. Having an overview of the present system, it is possible then to construct an essential model. The system essential model is a model of what the system must do in order to satisfy the users' requirements. It does not say anything about how this system is to be implemented. Thus it is the new logical model. In the 1993 version of YSM, essential modelling itself also has two major components:

- Environmental model building
- Behavioural model building.

In some descriptions, the creation of the entity-relationship diagram is seen as part of a third parallel component, referred to as the information model. In the following description, these aspects form part of the environmental model and behavioural model building phases. The activities are the same in either case. The key difference is, perhaps, more subtle. It represents the change in the approach from one which emphasises processes to one which emphasises both process and data aspects. It also enables stress to be made on the importance of comparing the data and process aspects to ensure consistency of models and therefore the integrity of the overall specification. This also separates YSM from STRADIS which was discussed in section 6.2.

The *environmental model* defines the boundary between the system and the environment in which the system exists. The data coming from and to the environment are identified. The model consists of a statement of purpose, context diagram and an event list.

The statement of purpose is a brief, concise, textual statement about the purpose of the system. It is provided for top management, user management and others who are not directly involved in the development of the system. It is only about a paragraph long.

The context diagram or level-0 data flow diagram (figure 4.37) represents the system in a circle in the middle of the page, along with the main sources and sinks of the data entering to and from it. It identifies the people, organisations and systems with which the system communicates. The data coming into the system that is processed in some way and then output in a different form is also identified along with any intermediary data stores. It also shows the boundary between the system and its environment.

The event list names the 'stimuli' that occur in the environment of the system to which the system must respond. An event may be flow-oriented, temporal or a control event. A flow-oriented event is one associated with a data flow. A temporal event is triggered by reaching a particular point in time. Control events occur at an unpredictable point in time and are therefore a special case of temporal event.

A first-cut data dictionary which describes the composition of each data element and a first-cut entity-relationship diagram highlighting the relationship between stores (the entities) may also be constructed at this time, but both are very early versions.

The *behavioural model* is a model of what the internal behaviour of the system must be in order to deal with the environment successfully. It includes a first-cut data flow diagram, entity-relationship diagram and state transition diagram and adds information to the data dictionary. A state transition diagram shows how the properties of an entity change over time and is therefore similar to an entity life cycle (section 4.12). Note that behaviour refers to the behaviour of the system and does not imply any emphasis on people-oriented aspects. The processing behaviour of the system, that is, how the system uses its inputs to produce the required output, is shown using data flow diagrams. The structure and use of the data in the system is shown using a data dictionary and a set of entity-relationship diagrams. The dynamic behaviour of the system, describing how events in time affect behaviour, is modelled by extending the data flow diagrams (which represent control) and state transition diagrams (which represent control behaviour).

From the event list obtained in the environmental model, a data flow diagram is constructed with one process representing the system's response to each event in the event list. Stores are then drawn as needed to enable the processes to access the required data and the input and output flows are connected to and from the processes. The data flow diagram or diagrams are then checked against the context diagram for consistency. In parallel, the control transformations are specified and the data relationships are modelled. By the end of this stage, the data flow diagrams are completed by a process of levelling out, and the data dictionary, process specifications, entity-relationship diagram and state transition diagram are also completed.

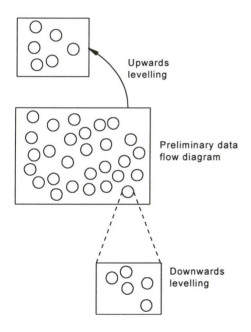

Upwards levelling

Preliminary data flow diagram

Downwards levelling

Fig. 6.5: Upwards and downwards levelling of YSM data flow diagramming

The process of levelling data flow diagrams involves restructuring so that there is a set of data flow diagrams, some the result of levelling upwards and some the result of levelling downwards. This is the key to the claim that YSM is middle-out rather than top-down. If the first-cut data flow diagram is too complicated with many processes, then related processes are grouped together into meaningful aggregates, each of which will represent a process in the higher-level data flow diagram. A rule-of-thumb is suggested that each data flow diagram will have around seven processes and stores in total. Downward levelling may be necessary where it is found that a process at the middle level is not a primitive process but needs to be expressed in more detail at a lower-level data flow diagram. This means that the initial process, which was a response to an event, is too complex for that middle level. The levelling process is seen in figure 6.5.

Note that as shown in figure 6.6, processes are illustrated by circles in the YSM standard. Other shapes are also used to represent sources and sinks (a simple rectangle with no shadowing) and data stores (two parallel lines). Further, Yourdon recognises two types of data flow. Discrete data flows arrive at their destination at discrete points of time (arrow as for Gane and Sarson) whereas continuous data flows are always available at their destination (arrow with two heads). This indicates that although many of the techniques are common to a number of methodologies, there are often variances in the way they are used, drawn and so on.

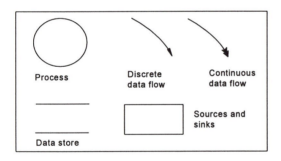

Fig. 6.6: YSM data flow diagram symbols

Process specifications are then drawn up for every 'functional primitive', that is, every process in the bottom level data flow diagram. These are referred to as minispecs, which are detailed specifications of each data process. Essentially, they state the rules that convert the inputs to the outputs. The process specification (see Chapter 4) may take the form of structured English, decision tables or any other method appropriate for the process which can be verified and communicated easily to the users. These will be cross-checked with the data dictionary and entity-relationship diagram and it might be necessary to modify the data flow diagrams as a result of this further detail.

The entity-relationship diagram also needs to be completed. YSM advocates iteration and this diagram will also be refined from its first cut form in stages. The knowledge gained when refining the data flow diagram will be used to help refine the entity-relationship diagram.

If the system being modelled has any real-time characteristics, then a state transition diagram is developed in addition to the entity-relationship diagram and data flow diagram. A state transition diagram specifies how a control process is to take account of its input control flows and how it is to output control flows. The effect which input control flows have depends on the 'state' of the system, and they may change the state of the system and cause control flows to be output. Moreover, the output control flows have an effect on the data flow model of system behaviour.

Like other structured approaches, 'methods' are emphasised in YSM rather than a 'methodology', indeed, many other techniques are described in detail. It is interesting that in introducing the modelling techniques, Yourdon (1993) suggests that there is no obvious sequence in which they are presented. Many methodology descriptions would describe them in the context of the particular phase or sub-phase in which they are used. Other techniques include entity-event matrices and function-entity matrices (section 4.15), the specification of entities, relationships, attributes, events and operations (figures 4.15 to 4.19) and normalisation (section 4.6).

The models together should describe 'what will the system do?' (for example, the data flow diagram), 'what happens when?' (for example, the event list) and 'what data is used by the system?' (for example, the entity-relationship diagram). Some models link these dimensions, for example, entity life cycles; entity-event matrices link time and information; and data flow diagrams link data and processes. Together, it is argued, the models provide a full description of the system.

3 Implementation modelling

This phase starts the systems design process. The limitations of such factors as the technology available, performance requirements and feasibility modify the essential model. Data flow diagrams and state transition diagrams are examined so that, for example, boundaries of computerisation are marked and within them groups of processes are bounded for particular devices and processes. This also includes allocating software environments to groups of processes. The entity-relationship diagram is examined to look for pointers as to which database management system might be suitable and how the data might be stored. Implementation modelling, in short, bridges the gap between specification and systems design.

6.4 INFORMATION ENGINEERING

The origins of Information Engineering (IE) differ according to which source is referenced. It appears that Clive Finkelstein first used the term to describe a data modelling methodology that he developed in Australia in the late 1970s. In early 1981 he renamed his consultancy company IE and wrote a series of articles on the methodology. In the same year he collaborated with James Martin on a two volume book entitled Information Engineering (Martin and Finkelstein, 1981). However, for some people the original ideas were developed by Ian Palmer and colleagues at CACI, Inc. International in London, also in the late 1970s. This methodology was originally known as the CACI methodology but was later called D2S2 (System Development in a Shared Data environment) (Palmer and Rock-Evans, 1981 and Macdonald and Palmer, 1982). A more up to date description is found in Martin (1989).

Since these early days there has evolved, rather confusingly, a number of versions of IE around the world which, whilst very similar in concept, have tended to develop along somewhat different lines. The reason for this is that James Martin, who is generally credited with evolving and popularising the methodology, set up a number of independent companies based on the methodology of IE. One such company, set up in the UK, was James Martin Associates (JMA), and a number of the original CACI people joined JMA. The result was the bringing together of James Martin's development approach and the ideas behind D2S2 into one methodology with the name IE.

Another of James Martin's commercial initiatives was an association with Texas Instruments (TI) in the US to develop the IEF (Information Engineering Facility), a CASE tool for the methodology. This was not the only co-operative arrangement to develop CASE tools and one was also entered into with KnowledgeWare to develop the IEW (Information Engineering Workbench) also based on IE. Things have simplified a little recently with TI becoming a more dominant partner and effectively taking over JMA which is now known as TI Information Engineering. This has meant a coming together of this particular 'brand' of IE. Nevertheless, there still appear to be a variety of different IE approaches and both Martin and Finkelstein still write independently on the topic. Martin says that IE 'should not be regarded as one rigid methodology but, rather, like software engineering, as a generic class of methodologies' (Martin, 1991).

The version of IE described here is based upon a number of sources and is sometimes termed 'classical' IE. There also exist a number of variants of IE for different development environments. These include a package-based approach and Rapid Application development (RAD) which is discussed later in this chapter (section 6.14).

IE is claimed to be a comprehensive methodology covering all aspects of the life cycle. It is viewed as a framework within which a variety of techniques are used to develop good quality information systems in an efficient way. The framework is argued to be relatively static, and includes the fundamental things which must be done in order to develop good information systems. The techniques currently used in IE are not part of those fundamentals, but are regarded as the best currently available to achieve the fundamentals. Thus the techniques can and do change as new and improved techniques emerge. The framework is also a project management mechanism, which reflects IE's philosophy of 'practicality and applicability'. It is not just a set of ideas but is argued to be a proven and practical approach. It is also said to be applicable in a wide range of industries and environments.

There are a number of philosophical beliefs underpinning IE. One of the original was the belief that data is at the heart of an information system and that the data, or rather the types of data, are considerably more stable than the processes or procedures that act upon the data. Thus a methodology that successfully identifies the underlying nature and structure of the organisation's data has a stable basis from which to build information systems. Methodologies which are based only upon processes are likely to fail due to the constantly shifting nature of this base as requirements change. This is the classic argument of the data modelling school of thought. However, IE also clearly recognises that processes have to be considered in detail in the development of an information system and balances the modelling of data and processes as appropriate.

A further aspect of the philosophy of IE is the belief that the most appropriate way of communication within the methodology is through the use of diagrams. Diagrams are very appealing to end users and end-user management and, it is argued, enable them to understand, participate and even construct for themselves the relevant IE diagrams. This helps to ensure that their requirements are truly understood and achieved. The diagrams are regarded as being rigorous enough on their own to ensure that all necessary information is captured and represented.

Each IE technique is orientated towards diagramming, and a diagram is a deliverable of each major stage in the methodology. One of the key elements in IE is the use of standard diagrams which initially are defined at high levels of abstraction and, as the methodology proceeds, they are gradually evolved, becoming more and more concrete and detailed until they ultimately form potentially executable applications. Standard symbols are used throughout, for example, boxes with square corners represent data and boxes with round corners represent activities. In IEF there are also standard colour codes, for example, red for data and blue for activities.

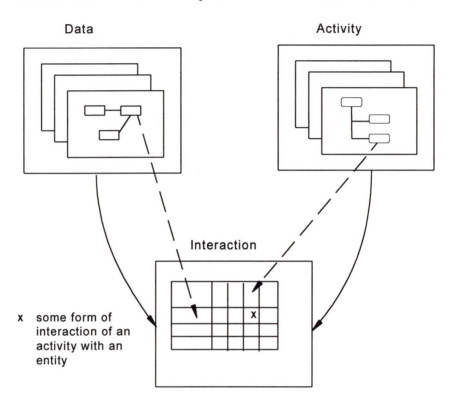

Fig. 6.7: Data, activity and interaction

The primary IE model consists of three components; data, activity and the interaction of the data and activities, as shown in figure 6.7. The interaction may be a matrix indicating at a high level which subject areas are used in which functions. At a lower level it may show which entity types are used by which processes. At an even lower level still, the interaction may be expressed as an action diagram (section 4.11) and finally as actual program code.

Automated support, that is, the use of an appropriate CASE tool, is identified as a basic requirement for the IE methodology. A description of IEF is given in section 5.7. Whilst IE can in theory be undertaken manually, the complexity of organisations and the need to develop systems quickly mean that in practice the support of CASE tools is imperative.

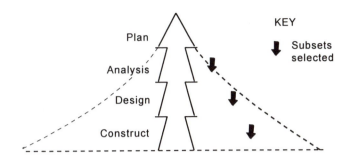

Fig. 6.8: Divide and conquer approach of IE

The methodology is top-down, and begins with a top management overview of the enterprise as a whole. In this way separate systems are potentially related and co-ordinated and not just treated as individual projects which enables an overall strategic approach to be adopted. As the steps of the methodology are carried out, more and more detail is derived and decisions concerning which areas to concentrate upon are made. Based on the overall plan, the business areas to be analysed first are selected and then a subset may be chosen for detailed design and construction. This approach to the management of the complexity of the information system requirements of an organisation is termed 'divide and conquer', and is illustrated in figure 6.8. The objectives and focus change as the methodology progresses with each stage having different objectives, although the overall objectives remain consistent. Progress is controlled by measuring whether the objectives have been achieved at each stage, not by how much detail has been generated.

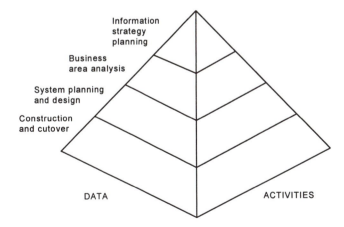

Fig. 6.9: The four levels of IE

The methodology is divided into four levels or layers. These levels are represented in figure 6.9. The four levels are:

- *Information strategy planning:* The objective here is to construct an information architecture and a strategy which supports the overall objectives and needs of the organisation. This is conducted at the enterprise level. One part of this planning is the identification of relevant business areas.
- *Business area analysis:* The objective here is to understand the individual business areas and determine their system requirements.
- *System planning and design:* The objective here is to establish the behaviour of the systems in a way that the user wants and that is achievable using technology.
- *Construction and cutover:* The objective here is to build and implement the systems as required by the three previous levels.

The first two levels are technology independent, whereas levels three and four are dependent on the proposed technical environment.

1 Information strategy planning (ISP)

Much of this level is really concerned with the overall corporate objectives. It may not always be part of the IE methodology, as it would normally be performed by corporate management and planners. However, it is recognised as a fundamental starting point for the methodology. It implies that the organisation's information system should be designed to help meet the requirements of the corporate plan and that information systems are of strategic importance to the organisation. The corporate or business plan should indicate the business goals and strategies, outline the major business functions and their objectives, and identify the organisational structure. The plan should ideally be in quantitative terms with priorities between objectives established.

ISP involves an overview analysis of the business objectives of the organisation and its major business functions and information needs. The result of this analysis is what is termed 'information architectures' which form the basis for subsequent developments and ensure consistency and coherence between different systems in the organisation. The resulting information strategy plan documents the business requirements and allocates priorities, which are the rationale for the development of the information systems. The plan enables these high-level requirements to be kept in view throughout the development of the project. In many other methodologies, it is argued, these needs get lost, if they are ever identified at all. It also provides a means of controlling changes to assumptions, priorities and objectives, should it become necessary. Apart from such changes, the information strategy plan should remain relatively static. Information strategy planning is a joint activity of senior general management, user management and information systems staff. It involves the performance of four tasks as follows:

1. *Current situation analysis:* This is an overview of the organisation and its current position, including a view of the strengths and weaknesses of the current systems. This overview will include an analysis of the business strategy, an analysis of the information systems organisation, an analysis of the technical environment, and a definition of the preliminary information architecture (data subject areas, such as customer or product, and major business functions).

2. *Executive requirements analysis:* Here, managers are provided with an opportunity to state their objectives, needs and perceptions. These factors will include information needs, priorities, responsibilities and problems. This also involves the identification of goals of the business and how technology can be used to help achieve these goals and the way in which technology might affect them. Critical success factors (CSFs) for the overall organisation are identified and these are also decomposed into CSFs for the individual parts of the organisation (see section 3.3).

3. *Architecture definition:* This is an overview of the area in terms of information (the identification of global entity types and the decomposition of functions within the subject areas described in the preliminary information architecture in the current situation analysis above), an analysis of distribution (the geographic requirements for the functions and the data), a definition of business systems architecture (a statement of the ideal systems required in the organisation), a definition of technical architecture (a statement of the technology direction required to support the systems including hardware, software, and communications facilities) and a definition of information system organisation (a proposal for the organisation of the information systems function to support the strategy).

4. *Information strategy plan:* This includes the determination of business areas (the division of the architectures into logical business groupings, each of which could form an analysis project in its own right), the preparation of business evaluations (strategies for achieving the architectures, including migration plans for moving from the current situation to the desired objective) and the preparation of the information strategy plan itself (a chosen strategy including priorities for development and work programs for high priority projects).

Recently IE has adapted the above approach to ISP to include elements of business process re-engineering (section 3.4). This is an example of the philosophy which suggests that emerging relevant techniques be adopted into IE.

2 Business area analysis

The business areas identified in the information strategy plan are now treated individually and a detailed data and function analysis is performed. Maximum involvement of end users is recommended at this stage. The tasks of business area analysis are as follows:

- *Entity and function analysis:* This is the major task of the stage. It involves the analysis of entity types and relationships, the analysis of processes and dependencies, the construction of diagrammatic representations of the above, such as entity models (section 3.8), function hierarchy diagrams, similar to that shown as figure 3.5, and process dependency diagrams (a kind of data flow diagram (section 4.7) but without data stores)) and the definition of attributes and information views.

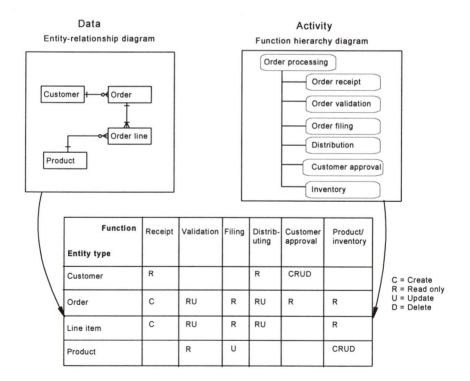

Fig. 6.10: Example function/entity type matrix

- *Interaction analysis:* This examines the relationship and interactions between the data and the functions, that is, the business dynamics. Figure 6.10 is an example of a function/entity type interaction matrix. In this example, the order processing function of a business is shown in the form of an entity-relationship model, a function/process hierarchy diagram and the matrix of interactions between the two. In this example, the matrix maps the interactions in terms of whether the function creates, reads, updates and/or deletes the entities. The example is somewhat simplified but it readily shows that orders and order lines are never deleted. This may be an error or it may indicate that an order archiving function is required. Interaction analysis also involves an analysis of entity life cycles (section 4.12), an analysis of process logic (section 4.8, 4.9 and 4.10), and the preparation of process action diagrams (section 4.11).

- *Current systems analysis:* This models the existing systems in the same way as for the entity and function analysis task in order that the models can be compared in the confirmation task (below), so that a smooth transition from one to the other can be achieved. The phase includes the construction of procedure data flow diagrams (section 4.7) and the preparation of a data model by canonical synthesis. Because this is a technique not described previously, an example of its use will be provided. Canonical synthesis is a technique for pulling together all the data identified in separate parts of the organisation, whether they be reports, screens, forms, diagrams and so on, in fact all sources, into a coherent structure, which is the entity model. The technique involves the drawing of bubble charts (user view analyses) and synthesising all the data into an entity model. A bubble chart is a graph of directed links between data-item types (figure 6.11). A double ellipse represents a key, an arrow represents a one-to-one dependency, and a double arrow represents a one-to-many dependency. In this case, the key completely determines (or identifies) the attributes, therefore the data is normalised. A separate bubble chart is constructed for each separate user view of the data. Figure 6.12 shows an example of three user views in a university environment. View 1 might be a secretary's view, view 2 a registrar's view, and view 3, a course manager's view. The process of canonical synthesis combines the separate views into one data model. Each view is normalised (section 4.6) and combined with another, and any duplications in the graph are eliminated. Figure 6.13 is the result of the combination of views 1 and 2 and figure 6.14 is the synthesis of all three views.

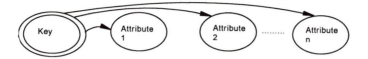

Fig. 6.11: A bubble chart

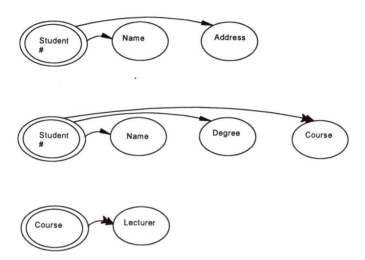

Fig. 6.12: Three user views (courtesy of James Martin Associates)

Fig. 6.13: Synthesis of views 1 and 2

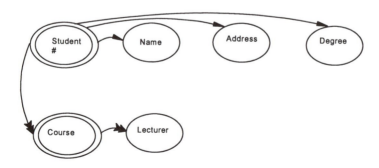

Fig. 6.14: Synthesis of all three views

- *Confirmation:* This is the cross-checking of the results of the above, in terms of completeness, correctness and stability. Hypotheses concerning business changes are also examined to see what effects these might have.

- *Planning for design:* This step includes the definition of design areas (which identify those parts of the model to be developed), the evaluation of implementation/transition sequences and the planning of design objects. This includes the identification of areas where existing re-usable objects (models and code) or components could be utilised. This may involve the re-use of objects generated internally or the purchase of objects externally. This is now an important area for IE, with the objective of speeding up the development process. In addition, areas where objects being designed for this particular area might be required themselves in the future for re-use should be designed with this in mind. Design for re-use involves additional requirements such as flexibility and ease of use (section 4.13). The effective identification of objects for re-use requires an appropriate repository (section 5.5).

The output from business area analysis is the business area description, which contains the business functions, and each function is broken down into its lower level processes and the process dependencies. On the data side, the entity types, relationships and attributes are described, along with their properties and usage patterns. The level of detail here is much greater than that arrived at during the construction of the architectures performed during the information strategy planning stage. This information provides the basis for the broad identification of business processes requiring computer support.

3 System planning and design

This level is divided into business system design and technical design. In some versions of IE, these are termed external design and internal design.

In the area of business systems design, for each design area identified, the facts gathered are used to design a system to fulfil the identified business requirements. The design is taken up to the point at which technical factors become involved, thus it is the logical design. The steps involved are as follows:

- *Preliminary data structure design:* In order to ensure integration and compatibility for all systems in the business area, this step is performed at the level of the whole business and not just the design area. It involves a first attempt at converting the entity model to the structure of the chosen database management system. This includes a summary of data model usage (basically an analysis of the way the data is used by the functions to produce a quantifiable view, sometimes referred to as volumetrics) and the preparation of the preliminary data structure.

- *System structure design:* This involves the mapping of business processes to procedures and the interactions are highlighted by the use of data flow diagrams. This phase involves the definition of procedures and the preparation of data flow diagrams.

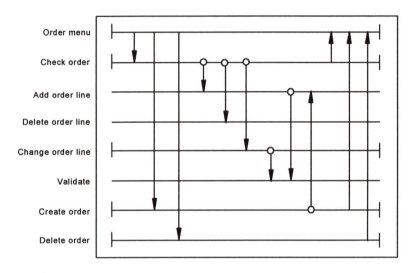

Fig. 6.15: A dialogue flow example

- *Procedure design:* This stage involves the development of data navigation diagrams (access path analysis, which examines the types and volumes of access required to particular entity types), the preparation of dialogue flows (that is, the various hierarchies of control of user interaction), and the drawing of action diagrams (section 4.11). Hierarchically structured dialogues and menus can be represented in action diagrams but non-hierarchical menus require an alternative diagrammatic representation. For this purpose, IE uses dialogue flow diagrams (sometimes called dialogue structure diagrams), a simple example being illustrated in figure 6.15. In overview, the horizontal lines represent screens and the vertical lines potential jumps or transfers between screens. This will depend on the choices made by the user. A horizontal line with bars at both ends represents a procedure, which could be a screen or a menu. Horizontal lines without bars represent a procedure step, that is, a part of a procedure. A vertical line between procedures indicates that control is transferred from the procedure at the tail of the arrow to the procedure or step pointed to by the arrow. If the tail of the arrow has a loop, this represents a link transfer which means that control may return to the originating procedure with the context retained. The example shown in figure 6.15 is the structure of a menu system for handling customer orders. The first procedure is the main order menu and the user can select three options: check order, create order or delete order. In check order, the customer order number is entered and the options of adding an order line, deleting an order line or changing an order line can be made. These are link transfers as the context, that is, the order details for the selected customer is retained on return. Note that add order line and change order line have a link transfer to validate, which is a procedure step of some other procedure. In an IE-based CASE tool, the dialogue flow diagrammer is usually linked to a screen design/paint function which provides the capability for a designer to click on a horizontal line and display the relevant screen layout, and flip backwards and forwards as necessary.

- *Confirmation:* Again, as part of business systems design there is a stage to confirm completeness, correctness and usability. Matrices are used to analyse completeness. For correctness, the question 'does it follow the IE rules?' is asked. For usability, verification is normally achieved by the users commenting on a prototype.
- *Planning for technical design:* The final phase of this stage involves the definition of implementation areas and the preparation of technical design plans.

At the end of this stage a business systems specification is produced which details, for each business process, the information flows and user procedures, and for each computer procedure, the consolidated and confirmed results of business area analysis, plus the dialogue design, screens, reports and other user interfaces. The scope of the proposed computer systems is defined along with the work programmes and resource estimates for the next stage.

The computerised aspects of the business systems identified above are designed at a technical level such that the final construction and operation of the systems can be planned and costed. The tasks of this technical design phase are as follows:

- *Data design:* which includes preparation of data load matrices, refinement of the database structure, design of data storage and the design of other files.

- *Software design:* which includes the definition of programs, modules, re-use templates, integration groups, the design of programs/modules and the definition of test conditions.
- *Cutover design:* which includes the design of software and procedures for bridging and conversion, the planning of system fanout (the phases in which it should be implemented by location) and the definition of user training.
- *Operations design:* which includes the design of the security and contingency procedures, the design of operating and performance monitoring procedures and the design of software for operations.
- *Verification of design:* which includes benchmark testing and performance assessment.
- *System test design:* which includes the definition of system tests and acceptance tests.
- *Implementation planning:* which includes a review of costs and the preparation of the implementation plan.

The output from this stage is the technical specification, including the hardware and software environment, its use, standards and conventions. It also includes the plan and resources for subsequent construction and cutover.

4 Construction and cutover

This level includes the stages of construction, cutover and production. Construction is the creation of each defined implementation unit and includes the following tasks:

- *System generation:* which includes the construction of the computing environment, preparation of development procedures, construction of database and files, generation of modules, generation of module test data, performance of integration tests and generation of documentation.
- *System verification:* which includes the generation of system test data, performance of system tests, generation of acceptance test data, performance of acceptance tests and obtaining approval. The use of test support tools is recommended.

Construction is completed once the acceptance criteria are satisfied.

Cutover is the controlled changeover from the existing systems and procedures to the new system. The tasks are:

- *Preparation:* which includes the preparation of cutover schedule, training of users and the installation of hardware.
- *Installation of new software:* which includes the conversion to the new software and execution of trial runs.

- *Final acceptance:* which includes agreement of the terms for acceptance and transferring fully to the new system.
- *Fanout:* which means the installation at all locations.
- *System variant development:* which is to identify requirements, revise analysis and design, and perform construction and cutover where a particular location requires a variance from the norm.

Cutover is regarded as complete when the system operates for a period at defined tolerances and standards, and passes its post-implementation review.

Following cutover, production is the continued successful operation of the system over the period of its life. The tasks are to ensure that service is maintained and that changes in the business requirements are addressed.

- *Evaluate system:* which includes performance measurement, comparing benefits and costs, user acceptability and making a comparison with the design objectives.
- *Tune:* which includes monitoring performance, tuning software and reorganising databases.
- *Maintenance:* which includes correcting bugs and modifying the system as required.

The levels, stages and tasks of IE outlined above are described in a sequence that would suggest that a top-down classic waterfall model is in operation. This is not necessarily the case, and much of the development after the information strategy planning level, and particularly after the business area analysis level, can be performed in parallel. To support parallel development a co-ordinating model is constructed. This is essentially a high-level model of data and processes, which identifies and highlights dependencies and necessary interfaces between systems and subsystems. This enables complex activities to be broken down into manageable components that can be developed independently. IE is also claimed to be able to support a variety of paths through the development layers. For example, reverse engineering starts at the bottom of the framework, that is, with an existing implementation, and deduces business rules from that system. This might be useful when existing legacy systems are to be included in an IE framework. It can support re-engineering which is a combination of forward and reverse paths through the framework. Developers may reverse engineer an existing application back to an appropriate point and then, when the design rules have been identified, combine these with some new requirements and then forward engineer it to implementation.

As mentioned above, IE is increasingly dependent on CASE tools and the ability to reverse engineer and re-engineer is predicated on the benefits of an integrated CASE tool with a sophisticated repository.

6.5 STRUCTURED SYSTEMS ANALYSIS AND DESIGN METHOD (SSADM)

SSADM is a methodology developed originally by UK consultants Learmonth and Burchett Management Systems (LBMS) and the Central Computing and Telecommunications Agency (CCTA) which is responsible for computer training and some procurement for the UK Civil Service. It has been used in a number of government applications since 1981 and its use has been mandatory in many Civil Service applications since 1983. It is said to be a data-driven methodology because of its history and emphasis on data modelling and the database, but in its later versions it has become more balanced, with, for example, importance attached to the role of the users. It is an important methodology, particularly in the UK, and version 4 was released in June 1990. A description of the methodology can be found in Downs *et al.* (1988), Weaver (1993), Eva (1994) and in its complete form in NCC (1995). The following represents only an outline of SSADM and is based on the much more detailed description found in Weaver (1993).

The methodology provides project development staff with very detailed rules and guidelines to work to. It is highly structured. Another reason for its success has been in the standards provided (often exercised by completing pre-printed documents). Documentation pervades all aspects of the information systems project. In many ways it is the true successor to the conventional approach described in Chapter 2, but includes the new techniques and tools developed since the 1970s.

SSADM version 4 has seven stages (numbered 0 to 6 below) within a five 'module' framework (the bullet points) with its own set of plans, timescales, controls and monitoring procedures. The activities of each stage are precisely defined as are their associated end-products (or deliverables), and this facilitates the use of project management techniques (the project management method PRINCE is recommended, see section 5.2).

- **Feasibility study**
    - 0   *Feasibility*
        - Prepare for the feasibility study
        - Define the problem

      - Select feasibility options
      - Create feasibility report

- **Requirements analysis**
  1 *Investigation of current environment*
        - Establish analysis framework
        - Investigate and define requirements
        - Investigate current processing
        - Investigate current data
        - Derive logical view of current services
        - Assemble investigation results
  2 *Business system options*
        - Define business system options
        - Select business system options
        - Define requirements

- **Requirements specification**
  3 *Definition of requirements*
        - Define required system processing
        - Develop required data model
        - Derive system functions
        - Enhance required data model
        - Develop specification prototypes
        - Develop processing specification
        - Confirm system objectives
        - Assemble requirements specification

- **Logical system specification**
  4 *Technical system options*
        - Define technical system options
        - Select technical system options
        - Define physical design module
  5 *Logical design*
        - Define user dialogues
        - Define update processes
        - Define enquiry process
        - Assemble logical design

- **Physical design**
  6 *Physical design*
        - Prepare for physical design
        - Create physical data design
        - Create function component implementation map
        - Optimise physical data design
        - Complete function specification

- Consolidate process data interface
- Assemble physical design.

These modules cover the life cycle from feasibility study to design, but not program design. Planning is therefore assumed to have been done, and the stages following design are presumably seen as installation-specific, and therefore not covered by the methodology. We will now look in outline at each of the seven stages of SSADM.

0 Feasibility

This stage is concerned with ensuring that the project which has been suggested in the planning phase is feasible, that is, it is technically possible and the benefits of the information system will outweigh the costs.

This phase has four steps: prepare for the study, which assesses the scope of the project; define the problem, which compares the requirements with the current position; select feasibility option, which considers alternatives and selects one; and assemble feasibility report.

Systems investigation techniques, such as interviewing, questionnaires and so on, discussed in section 2.2 in the context of the traditional life cycle, are used in this stage as are the 'newer' techniques of data flow diagramming (referred to as data flow models), drawn from an analysis of the flow of documents. The latter have different symbols (see figure 6.16), but essentially the technique is the same as described in section 4.7. Entity models (referred to as logical data structures) similar to that described in section 4.5 are drawn. As one would expect at the feasibility stage, these are all done in outline and in not too great a detail. This detail will come in later stages.

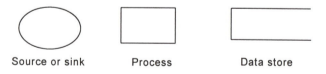

Source or sink   Process   Data store

Fig 6.16: Data flow diagram symbols in SSADM

The requirements of the new system, in terms of what the system will do and constraints on the system, are partly defined by considering the weaknesses of the present system.

Once the problem has been defined in this way, it is possible to consider the various alternatives (there might be up to five business options and a similar number of technical solutions) and recommend the best option from both the business and technical points of view. All this information is then published in the feasibility report.

1 Investigation of current environment

The second module, requirements analysis, has two stages: investigation of current requirements and business system options. This module sets the scene for the later stages, because it enables a full understanding of the requirements of the new system to be gained and establishes the direction of the rest of the project.

The first of these stages repeats much of the work carried out at the feasibility study stage but in more detail. For example, at the feasibility stage, the data flow diagrams may not have included much of the processing which is not related to the major tasks nor decomposed to more than two levels of detail (level 3 diagrams would be the norm at this stage). Further, the conflicts and ambiguities of the entity model need to be resolved. Indeed, in some projects, the feasibility stage is carried out very much in outline and the investigation of current environment may have much less of a basis for the tasks of this stage.

The results of the feasibility study are examined and the scope of the project reassessed and the overall plan agreed with management. The requirements of the new system are examined along with investigating the current processing methods and data of the current system, again in more detail than that carried out at the feasibility stage. The present physical data flow model is mapped onto a logical data flow model and this helps to assess the present functionality required in the new system. Matrices (section 4.15) might be constructed which, for example, show the relationship between processes and entities (that is, which processes access the information in the various entities). Catalogues will be created, such as the user catalogue, which lists the activities carried out in each job, and the requirements catalogue, which lists the functional and non-functional requirements. Again, there is a complete description of the results of this stage assembled and reviewed as the deliverable.

2 Business systems options

It is at this stage that the functionality of the new system is determined and agreed. The user requirements were set out in stage 1, but it is at stage 2 that only those requirements which are cost-justified are carried forward (using standard cost-benefit analysis techniques) and these requirements are specified in greater detail. A number of business system options are outlined, all satisfying this minimum set of user requirements, and a few of these are presented to management so that one can be chosen (or a hybrid option chosen, taken from a number of the options presented). Each of these will have an outline of its cost, development time scale, technical constraints, physical organisation, volumes, training requirements, benefits and impacts on the organisation. The option chosen is documented in detail and agreed as the basis of the system specification which is the next stage of SSADM. Data flow diagrams and entity models are developed, but this stage is largely a specification in narrative.

3 Definition of requirements

This stage leads to the full requirements specification and provides clear guidance to the design stages which follow. Weaver (1993) describes this stage as the 'engine room' of SSADM where investigation and analysis are replaced by specification and design. For example, stress is placed on the required system design rather than the functionality of the current system. The requirements catalogue will be consulted and updated and the logical entity model extended followed by normalisation (section 4.6) of the relations (to third normal form). The data flow model is also extended and used as a communication tool with users, with the definition of user roles in the new system, but it is the entity model which is emphasised at this stage and is the essential basis of the logical design of the new system. Documentation forms for all the entities and attributes (similar to that shown in figures 4.15 and 4.16) are completed.

Although the data model is emphasised at this stage, the components of each function (in terms of inputs, outputs and events or enquiry triggers) are defined. Each function is documented in detail and a form is used which includes space for function name, description, error handling, data flow diagram processes, events and input and output descriptions. Jackson structure diagrams (section 6.7) are used to show the input and output structures. Further documentation shows other detail, such as the relationship between user roles and functions (via a user role/function matrix).

This stage in SSADM also has an optional prototyping phase. The methodology suggests demonstrating prototypes of critical dialogues and menu structures to users and this will verify the analysts' understanding of the users' requirements and their preferences for interface design. As well as verifying the specification, this phase can have other benefits described in section 3.11, such as increased user commitment.

Entity life histories (called entity life cycles in section 4.12) are also constructed during this phase. These document all the events which can affect an entity type and model the applicable business rules. Events affecting each entity may have been identified previously by constructing an event/entity matrix (similar to the CRUD matrix, figure 4.64). Again, this is useful for verification purposes, as an entity should normally have at least one creation and one deletion event, and every event should lead to the update of at least one entity. Finally, at this stage, the system objectives are verified, the functions checked for completeness of definition and the full requirements specification documented.

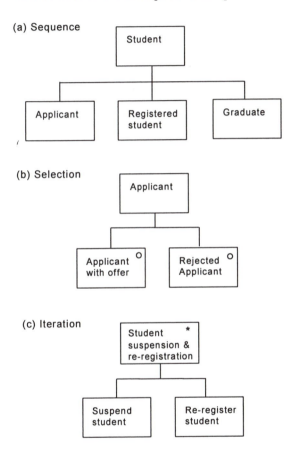

Fig. 6.17: SSADM entity life history constructs

The diagramming conventions used in drawing entity life cycles (called in SSADM entity life histories) are very similar to the entity structure step conventions of Jackson System Development (JSD) (section 6.7) and the technique is described in more detail there because it is of such crucial importance in JSD. In SSADM, the diagrams look like hierarchies, but they are meant to be read from left to right, and, in so doing, progressively suggest the different states of the entity. Using an example from the academic world, figure 6.17(a) shows how the entity 'student' changes over time, as an applicant, registered student and graduate (there will be other intermediary states). Figure 6.17(b) shows the use of the selection construct, whereby the 'o' in the 'applicant with offer' and 'rejected applicant' boxes denote alternative conditions (these are mutually exclusive). Figure 6.17(c) shows the iteration construct, marked with an asterisk, which shows an event that may repeat (in this example, the possible repeated suspension and re-registration of a student who might regularly pay fees late).

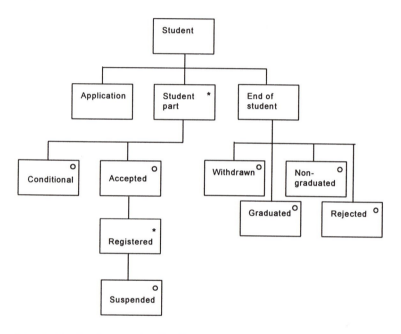

Fig. 6.18: Student entity life history

Figure 6.18 presents an SSADM entity life history. The first level contains the events that cause an entity to be initiated into the system and those events that terminate the entity from the system. There is an iteration construct relating to whether the student is accepted conditionally or not, and to reflect suspended or registered states. There are four states for the end condition: withdrawn, graduated, non-graduated or rejected. These are all mutually exclusive (the selection construct). Notice that in this model it is not possible to show that 'graduated' can only happen from registered and that 'suspended' can only terminate with non-graduated. Thus, some information is lost when compared to the earlier entity life cycle representation described in section 4.12, and seen in figure 4.55.

4 Technical system options

This stage and the following stage are carried out in parallel. In the technical system options stage, the environment in which the system will operate, in terms of the hardware and software configurement, development strategy, organisational impact and system functionality, is determined.

The definition of technical options will be implementation specific, because there are so many alternative hardware, software and implementation strategies. The analysts need to identify constraints, for example, the hardware platform may be 'given' along with time and cost maxima and minima. System constraints might include performance, security and service level requirements that must be met and these will limit choice. Technical system options need to meet all these constraints and a chosen option has to be agreed with management.

5 Logical design

This is a statement of what the system is required to do rather than a statement about the procedures or program specifications to do it. The latter is the realm of the final stage 6, the physical design. In stage 5 the dialogue structures and menu structures and designs are defined for particular users or user roles. User involvement is recommended at this stage and the prototypes developed in stage 3 are referred to. Furthermore, following the entity life cycles designed in stage 3 (which are developed further), the update processes and operations are defined along with the processing of enquiries, including the sequence of processing. In other words, it is at this stage that further detail about how the system will apply and control the operations following each event will be defined. Detail such as the rules of validating data entered into the system, will be specified. All the requirements to start designing the physical solution are now in place.

6 Physical design

It is at this final stage that the logical design is mapped onto a particular physical environment. A function component implementation map (FCIM) documents this mapping. The phase provides guidelines regarding physical implementation and these should be applicable to most hardware and software configurations. However, this stage will be carried out with the actual configuration in mind. The role of the technologist, the programmer and database designers in particular, is stressed in this phase, although the analyst and user need to be available to verify that the final design satisfies user requirements.

The logical data model will be converted into a design appropriate for the database management system available. The database mapping will be a key aspect of final implementation and include not only the way data and data relationships are held on the database, but also key handling and access methods. Much will depend on performance measurement so that database access is efficient, and again this will depend on the actual hardware and software configuration (including database management system).

The function component implementation map lists the components of each logical function and their mapping onto the physical components of the operational system. The principles of the FCIM are well specified in SSADM, although the form of the FCIM is somewhat ambiguously expressed. Presumably, this is seen as dependent on the standards of the particular organisation. Designs are optimised according to storage and timing objectives. From this stage it should be possible to design and develop the programs necessary to provide the required functionality. It is at that point the SSADM stops and detailed software design and testing starts.

The well-defined structure of SSADM make it teachable and many UK university courses in information systems have used this methodology for in-depth treatment and discuss other methodologies in overview only for comparative purposes. Its three basic techniques, entity models, data flow diagrams and entity life histories, are common to a number of methodologies and they ensure that there has been a detailed analysis of the target system. Along with the well-defined tasks, and guidance with the techniques, the methodology defines the outputs expected from the stage, and gives time and resource management guidelines.

SSADM is expected to be used along with computer tools and there are many tools designed specifically for users of SSADM as well as those designed for other methodologies which are also useful to followers of SSADM. There are, for example, a number of CASE tools (section 5.7), supported by data dictionaries and systems repositories (sections 5.4 and 5.5) which help analysis and design, some generating code from the SSADM design, and drawing tools (section 5.6), to help draw entity-relationship diagrams, entity life histories and data flow diagrams, and all these can be very supportive of the information systems development process.

The proponents of the methodology also recommend 'Quality Assurance Reviews' based on structured walkthroughs (which were described in section 3.7). They are meetings held to review identifiable end products of the various phases of the methodology, such as entity models, data flow diagrams, entity life histories and process details. Usually, the end product is presented by the authors and reviewed by personnel from related project teams (helping good communication between project teams and ensuring a common standard of work), specialist quality assurance teams or groups of users. The purpose of the meetings is to identify errors in the product. Solutions are resolved outside the meeting. Post implementation feedback is also encouraged and there is an audit at this time.

The successful implementation of the methodology relies on the skills of key personnel being available, though the techniques and tools are widely known and the project team method of working, along with systems walkthroughs, encourages good training procedures and participation. Analysts trained in the conventional approach, discussed in section 2.2, will recognise many features in SSADM, particularly the emphasis on documentation standards, clear and detailed guidelines and thorough quality assurance.

6.6 MERISE

Merise is the most widely used methodology for developing information systems in France. It is used in both in the public and private sectors. Its influence has spread outside France, and it is now being used in Spain, Switzerland and North America. Like SSADM, Merise may become very influential in any future European standard. The approach is appropriate for data processing applications that use databases and those in real-time and batch processing environments.

The essentials of the approach lie in its three cycles: the decision cycle, the life cycle and the abstraction cycle, which cover data and process elements with equal emphasis. Although it is prescriptive to some extent, Merise permits the participation of end users and senior management as well as data processing professionals, in its decision cycle. There are a number of software tools for use with Merise.

The project which led to Merise was launched in 1977 by the French Ministry of Industry, and included research groups, consultancy and engineering firms, and academics, the inspiration coming from Hubert Tardieu. Merise has since developed into a very thorough and comprehensive methodology.

The core of the Merise approach lies in its three cycles: the decision cycle, which relates to the various decision mechanisms; the life cycle, which reflects the chronological process of a Merise project from start to finish; and the abstraction cycle, the key to Merise, which describes the various models for processes and data in each of three stages. Each of these three cycles will be considered in turn, with the major emphasis being placed on the abstraction cycle.

1 Decision cycle

The decision cycle, sometimes referred to as the approval cycle, consists of all the decision mechanisms, including those for choosing options, during the development of the information system. Decision making is a joint process concerning senior management, users and systems developers. Decisions will include:

- Technical choices regarding hardware and software
- Processing choices, such as real-time or batch
- User-oriented choices relating to the user interface
- Identification decisions regarding the major actors of the information system and the organisation
- Financial decisions relating to costs and benefits
- Management decisions concerning the functionality of the information systems.

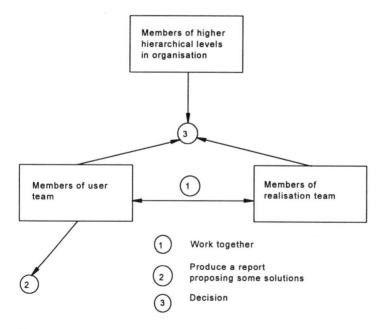

Fig. 6.19: Schema of decision-making process at each step

Each decision point during the development of an information system is identified by Merise. It is essential to know who takes the decisions, particularly those relating to the validation of the various models used by the method, and when to complete one stage to start the next. The Merise authors suggest that the decision-making process will follow the scheme as shown in figure 6.19. The groups of users and systems developers will together discuss various options (1) and it is the responsibility of the user team to produce a report reflecting these deliberations (2). This is then discussed at a joint meeting (3) of senior management, users and application developers, and the decision made at this point.

It is necessary to specify how a compromise should be reached in the case of conflicting views. This will depend on the norms of the specific organisation, but there is a strong user element suggested in the decision-making process and this will influence the acceptance of the final system, from the point of view of operational and technical criteria and usability (Eason, 1984).

Thus, in Merise, there are opportunities for user influence and participation, but this is not spelt out in detail as it will depend largely on the norms of the organisation.

2 Life cycle
The life cycle shows the chronological progress of the information system from its creation, through its development, until its final review and obsolescence. Each of these stages is well defined in Merise. The main phases of the life cycle are:

1. *Strategic planning (at the corporate level):* which maps the goals of the organisation to its information needs, and partitions the organisation into 'domains' for further analysis (such as purchasing, manufacture, finance and personnel). For each of these a schedule of applications is devised to include a policy for human resources, software and hardware products, and system development methodology implementation. Within the frame of the strategic plan, the analysis that has just been carried out for one domain should be done for all the others, and then it will be possible to understand better and more coherently the connection between them.

2. *Preliminary study (for the domain of interest):* which describes the proposed information systems, discusses their likely impact, and details the associated costs and benefits, which should be consistent with the strategic plans.

3. *Detailed study (for a particular project):* of only those aspects which will be automated, including detailed specifications for the functional design (the requirements specification) and the technical design (the technical architecture of programs and files).

4.   *Schedules   and   other   documentation:*   for   development, implementation and maintenance (all three at the application level).

Sometimes the last stage is defined as consisting of three separate stages: development, implementation and maintenance. This is little different from conventional systems development as described in section 2.2. In practice it will differ through its use of techniques and tools which were developed since the 1970s.

The whole of this second cycle is similar to the conventional life cycle as found, for example, in SSADM and other methodologies, and for this reason will not be discussed further. It should be pointed out, however, that unlike many alternative approaches, Merise includes a strategic planning phase, and in this respect Merise is similar to Information Engineering. The objective of this stage is to link the goals of the business with the information systems needs.

Nevertheless, like SSADM, the emphasis is on the analysis and design of the database and corresponding transactions. The reference to SSADM made earlier is apposite, for the nearest UK equivalent to Merise is SSADM, being the most used methodology in the UK and widely adopted by the UK Civil Service and other public and private sector organisations.

3 Abstraction cycle

The abstraction cycle is the key to Merise. Unlike many alternative approaches, the separate treatment of data and processes is equally thorough and both are taken into account from the start. The data view is modelled in three stages: the conceptual, the logical, through to the physical (a framework borrowed from the database approach as originally specified in ANSI-SPARC, 1975, as described in section 5.3). Similarly, the process-oriented view is modelled through the equivalent three stages of conceptual, organisational and operational. Each of these six abstraction levels in the abstraction cycle is a representation - albeit a partial one - of the information system, and they should be consistent.

The abstraction cycle is a gradually descending approach which goes from the knowledge of the problem area (conceptual); to making decisions relating to resources and tasks; through to the technical means on which to implement it. The conceptual stage looks at the organisation as a whole; the logical stage addresses questions, such as, who must do what, where, when and how; whereas the physical stage looks at resources and technical constraints surrounding what will be the operational system. Merise is therefore independent of the technology until the later phases.

The modelling logic of Merise, outlined above, is shown in figure 6.20. At the conceptual level the group of entities dealt with by the information system will be represented in a totally independent way from the organisation and from the existing or future technical means for developing the project. At this level it is necessary to find out what the business does and the essence of the problem situation. At the logical level it is necessary to make choices (using methods developed at the conceptual level) in terms of the organisation for the processing and with regard to the database models for the data, which will be part of the automated system. The physical level is the level at which constraints related to the operating system, database management system and programming languages are going to be introduced.

LEVEL	CONCERN	DATA	PROCESSING
CONCEPTUAL	What do you want to do?	Conceptual data model	Conceptual processing model
LOGICAL OR ORGANISATIONAL	Who does what, when, where, how?	Logical data model	Logical or organisational data model?
PHYSICAL OR OPERATIONAL	By what means?	Physical data model	Operational processing model

Fig. 6.20: Merise by levels of data and processing

An initial overall view of the system is given in the Merise flow diagram, and the construction of this precedes the conceptual models (both data and processing). The Merise flow diagram is not to be confused with the more conventional data flow diagrams. The Merise flow diagrams bring to light the information flows between the various actors in the domain studied, together with the environment. They serve as a base for developing the conceptual data model and the conceptual processing model. Data flow diagrams relate to the processing side. The actors are described in the ellipses and arrows represent the information flows between them. Thus the flow diagram showing the accounts, suppliers and customers might be as shown in figure 6.21, where the actors which are hatched are external to the information system. From it, we can see directions for the conceptual data model (concerning customers, accounts and suppliers) and for the conceptual processing model (concerning the settlement of invoices) in this example.

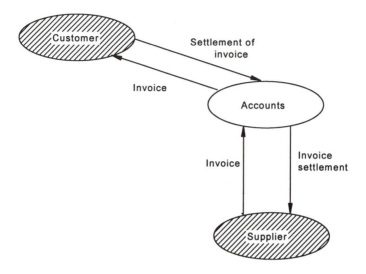

Fig. 6.21: A Merise flow diagram

At the conceptual level, the information system is represented independently of its organisation and of the physical and computing means that it could use. The objective of the conceptual level is to answer the question 'what?' and to understand the essence of the problem. The rules evidenced at the conceptual level are the 'management rules' of the domain under analysis.

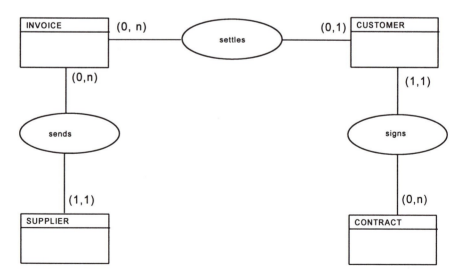

Fig. 6.22: A conceptual data model (part)

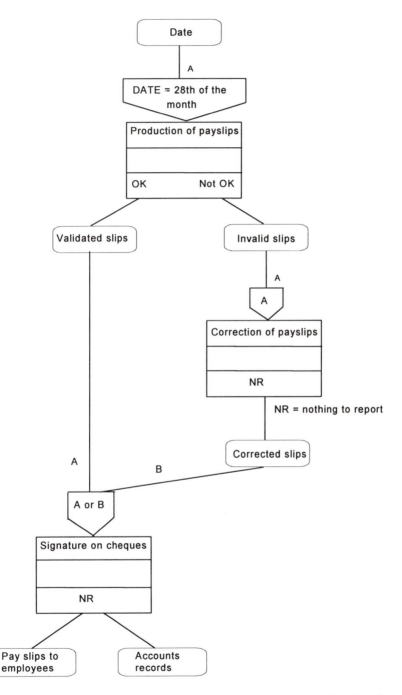

Fig. 6.23: Conceptual processing model of the process 'production of payslips'

The graphical representation of the conceptual data model is the entity-attribute-relationship model developed from Chen (1976), and similar to that described in section 4.5. An entity is represented by a box in Merise; a relationship is represented by an ellipse. Merise has a number of rules which enable the verification of the model.

The conceptual processing model describes the activities of the organisation. The concern of the conceptual processing model is with events, operations and their synchronisation.

Many other approaches which include the concepts of operations and events do not include that of the synchronisation of an operation. This is the condition or conditions (events) which must have occurred to trigger the operation, and a rule or set of rules regarding the necessary condition for the operation to be triggered. For example, payslips should be produced if it is the 28th of the month. The conceptual processing model related to the production of payslips might include a synchronisation (28th of the month when payslips are produced) following the event that a new day has dawned (see figure 6.23). This will trigger the process to produce payslips, and routines which are dependent on whether the payslips are valid or invalid. In the figure, we also see an issuing rule, related to whether the payslips are OK or not OK and, depending on this state, the processes that follow are different. Cheques are signed, however, whether the event is A or B, that is, a validated payslip is received or a corrected invalid payslip is received. Merise also provides a series of rules to enable the verification of the conceptual processing model.

Having established what to do at the conceptual level, at the logical level all the organisational alternatives are identified in order to discover who will do what, where and when, and how the processing will be carried out. The information system is represented by taking into account the constraints imposed by these alternatives. The rules brought to light at the logical/organisational level are the 'organisational rules' of the domain under analysis. The organisational processing model is used again to clarify all the concepts described in the conceptual processing model. It is therefore a question of describing how the processing methods are executed within the organisation, which could be manual (where the procedure is carried out without computing resources), conversational automatic (where the procedures are carried out by computer but with the intervention of people) or automatic (where the procedures will, once started, run without human intervention). The organisational processing model is used to define who carried out the processing, and when and how it is achieved. The organisational processing model will be based on the conceptual processing model with some changes, such as the names of departments where the processing will take place. Figure 6.23 might be amended so that processing is allocated to the personnel department (signing cheques), computing department (producing payslips) and accounts (records).

The logical data model is situated between the conceptual data model and the physical data model. It represents the world of data, described in the conceptual data model, but which takes into account the type of database management system chosen. In other words, the logical data model transforms the conceptual data model into a form that is suitable for computerisation. Merise also offers a full description of the normalisation process (section 4.6).

CONCEPTUAL DATA MODEL		LOGICAL DATA MODEL IN BCNF
Identifier of entity	becomes a	Key
Property of identifier	becomes an	Attribute
Entity	becomes a	Relation
Relationship not of cardinality (1,1)	becomes a	Relation
Relationship of cardinality (1,1)	disappears	

Fig. 6.24: Rules for mapping the conceptual data model to normalised relational model

One of the most important aspects of Merise is the detailed rules for converting from one model to the next (as well as specifying rules for creating each model, for example, there are ten verification rules for the conceptual processing model). Thus the rules for mapping the conceptual data model to the logical model in relational Boyce-Codd Normal Form (BCNF) is shown in Figure 6.24 and the rules for special cases, such as those of binary and n-ary relationships, are described in detail. Rules for mapping to other data models and optimising the relational and other data models, taking into account volume and activity, are also provided.

At the physical level, all the technical alternatives are identified which make it possible to define the computing needs, and this definition represents the last stage before the development of the software. The objective of the physical level is to answer the question 'with what means?'. The rules brought to light at the physical level are the 'technical rules'. It therefore takes into account the physical resources (database management system, hardware, support tools and so on). Typically, the physical data model might be represented as a series of SQL definitions and the operational processing model as structured English, along with equivalent SQL queries to the database (section 5.3). The diagrammatic representation of the database processes and queries might be in the form of data flow diagrams and mapped onto structure diagrams. Again rules of mapping are provided.

Another important aspect of Merise is that the methodology has been designed to reflect the world where change is common, and new needs and directions can be incorporated into the design as the information systems develop. For example, guidelines are given to show how the conceptual data model and the conceptual processing model can be modified to take account of new management rules for data and processing. Further, unlike a conventional database methodology which is directed towards the static, data-oriented aspects of information systems, Merise, as we have seen, includes a thorough analysis of events, operations and synchronisation, all dynamic aspects of an information system.

Each of the six models of the abstraction cycle has a graphic formalism, with the possible exception of the physical data model, hence the methodology lends itself to the use of support tools. Many tools support all three levels of the abstraction cycle, both for data and processes, and thus ease the task of drawing the models. Some will validate, or partially validate, each of the models, help generate the required documentation at each stage, and may incorporate an applications generator, query language and data dictionary interface, which again will lighten the task of developing information systems.

Significantly, there are a number of support tools which help the user of Merise. These tools are varied: some are design tools (for example, to develop the various conceptual and logical models and diagrams), others are modelling and prototyping tools (to give alternative views of the final system) and yet others, execution and code generation tools (to generate the future application). Many purchasers of methodologies look as much at the quality of the tool support as to the basics of the methodology itself.

One of the important reasons for adopting an information systems development methodology is that of common standards, and as well as the graphical support tools for each of the six models which are part of the abstraction cycle, there are standards and tools supporting strategic planning, project planning, requirements specification and the file of options, and the documentation of each entity, relationship, attribute, event and operation defined. The strategic plan, for example, is likely to contain a diagnosis of the present situation, perspectives on the evolution of the organisation, a description of the conceptual solution (what we want to do) and plans for development regarding organisational and technical solutions. It will also include, for the adopted solution, a description and information about its impact (including advantages, risks and means), as well as reasons for rejecting other solutions.

For the processing, the overall specification is likely to include the organisational processing model, and a detailed description of processes and the operational processing model, with a list of applications, transactions and batch chains and their arrangement into computer programs. For the data, it will include the conceptual, logical and physical data models. It will also include a study of constraints (security and control policy), details about interfacing with existing applications and responsibilities. Appendices are likely to include definitions of relations (depending on the eventual database approach chosen), a list of states and screens for each process and their sequence, and a physical description of records or relations.

Of course there are problems associated with the adoption of any methodology, in particular training and motivating staff to use it, but Merise does have many aspects which commend it:

- It has been well tried and tested on a number of projects in France, elsewhere in Europe and in North America
- In its life cycle it covers a strategic plan as well as the conventional life cycle, from conception and preliminary study, through its development, implementation, maintenance, and decline and replacement
- In its three-level abstraction cycle it covers both data and process elements with equal emphasis
- Although it is prescriptive to some extent, it enables the participation of end users and senior management as well as data processing professionals in its decision cycle
- It attempts to take into account any new needs and directions identified both when compared to the present operational system and as the information system project develops
- There are now many tools available which support Merise.

A fuller description of Merise in English can be found in Quang and Chartier-Kastler (1991).

6.7 JACKSON SYSTEMS DEVELOPMENT (JSD)

Michael Jackson's program design methodology, Jackson Structured Programming (JSP), which is described in Jackson (1975), has had a profound effect on the teaching and practice of commercial computer programming. Jackson (1983) on Jackson Systems Development (JSD), argues that system design is an extension of the program design task, and that the same techniques can be usefully applied to both. Aspects of JSP are diffused throughout JSD, so that the JSD methodology is a significant development on its precursor, and therefore should not be seen as a 'front end' to JSP but an extension of it, where JSP is the core. 'In principle', says Jackson, 'we may think of a system as a large program'. However, the primary purpose of JSD is to produce maintainable software, and its emphasis is on developing software systems. This leads to a potential criticism of JSD in that, in the context of this text, it is orientated towards software and not to organisational need.

Given this comment, therefore, it is not surprising that Jackson's text does not address the topics of project selection, cost justification, requirements analysis, project management, user interface, procedure design or user participation. Further, JSD does not deal in detail with database design or file design. At least as described in his book, Jackson's methodology is not comprehensive in the sense that it does not cover all aspects of the life cycle.

However, the commercial version of JSD, because of practical necessity, has now been extended to include some of these aspects. Later texts about JSD, for example, Sutcliffe (1988), Cameron (1989) and Davies and Layzell (1993), inform of the broader version of JSD, indeed, the latter merges JSP and JSD into one overall approach. We will use Davies and Layzell and Sutcliffe as well as the original Jackson version of 1983 as the basis of this section.

The emphasis in the methodology is solving what Jackson terms the hidden path problem, that is, the path between the presentation of a specification to the design/programming group and the completed implemented system, which could be described as a 'bundle' of documentation, listings and executable programs. Jackson asks, 'What reasons do we have to support the claim that we have delivered what is required in the specification?' The traditional response is that the answer is found in the processes of testing and checking. But there are two problems here. We cannot be sure that the tests are complete and, in any case, when testing is possible, the system is already complete and it is usually rather too late and too costly to repair the damage. A second answer is to apply formal methods, and the JSD methodology does have some links with formal methods (see section 3.12). However, Jackson is aware of the problem of their inaccessibility, and the difficulty of communication.

JSD uses transformation through process scheduling as the answer to the hidden path problem and a major contribution of JSD lies in the areas of process scheduling and real-world modelling. Further, JSD deals with the problem of time in systems modelling and systems design in a way that most other information systems design methodologies do not, as the latter tend to model static elements in the system.

There are three major phases in JSD: the modelling phase, the network phase and the implementation phase. In the modelling phase, events and entities are identified and entity structures and entity life cycles formed. In other words, analysts ask what is happening in the real world and how might this be connected to the computer world. In the network phase, the inputs and outputs are added to the model so far derived so that the input and output subsystems can be analysed. In other words, the analysts ask what outputs are needed from the system, and what processes and operations must be added to produce these outputs. The implementation phase is concerned with detailed design and coding, that is, how can the specification (model plus function) be transformed to run on the hardware:

- **Modelling phase**
  1. Entity action step
  2. Entity structure step.
- **Network phase**
  3. Initial model step
  4. Function step
  5. System timing step.
- **Implementation phase**
  6. Physical system specification step.

In the entity action step the systems developer defines the real world area of interest by listing the entities and actions with which the system will be concerned. In the entity structure step the actions performed or suffered by each entity are ordered by time. In the initial model step communications between entities are depicted in a process model linked to the real world. In the function step functions are specified to produce the outputs of the system, and this may give rise to new processes. In the system timing step some aspects of process scheduling are considered which might affect the correctness or timeliness of the system's functional outputs. In the physical system specification step the system developer applies techniques of transformation and scheduling that take account of the hardware and software available for running the system. JSD is applied iteratively, and as increasing detail is revealed, data and functions will also be revealed. Each of these stages will be looked at in turn.

1 Entity action step

JSD aims to model the real world. In the entity action step, real-world entities are defined. These might include SUPPLIER, CUSTOMER or PART, but unlike the data analysis approaches, JSD is more concerned with the behaviour of the entity, than its attributes or its relationships with other entities. Conventional entity modelling presents a static view of the real world, whereas JSD is concerned with modelling system dynamics.

To be defined as an entity in JSD, an object must meet the following criteria:

1.   It must perform or suffer actions in a significant time ordering.
2.   It must exist in the real world outside the system that models the real world.
3.   It must be capable of individual instantiation with a unique name.

Entities may also be collective (for example, BOARD OF DIRECTORS) if the instantiation has objective reality without considering its component objects. Entities may be generic (for example, SPARE-PART) thus supporting the abstraction of classification, or specific (for example, INNER-FAN-SHAFT). Entities that exist in the world may be ignored if it is impossible or unnecessary to model their behaviour. Thus, only a relevant subset of the real world is modelled.

An action describes what an entity does within a system. Since the distinctive feature of JSD entities is that they perform or suffer actions, it is necessary to specify the criteria for something to be an action. These are as follows:

1.   An action must be regarded as taking place at a point in time, rather than extending over a period of time.
2.   An action must take place in the real world outside the system and not be an action of the system itself.
3.   An action is regarded as atomic and cannot be decomposed into sub-actions.

Since the original version of JSD, more emphasis has been placed on the process of eliciting attributes, both action attributes and entity attributes. Whereas action attributes come from outside the system and trigger the action, entity attributes add information about the entity and will be updated by its entity actions. The actions and changes to the entity attributes form the entity life history. It is important to analyse when these changes occur so that the entity life history will have the correct time ordering. But we are now discussing the beginnings of the next step, the entity structure step.

The end result of the entity action step is a list of entities and their attributes and a list of actions and their attributes. The list of entities is liable to be much shorter than that produced by an equivalent data analysis process, particularly if the latter normalises the entities, because the functional components of the system are excluded at this stage.

2 Entity structure step

The actions of an entity are ordered in time and are expressed diagrammatically in JSD. This is similar to the technique of entity life cycles (section 4.12), although there are differences. They show the structure of a process in terms of sequence, selection, and iteration. The diagram shown as figure 6.25 is read from top to bottom as a hierarchical decomposition. Actions are shown as the leaves, whereas components higher up represent aggregations of actions. Each structure diagram is intended to span the whole lifetime of an entity, including therefore an action that causes the entity to come into existence and one that causes it to cease to exist. The model must illustrate time ordering of these elements. The lifetime of an entity may span many years in the real world.

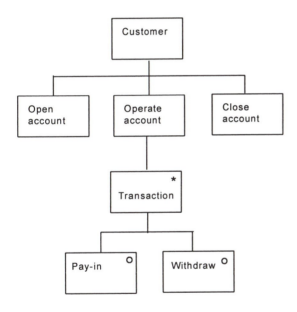

Fig. 6.25: JSD structure diagram

JSD structure diagrams do not support concurrency. For example, the entity CUSTOMER in a banking system, an example discussed fully in Jackson (1983), might have been specified as a sequence (as in figure 6.25) of OPEN-ACCOUNT, OPERATE-ACCOUNT and CLOSE-ACCOUNT. OPERATE-ACCOUNT is an iteration of TRANSACTIONs (hence the asterisk (*) which illustrates iteration, each of which is a selection of either PAY-IN or WITHDRAW (hence the small circle which indicates selection). Such a structure would constrain a customer to having only one account. To relax this constraint, the systems developer may be tempted to re-draw the

structure diagram as in figure 6.26. It now appears that the customer may
have many accounts each being operated as in figure 6.25. The diagram
now specifies that the customer can have more than one account, but not
more than one at the same time. A customer may only open a new account
after an existing account has been closed. The JSD structure diagram
cannot show the simultaneous operation of many accounts. The answer to
this problem is to specify a new entity ACCOUNT whose life history
proceeds in parallel to the life history of the CUSTOMER entity. CUSTOMER
now appears as in figure 6.27(a) and ACCOUNT as in figure 6.27(b).

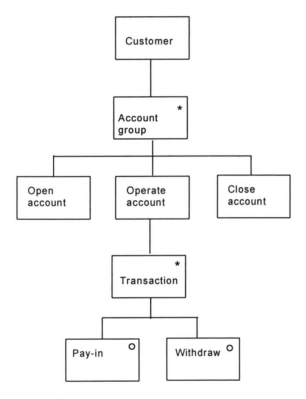

Fig. 6.26: JSD structure diagrams - customer with more than one account

In JSD, discrimination between entity roles is necessary if an entity can
play more than one role simultaneously. Jackson provides an example
using the entity SOLDIER. A soldier enlists in the army and may be
promoted to a higher rank at various points in his career. Soldiers are also
given training and may attend training courses, which they may or may not
complete successfully. If successful completion of a course always leads
to promotion, then these facts can be accommodated in one structure

diagram. If there is no necessary connection between training and a career, then two structure diagrams are required, one for the soldier's promotion career and one for his training career. The soldier in this example is playing two roles, one as a person being trained and one as a person being promoted to a higher rank. Multiple roles may be synthesised into another structure diagram showing a selection of the possible activities in the possible roles that can be played.

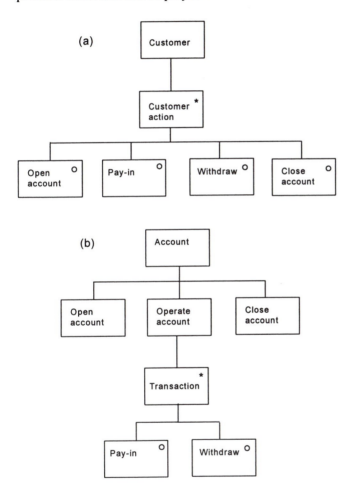

Fig. 6.27: JSD structure diagrams - simultaneous operation of many accounts

Entity structure diagrams represent a sequence of activities ordered in time, without concurrency, from the 'birth' of an entity to its 'death'. One final problem addressed by the methodology in this step is that of the premature termination of the life cycle. In the real world, events may occur that prevent an entity making an orderly progression through its life cycle. For example, a soldier may be killed in battle without proceeding to retirement. It may not be feasible to draw a structure diagram for every possible variation on a prematurely terminated life cycle. JSD allows for a general specification of premature termination. This recognition of such a circumstance is an example of 'backtracking', in JSP.

It is possible to use tools which support the methodology. Drawing and re-drawing design structures can be somewhat tedious, and the Program Development Facility (PDF) is a Jackson diagram editor which eases this process.

The end result of the entity structure step in JSD is a set of structure diagrams. New entities and multiple roles for the same entity may have been generated during this phase.

3 Initial model step

In this third step the systems developer creates a model that is a simulation of the real world. For each entity defined in the preceding two phases a sequential process is defined in the model that simulates the activities of the entity in such a way that it could be implemented on a computer. This is not to say that the implementation necessarily has to be computerised, merely that it could be if this were required.

In the model there will be a sequential process for each instance of an entity type, not one process for all instances. Thus if there are a hundred instances of entity type CUSTOMER, there will be one hundred sequential processes in the model. Moreover, the processes notionally execute at exactly the same speed as the real-world processes. Thus if a customer has a bank account for fifty years, the matching processes will also execute for fifty years.

The sequential processes specified in the initial model step are documented both by a diagram showing the interconnection of processes and by a pseudo code definition of each model process. Pseudo code is a language similar to structured English, which was described in section 4.10, but nearer to a programming language in type. The pseudo code is known as structure text in JSD, and resembles a high-level Algol-like programming language. Structure text exactly matches a corresponding entity structure diagram and major constructs are sequence, selection and iteration. The value of structure text is that it may be elaborated in later phases of the JSD methodology in a manner similar to the program design technique of stepwise refinement used in JSP. This process should be straightforward. An example of structure text for the ACCOUNT entity is provided in figure 6.28.

ACCOUNT-1 *seq*
    read data-stream
    OPEN-ACCOUNT; read data-stream
    OPERATE ACCOUNT *itr while* (PAY-IN or WITHDRAW)
        TRANSACTION *sel* (PAY-IN)
            PAY-IN; read data-stream
        TRANSACTION *att* (WITHDRAW)
            WITHDRAW; read data-stream
        TRANSACTION *end*
    OPERATE ACCOUNT *end*
    CLOSE-ACCOUNT;
ACCOUNT-1 *end*

Fig. 6.28: JSD structure text

Process connection in JSD is achieved by either data stream connection or state vector connection. In data stream connection one process writes a sequential data stream, consisting of an ordered set of messages, and the other process reads this stream. This is similar to process connection in a data flow diagram.

The JSD system specification diagram (SSD) models the system as a network of interconnected processes showing how they communicate with each other. In the diagram (figure 6.29) processes are shown in boxes, with the data streams that connect processes shown as a circle with its identification given (in the example, the identification is 'C'). Arrows show the direction of data stream movement.

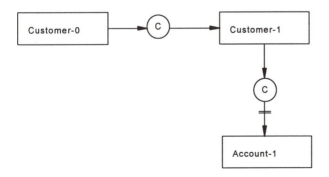

Fig. 6.29: JSD system specification diagram

In figure 6.29, CUSTOMER-0 is intended to represent a real-world instance of a customer, sending messages about his actions to a process that simulates this behaviour (CUSTOMER-1). A circle in an SSD indicates data stream connection. CUSTOMER-1 is sending a stream of messages to ACCOUNT-1. Since a customer can have many accounts, a double bar is used on the diagram to represent this multiplicity. Data stream connection is appropriate in the banking example, as it is not practical to telephone the customer every ten minutes to find out if he or she has paid in or withdrawn money.

Jackson (1983) also gives the example of a lift system that finds out whether a button has been pressed in a lift by linking the button via a state vector to a process that models the button's behaviour. The alternative state vector connection is appropriate here because the button is essentially a switch, denoting an on or off state.

In the state vector connection, one process inspects the state vector of another process. A state vector is the internal local variable which describes and is owned by a particular process. State vector connection has no equivalent in data flow diagramming because the data flow technique permits process connection via logical files. There are no logical files in a JSD system specification diagram. State vectors are shown as a diamond on the SSD. In figure 6.30, the data relating to the account is used to produce a report. Normally a data stream is used where a long-term view of events is required and a state vector used where a short-term snapshot is required.

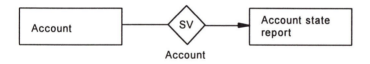

Fig. 6.30: State vector connection between processes

Data stream connection is considered to be buffered (a data stream will be read on a first in, first out basis), so writing processes are never blocked; reading processes may lag behind writing processes. State vector connection is also unbuffered and again no blocking occurs. State vector inspection therefore depends on the relative speeds of the processes involved. This is not true of data stream connection. If there is more than one input to a process, rules must be specified for determining which input is to be taken next. The determination may be made by fixed rules (fixed merge), or specified as part of the message stream (data merge), or determined simply by the relative availability of messages (rough merge). Such careful attention to synchronisation details is absent from most other methodologies. JSD also allows for time grain markers to indicate the arrival of particular points in real-world time (see step 5).

The end result of the initial model step is a systems specification diagram depicting a set of communication processes each of which is specified by a pseudo code structure text.

4 Function step

The model created in the first three phases of JSD has no outputs; it models the dynamic behaviour of the real world. In the function step, further elaboration takes place, and functions are added to the model to ensure that the required outputs are produced when certain combinations of events occur. The addition of functions may require no change to the SSD, in which case structure text is elaborated to specify the functions required. Alternatively, it may be necessary to create new processes, which are added to the SSD and specified with new structure text.

```
ACCOUNT-1 seq
        read amount-deposited
        OPEN-ACCOUNT seq
                balance:= amount-deposited
        OPEN-ACCOUNT end
        read transaction;
        OPERATE ACCOUNT itr while (PAY-IN or WITHDRAW)
                TRANSACTION sel (PAY-IN)
                        PAY-IN seq
                                balance:= balance + amount;
                        PAY-IN end
                        read transaction;
                TRANSACTION alt (WITHDRAW)
                        WITHDRAW seq
                                balance:= balance - amount;
                        WITHDRAW end
                        read transaction;
                TRANSACTION end
        OPERATE ACCOUNT end
        CLOSE-ACCOUNT;
ACCOUNT-1 end
```

Fig. 6.31: Addition of functions to structure text of figure 6.28

To give an example, we may wish to provide the facility in the banking application to interrogate customer balances on demand. Thus functions must be added to the existing SSD and structure text that record and display account balances. The elaboration of the ACCOUNT text is shown in figure 6.31. Clearly, the state vector of ACCOUNT now includes knowledge of the customer's balance, because the structure text has been elaborated to update that balance. The SSD can now be amended (as in figure 6.32) to show the new interrogation process. INTERROGATE can inspect the state vector of any ACCOUNT-1 process (as indicated by a diamond symbol). It will do so when it receives a message specifying an account enquiry, and it will produce an output showing the balance of the customer's account. Thus, the addition of function to an initial model may cause the elaboration of existing structure texts and/or lead to the specification of new processes with their own structure texts.

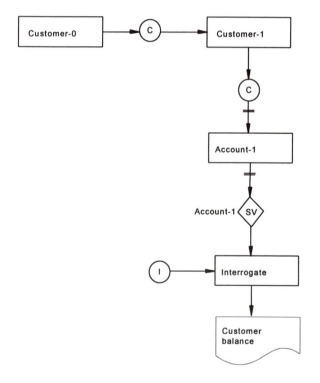

Fig. 6.32: System specification diagram with interrogation process

Whereas the model of processes and their input and output data streams correspond to the basic system and are fairly stable, the functions represent ancillary processes and are likely to be less stable. They relate to reports, queries and the user interface, which might change much more frequently, but the partitioning of parts of the system that change frequently into separate functions is relatively easy to carry out.

Whilst the earlier versions of JSD did not provide many guidelines regarding data analysis, later versions do require the identification of attributes relating to entities and these are referred to as fields of the state vector record. Update procedures and integrity constraints are also defined. However, data analysis is not as complete as in other methodologies in that, for example, it does not suggest normalisation.

The end result of the function step is an amended system specification diagram with associated structure texts.

5 System timing step

The JSD modelling process so far described has not yet explicitly raised the question of speed of execution of processes and their timing. Implicitly, the model must lag to some extent behind the real world because input must take some time to arrive. In the system timing step (sometimes included within the function step) explicit consideration is given to permissible delays between receipt of inputs and production of outputs. Different parts of the system may be subject to different time lags. Timing is both important within a process and between processes. JSD uses messages known as time grain markers (TGM) which act like data streams but which contain timing information. They are rough-merged with other data streams to control the arrival of messages and the timing of the execution of processes. They are used to trigger actions within processes, start and stop processes and generally aid the synchronisation of processes.

Time constraints will derive either from user requirements (for example, for a monthly report or for an immediate response to an enquiry) or from technical considerations. Examples of the latter are state vector retrievals that must be sufficiently frequent to capture changes of state (as in a process control application) but not so frequent that they capture too many instances of the same state. The system timing step will gather information usable in the next phase when decisions are made. These decisions may concern questions relating to on-line, real-time, or batch implementation of aspects of the model system.

The end result of the system timing step is a specification of timing constraints using the TGMs associated with processes. The step does allow for the addition to the SSD of synchronisation processes whose sole function is to ensure that certain actions have been completed satisfactorily before a further process is initiated.

6 Implementation step

Jackson's account of systems implementation is not a comprehensive treatment of all implementation considerations. Moreover 'implementation' in JSD includes activities that would be regarded in other methodologies as 'systems design', for example, file and database design, although JSD does not describe these processes in any depth.

The JSD implementation step concentrates on one major issue, the sharing of processors among processes. A system specification diagram can be directly implemented by providing one processor for each sequential process. Since there is one sequential process for each instance of an entity type (for example, one for each customer of a bank) this might imply many thousands of processors. If this is an unacceptable implementation of the model, then the implementation step provides techniques for sharing processors among processes.

The direct opposite of providing one processor for each process is to provide one processor for all processes, which could be provided by a centralised mainframe computer. In this case, JSD provides for a transformation of the model into a set of sub-routines. JSD is not recommending computer users to write their own operating systems and teleprocessing monitors, however. If these items of software are available on a machine and match the process scheduling requirements of a system, then Jackson would recommend using them. In fact, most computer-based systems are scheduled by a mixture of administrative, clerical and software action. The structure diagram for a scheduler can be drawn to alternatively represent an on-line system, a batch system or a mixture of these, together with actions that may, in fact, be performed by human beings.

The JSD systems implementation diagram (or SID) is, in a way, an abstraction of all these real-world scheduling possibilities. Figure 6.33 shows the sequence of processes hierarchically, with the scheduler at the top (drawn as a vertical bar); with the processes (contained in the SSD) shown also in boxes; inversion sequences, that is, the hierarchy of processes in terms of a main program calling sub-programs shown as parallel lines representing the data stream as a pipe; using data contained in a state vector file (rounded box); and data streams as buffers. Thus, if we return to the banking example, the buffer could represent a data stream of credit amounts; with the processes arranged hierarchically as deposit, interest being added and deposit account; and the state vector file providing data about the rate of interest. The scheduler handles the overall sequencing and this can be represented as a process structure diagram and as structure text.

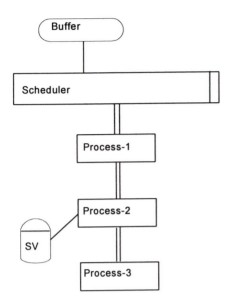

Fig. 6.33: System implementation diagram

The system may be implemented by one or more real processors, thus giving rise to possible implementations that range from completely centralised to completely distributed. In general, multiple instances of common processes would share process texts (that is, programs) as well as processors. To make process text sharing possible, it is necessary to separate the state vectors of processes from the shared process text. A concatenation of state vectors is then transformed into a file or a database. Thus the implementation step in JSD can give rise to perfectly conventional data processing solutions and the information provided in the above steps gives the programmer all that is necessary to code the system using JSP procedures.

JSD's strength as a methodology lies in its determined and detailed attempt to model the dynamic aspects of real-world systems. Its treatment of concurrency, timing and process scheduling is far more comprehensive than any other methodology. The model is, as far as possible, kept up to date so that it is always a fair refection of the real world that this abstract model represents. In this way, it is possible to see what is happening in the real world and address it in the decision-making process. Entity modelling, it is argued, only represents a static view of the real world, whereas this process-orientated approach is dynamic.

It is, however, self consciously incomplete as a methodology. Jackson wishes to say nothing about areas that are satisfactorily treated by other methodologies, only those that are not. Jackson is critical of methodologies that rely on structured decomposition, on the grounds that they confuse a method of documenting a design with the design process itself. JSD is not top-down design. Jackson is similarly critical of data modelling approaches: 'It is not much more sensible to set about designing a database before specifying the system processes than it would be to declare all the local variables of a program before specifying the executable text: the two are inextricably intertwined.' (Jackson, 1983).

6.8 OBJECT-ORIENTED ANALYSIS (OOA)

There are many different approaches to the analysis and design of object-oriented systems and, indeed, despite the relative newness of object orientation, there are already a number of object-oriented methodologies and approaches (for example, Booch 1991, Coad and Yourdon 1991, Martin and Odell 1992 and Rumbaugh *et al.*, 1991).

Of these competing methodologies, possibly the most well known is what has become known as the Coad and Yourdon Object-oriented Analysis (OOA) Methodology. It was first described in 1990 and has in the past few years become one of the major object-oriented approaches. However, even Coad and Yourdon recognise that it is early days to be defining a methodology in this area and they warn that it is still evolving and that it will change and be modified in the light of continuing practical experience. In the second edition of their book, Coad and Yourdon implore people not to say that they are 'developing systems compliant with the Coad/Yourdon OOA standard'. They argue that their book should be used as a starting point for applying OOA and the method should be tailored and expanded to suit the specific organisation or project needs. By 1994, however, Yourdon claims that he is the 'widely-known developer of the Coad/Yourdon Object-Oriented Analysis Methodology'. So we see just how rapidly things are changing in the object-oriented world.

OOA consists of five major activities:
1.   Finding class-&-objects
2.   Identifying structures
3.   Identifying subjects
4.   Defining attributes
5.   Defining services.

Coad and Yourdon emphasise that these are activities that need to be performed. They should not necessarily be seen as stages or sequential steps. They point out that many analysts prefer to iterate around the various activities in a variety of sequences. Nevertheless, we shall describe them in this order which progresses from a high level to increasingly lower levels of abstraction.

1 Finding class-&-objects

This activity is about increasing the analyst's understanding of the problem domain and, as a result, identifying relevant and stable classes and objects that will form the core of the application. Coad and Yourdon describe this as the 'system's responsibilities'. The problem domain is the general area under consideration, and the system's responsibilities are an abstraction of those elements that are required for the system that is conceived. It is the system's responsibilities which are modelled. The analysis of the problem domain is not particularly original nor examined in great detail by Coad and Yourdon. The approach recommended is first-hand observation, talking (or rather listening) to 'domain experts', reading (or 'read, read, read' as they suggest), gathering experience from previous, and related, systems, and finally prototyping.

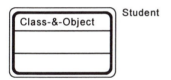

Fig. 6.34: Objects in a class

The relevant classes and their associated objects are filtered out from the problem domain. The specific term class-&-objects (represented by a particular symbol as shown in figure 6.34) includes a class (the bold inner box) and the objects in that class (the outer box). In the symbol, the class is divided into three parts. The top part is for the name of the class-&-object, the middle part for the attribute names, and the lower part for the services. An object in OOA is an abstraction from the problem domain, about which we wish to keep information (attributes of the object) and with which we can interact (the services). A class is a description of one or more objects with a common set of information and interactions.

An example in the domain of university student administration might be the classes of registration, student, course, registration-clerk and so on. For the class registration, attributes might include date, number and fee. Services might include create, renew, terminate, suspend, approve and check-qualifications. An object might be an instance of the class Student, for example, the attributes and processing for student Smith. An object embodies the notion of encapsulation (see section 4.13).

Coad and Yourdon offer a set of helpful hints in order to find relevant class-&-objects:

1.   Most importantly, look for structures which is the second activity of OOA as is discussed later.
2.   Look at other systems with which the system under consideration interacts as a way of prompting potential class-&-objects.
3.   Ask what physical devices the system interacts with. In our student administration system it is difficult to think of any example, although perhaps the photo booth might qualify. In a manufacturing system it is more obvious, for example, a weighing platform or a bottling machine. (It should be noted that these devices are not the technology with which the system might be implemented).
4.   Examine the events that must be remembered and recorded, for example, the date of registration, then the roles that people play, for example, the owner, manager and client.
5.   Examine the physical or geographical locations of relevance and also the organisational units, for example, divisions and teams.

An examination of all these factors will help to reveal relevant class-&-objects. This is by way of a checklist and may or may not lead to the identification of all the relevant ones, but it is argued that it is a useful starting point. However, even with object orientation, the traditional problems of systems analysis remain, including users and stakeholders not really knowing what they require. Yourdon suggests that a common problem in OOA is the identification of too many objects and so a criterion for evaluating objects is provided. This Coad and Yourdon term 'what to consider and challenge' and is somewhat similar to the criteria that are applied when building entity models:

- *Needed remembrance:* Is there anything, that is, any data, that must be kept by the system for this object? If there is no data, it probably means it is not an object.
- *Needed behaviour:* Is there any behaviour, that is, processing or functionality, that must be kept by the system for this object? If there is no behaviour, it probably means it is not an object, and it will certainly not be an object if there is no needed remembrance and no behaviour.
- *More than one attribute:* An object is likely to have more than a single attribute and it should be reviewed if it has only one.
- *More than one object in a class:* If there is only one object in a class, then it should be seriously challenged, indeed Coad and Yourdon suggest that this is a 'suspect' object.

- *Always applicable attributes:* Are the attributes common, that is, applicable to each object in the class? If not, it is probable that the model should contain a class hierarchy. If the object is 'student' and the attribute 'employer', but this does not apply to full-time students, then it is likely that we have two sub-types of student, full-time and part-time. (This is examined in a subsequent activity.)
- *Always applicable behaviour:* Similarly, we apply the same test to the behaviour, or in Coad and Yourdon terminology, services. If certain services do not apply to all instances of the object, then we should consider sub-types or breaking down the structure.
- *Domain-based requirements:* Ensure that all the objects are derived from the domain and not from implementation considerations. For example, 'student' is clearly derived from the domain, whereas 'registration-card' or 'application-form' are about a particular design of implementation. It is recommended that the model is kept at the highest possible level of abstraction, because the concept of a registration-card may not exist in some possible implementations, that is, it is a design consideration. Application-form is a similar case. It might preclude a design which enables a direct application via the telephone. In this case, no application form is completed and therefore the object should be less specific and focus on the logical requirement or event rather than the document or the implementation. In the examples we might prefer to consider a registration-complete object or an application-event object rather than the registration-card or application-form respectively.
- *Not merely derived results:* Derived results, that is, things that can be derived or calculated or implied from other attributes, should be avoided. For example, holding a student's examination grade (A, B, or C) as well as the percentage mark, is not relevant, as the grade can always be derived from the mark. Such consideration avoids duplication and helps to simplify things.

The end result of the class-&-objects activity is a set of relevant classes and, for each class, the associated objects modelled using the appropriate conventions. These classes and objects should have been challenged and accepted or modified according to the guidelines outlined above. These class-&-objects will form the basic structure of the system under consideration.

2 Identifying structures

The next activity is to organise the basic classes and objects into hierarchies that will enable the benefits of inheritance to be realised (see section 4.13). This involves the identification of those aspects or objects that are common or generalised, and separating them from those that are specific. (Yourdon points out that some analysis of structure will probably already have occurred in the class-&-objects activity.)

Coad and Yourdon use the terms generalisation and specialisation (which they shorten to 'gen-spec') for what is otherwise known as the identification of superclasses and subclasses. First, each class-&-object is examined to identify the gen-spec structure for each class. In other words, the generalised form of the class is examined and any specific subclasses are identified. There may be many ways of breaking down the generalised elements into specific elements but what should ideally be identified are those that will lead to the greatest degree of inheritance.

(1) Gen-spec structure

The gen-spec structure is graphically modelled as in figure 6.35 and usually reflects a hierarchy of classes. Thus, for example, we may break our student-class into full-time-student and part-time-student, keeping as much as possible that is common to both the lower-level classes in the higher-level class. This will enable all the common aspects (data and behaviour) to be inherited from the higher level to the lower levels. The benefits of this were seen in section 3.9. Any specific factors or requirements (known as specialisations) for the lower levels can then be added to the general ones inherited from the higher level. Thus, for example, all those aspects common to all types of students will be included in the high-level class of student, such as registration and qualifications. Those attributes and services specific to part-time-student, possibly employer details (full-time students would not have employers) and processing of student progress reports to employers, would be added at the lower level. This would mean that in implementation, the code for part-time-student could mostly be inherited from student and only a small addition need specifically be added for part-timers.

As a way of testing the gen-spec structures, Coad and Yourdon suggest asking the following questions of each specialisation, that is, the lower level classes:

- *Is it in the problem domain?* Does it make sense in terms of the business or organisation? In the example, do we really have a distinction between part-time and full-time students that is of relevance in this context? Objects should not be broken down just for the sake of it. Some students are male and some are female, but we must ask whether this is of importance and relevance in the problem domain. In most cases it probably is not and we would distinguish between male and female simply by an attribute of gender at the generalisation level rather than identifying separate classes of male-student and female-student.

- *Is it within the system's responsibilities?* Again, if the system does not need to make a distinction then it should not be broken down.
- *Will there be inheritance?* Are there some attributes and/or services that are common (shared) and some that are specific? If there are not, then there is not much point in breaking it out.
- *Will the specialisations meet the 'what to consider and challenge' criteria?* This will be detected from the class-&-objects activity above.

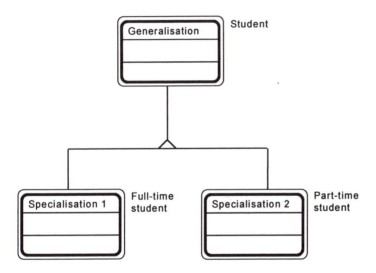

Fig. 6.35: Gen-spec structure

At this stage, the diagrams are drawn to indicate the class hierarchies. They are usually hierarchies, although multiple inheritance is allowed. This is where a specialisation (lower level) inherits elements from more than one generalisation (higher level). An example might be a specialisation course-exam which inherits some aspects from exam, such as common examination standards and procedures which apply to all courses in the university, and some aspects from course, such as examination weightings and course-specific data and procedures, for example, that it is laboratory-based or the requirements for practicals. If there is multiple inheritance, then the gen-spec structure becomes a lattice rather than a hierarchy (see figure 6.36). It should be noted that at this stage the model simply indicates that there are some attributes and behaviour that are general and some that are specific, they do not specify the detail.

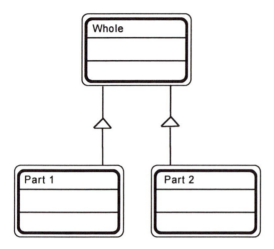

Fig 6.36: Gen-spec structure, multiple inheritance

(2) Whole-part structure
OOA also identifies what are termed whole-part structures. These are hierarchies of objects which indicate that one object is composed or made up of a series of sub-objects. The notation is illustrated in figure 6.37. The distinction between the gen-spec structure and the whole-part structures is indicated by the triangle. The cardinality of the relationship may also be indicated on the model, for example, that a course may be composed of a minimum of one module and a maximum of six modules. A module, on the other hand, may not necessarily be part of any course (figure 6.37). Coad and Yourdon suggest that there are three types of whole-part structures that might be considered.

1.   The 'assembly and its constituent parts' type, for example, an organisation is composed of various departments.
2.   The 'container and its contents' type, for example, a lecture hall and its seats
3.   The 'collection and its members' type, for example, the football club and its players and helpers.

A set of criteria for considering and challenging the identification of whole-part structures are similar to those used for gen-spec structures as outlined above, with the exception of the inheritance test.

Whole-part structures often present people with difficulties, and this is usually to do with their purpose in relation to an object-oriented approach.

It seems that they have been introduced to capture elements that have been found to be needed but not captured in the traditional object-oriented methods, for example, in gen-spec structures. Whole-part structures do not imply any notion of inheritance by the parts from the whole, they simply indicate that an object is composed of various other objects or parts. Coad and Yourdon suggest that whole-part structures are particularly useful for identifying class-&-objects at the edges of the problem domain, and these objects are dealt with by other systems.

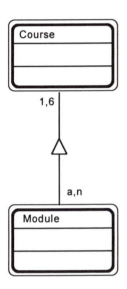

Fig 6.37: Whole-part structure

Diagrams may be constructed that include both gen-spec structure and whole-part structure together.

3 Identifying subjects
The third activity of the OOA methodology is the identification of subjects. The purpose of this is to reduce the complexity of the model produced so far by dividing or grouping it into more manageable and understandable subject areas. This is somewhat analogous to the levelling of a DFD in other approaches and is about presenting relevant chunks of the model to users or designers that are understandable on their own but are also set in context as part of a larger whole. Obviously in a small system there may be no need for this, but in larger systems with more than about 20-30 classes it becomes important. Guidelines are provided for the grouping of related classes together and it is a bottom-up process which

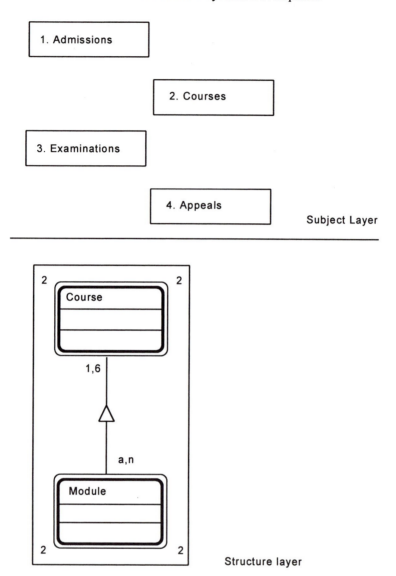

Fig 6.38: Subject and structure layers

produces a top-down view. The groupings may be based on any criteria that are relevant to the area of concern and it might be a traditional functional type of decomposition but it could also be based on problems or issues that emerge from the problem domain. For example, in a university problem domain, the subject layer might be admissions, courses, exams, appeals and so on, where admissions might be composed of classes concerning applications, criteria, acceptance, references and payments. Figure 6.38 illustrates the notation used, and shows the structure layer for the subject courses. The subject identification provides a particular view or picture of the system and there may be a number of relevant, and overlapping views. At any particular point, the most useful view is used depending on the objectives, which might be explanation to senior management, or verification by a user or the creation of a work-package for an analyst or designer.

4 Defining attributes

In this activity, the attributes of the class-&-objects are defined. This is very similar to the identification of attributes for entity models (section 4.5). It is the data elements of the object that are defined. The only difference is that attributes that define the state of the object are perhaps given more prominence, for example, things that might be defined using an entity life history diagram are emphasised (section 4.12). Examples of attributes for an object student might be student-number, name, address, date-of-birth, suspended or current. Attributes are normally listed by name in the middle part of the class-&-objects box (figure 6.39). Attributes that 'point' to other objects are included. In relational database terms this means that foreign keys are included. For example, the attribute course above is in effect a pointer from the student object to another object called course. In the model these objects would be connected with a line to indicate a relationship and the degree and cardinality of the relationship expressed. This part of the identification of attributes is termed 'instance connection' by Coad and Yourdon and indicates that the connection is between instances of the individual objects rather than between classes. This is the same for entity models, where the relationships connect occurrences of the entity type rather than the entities themselves.

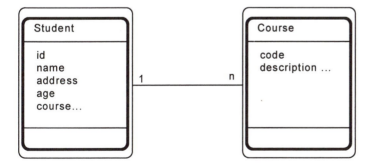

Fig. 6.39: Defining attributes

For each named attribute that has been identified, a short description is specified along with various constraints that apply to the attribute. These may include domain information, the allowable range, any default value, the various states that may apply and, any constraints implied by other attributes. An example would be: 'if in object student the attribute registration-suspended is set, then attribute fee must be zero'.

All implementation considerations associated with attributes are deferred to design. Therefore the attributes should not be normalised nor performance considered, and specific identification keys are not defined.

5 Defining services

In this, the final activity of the OOA methodology, the services of the class-&-objects are defined. In the object-oriented approach, as we have seen, the objects are composed of data and processing. The previous activity defined the data and this activity defines the processing, or in Coad and Yourdon's terminology, the services. The terms 'method' or 'behaviour' are also used to mean the same thing. A service is the operation or processes performed by the object in response to the receipt of a message.

In the previous activity, attributes that defined the state of an object were identified. In this activity, the services required to change or modify states are identified. In the student object, we might have identified a state of suspended-registration. This would imply that we need to define the services to suspend a student's registration and perhaps to unsuspend, or reinstate, the registration of that student. All the services needed to achieve the changes of state identified in the entity life history diagrams should be fully defined.

Next, what are termed the 'algorithmically-simple' services are defined. Yourdon later changes the terminology and calls these 'implicit' services. These are the ones that are likely to appear in some form for each class-&-object in the model, they are create, release (that is, delete), connect and access. The services required for create, delete, and access are fairly obvious. For example, what is required to create an object student might be a check for a valid registration form, followed by the allocation of the next available registration number, the creation of a new object and then the return of a 'successful creation' message. The connect is perhaps less obvious, but means a service that creates or terminates a mapping between objects, that is, the establishment of a relationship between objects. An example might be the allocation of a new student to a personal tutor as this would require the creation of a mapping from object student to object tutor. These algorithmically-simple services are usually not included on the OOA diagrams.

As can be imagined, if the algorithmically-simple services are identified, this will be followed by the identification of any algorithmically-complex services. These are classified into two types:
1.   Services concerned with calculations.
2.   Services concerned with monitoring the external environment, that is, the services required to detect and respond to events.

Finally, the services required for processing a message received from another object, and any processing triggered by the message, are defined.

Once the required services have been identified they are specified in detail using either a form of structured English notation (section 4.10) or via a service chart, which is a kind of program flowchart. Figure 6.40 provides an example of a service chart.

Fig. 6.40: Service chart

It should be emphasised again that the description of the methodology as a linear series of activities is not necessarily how it would actually be approached. Some analysts might identify a few key class-&-objects, then drive down through the activities and then iterate the process with other objects. It does not really matter as long as the outcome is a complete set of OOA models and diagrams. A further aspect that needs highlighting is the importance of the activities of identifying reusable objects, classes and services, and therefore, ultimately, reusable code.

The methodology of OOA, as its name implies, does not include design and implementation phases, although the authors address design in some detail in other sources. In this book we will not extend the description into detailed design because that is not our focus. However, an important aspect of the transition from analysis to design in object-oriented methods is that it is not a question of changing, or introducing, new concepts. The transition is simply a matter of extending the detail of the object-oriented models and specifications, and adding components concerning human interactions (such as dialogue design), task management (such as real-time tasks, communication and hardware considerations) and data management (for example, designing the database). The detailed design stage slowly becomes program language-dependent, that is, we need to know what the target program language is, and the actual implementation will normally take place in an object-oriented programming language in order to utilise most easily the object-oriented concepts. Coad and Yourdon point out that the results of the OOA can be implemented in a non-object oriented language, although it would be much more difficult.

6.9 INFORMATION SYSTEMS WORK AND ANALYSIS OF CHANGES (ISAC)

The ISAC methodology has been developed since 1971 by a research group at the Department of Administrative Information Processing at the Swedish Royal Institute of Technology and at the University of Stockholm. The methodology has been developed by use and experience in a number of commercial organisations and Swedish government agencies. Most users of the methodology are Scandinavian, although a number of users are claimed in other parts of Europe and North America. The methodology is closely associated with Mats Lundeberg (and is described in Lundeberg et al., 1982).

The methodology covers all aspects of information systems development, although some users only apply the analysis and design parts of the methodology, which are probably its best known aspects. ISAC is a problem-orientated methodology and seeks to identify the fundamental causes of users' problems. The approach is designed to analyse users' problems and to solve aspects of them where appropriate. The methodology begins at an earlier stage than most methodologies and does not assume that the development of an information system is necessarily the solution to the problem. If a need for an information system is not identified, then the role of the methodology terminates. Need is established only if it is seen that an information system benefits people in their work, so that pure financial benefit to the organisation, or some other benefit, is not thought to be an indication of need for an information system. An information system is thought to have no value in its own right and without benefiting people should not be developed. ISAC is a people-orientated approach.

People, and the problems they have, are seen as the important factors in organisations. The term 'people' includes all people in an organisation: users, managers, workers, as well as people usually thought of as outside of the organisation, such as customers and funders. People in an organisation may have problems concerning the activities that they perform. These problems may be overcome, or the situation improved, by analysis of these activities and the initiation of various changes. The ISAC authors believe that the people best equipped to do this analysis, in terms of their knowledge, interest, and motivation, are the users themselves. The methodology attempts to facilitate this by providing a series of work or method steps and a series of rules and techniques which, it is claimed, can be performed by these users. For ISAC, an important part of this process is the education of the users to understand the organisation better and to improve the communication between people in the organisation.

If the need for an information system is established, then the methodology emphasises the development of a number of specific information subsystems rather than one 'total' system. The subsystems are local systems tailored to groups of individual needs, and these subsystems may well overlap in content and function. However, it is argued that the benefit that accrues is in the specific relationship to the local needs. The assumption is therefore that solutions to sub-problems will give solutions to the organisation's problems as a whole (which perhaps conflicts with the holistic, systems view described in section 3.2).

In ISAC terms, an information system is an organised co-operation between human beings in order to process and convey information to each other, it does not necessarily involve any form of automation.

The major phases of ISAC are:
1.   Change analysis.
2.   Activity studies.
3.   Information analysis.
4.   Data system design.
5.   Equipment adaptation.

The first three phases are classified as problem-orientated work, and focus on users and their problems, whereas the latter two phases focus on data processing-orientated work. Within each phase a number of work steps are identified and within these work steps various techniques concerning documentation are employed.

1 Change analysis
Change analysis seeks to specify the changes that need to be made in order to overcome the identified problems. Change analysis begins with the analysis of problems, the current situation and needs. The following method steps are used:

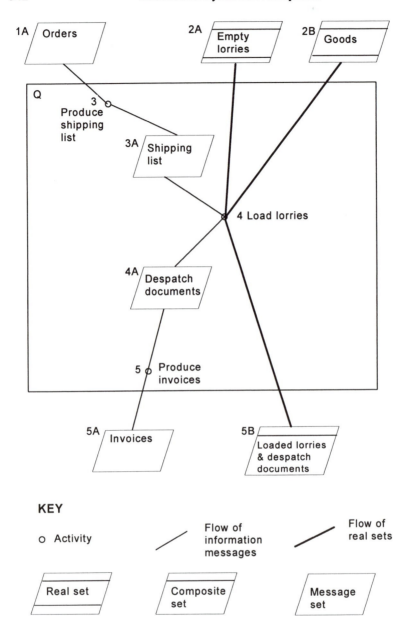

Fig. 6.41: A-graph

- *Problem listing:* This is a relatively unstructured, first attempt, look at current problems and any anticipated future problems.
- *Analysis of interest groups:* ISAC acknowledges that problems do not usually have objective status. They are relative to the viewpoint

of the participants in the system. At this stage, the different interest groups are identified. These interest groups may be system end users, public users, departmental managers and so on.

- *Problem Grouping:* Here, the identified problems are grouped into related sets or categories.
- *Description of current activities:* The activities that the identified problems relate to, plus the activities undertaken by the concerned interest groups, are now modelled. This activity model is a functional view and shows processes performed on inputs to produce outputs. These aspects are not just concerned with information, but include physical activities, inputs and outputs, such as the loading of a lorry. The activity model is documented in the form of an A-graph (an activity-graph).

An A-graph depicts three things:

1. Sets which can be real or physical sets, concerning, for example, people or lorries or stock, or they can be message sets, containing only information or they can be a combination of both.
2. A-graphs, which depict activities.
3. Flows, which can be shown in detail or in overview.

Activities are transformations of sets into new sets. Flows represent the movement of sets to and from activities. They are very similar to data flow diagrams (section 4.7), except that they also represent physical objects as well as data flows. Figure 6.41 shows an example of an overview A-graph concerned with the despatch of goods. The message set 'orders' flows into the activity 'produce shipping list' which results in a message set 'shipping list'. This itself flows to an activity called 'load lorries' which has input flows of real sets 'empty lorries' and 'goods'. A-graphs exhibit a hierarchical structure capable of showing an overview picture which can then be broken down to show the detail at lower levels. The A-graph is supplemented by narrative or descriptive text and property tables. The property tables show quantitative information such as volumes.

- *Description of objectives:* The previously identified interest groups are perceived to have a variety of different general objectives and desires. Here, firmer and more specific objectives are defined and these are unified into a single set of objectives, via a process of negotiation and compromise. An attempt is made to reach a situation where the achievement of a set of agreed objectives solves the problems that have been identified.
- *Evaluation of current situation and analysis of needs for changes:* This is where all the previous work comes together and enables the methodology to progress. What is wanted (the objectives) is

compared to what is available. What is available is described by the activity model, but this is tempered by the problems that have been identified. The differences between what is wanted and what is available are defined as the needs for changes. These needs are then evaluated and prioritised according to the values of the different interest groups involved. This evaluation of the importance of the various needs for changes leads directly into the next stage which is the generation and study of different change alternatives.

ISAC gives no guidance on how to generate ideas for changes, since this requires creativity in the context of the situation rather than the use of techniques, except to say that an analysis of flows and activities might be helpful. ISAC does not presuppose that the solution to the problems lies in automation or indeed in the construction of any form of information system. The type of solution is not constrained and may involve purely organisational and physical changes that do not result in the generation or modification of information systems. Once possible changes have been generated they are described through a new A-graph. The changes and the models are then analysed and evaluated from human, social and economic feasibility viewpoints.

The final part of change analysis is to choose the most appropriate change approach based on the previous evaluations, and to document the reasons for the choices made. If the recommended changes do not involve information systems then the role of ISAC ceases. More likely however is that the recommendations involve a combination of types of change, and a plan is made concerning the necessary development measures for each type. An analysis of the effect and consequences of these parallel development measures is also made.

2 Activity studies

The starting point for activity studies is a proposal for a new system modelled and described in a number of ways, in particular, as an A-graph. The activity models that were produced in change analysis for the purpose of identifying needs for changes were at a relatively high, overview level, and these need to be expanded and investigated at a more detailed level. This is shown as figure 6.42, which is a decomposition of process 4 on figure 6.41. The object is to get to the level at which the information system is separated from the human activities which it supports, such that each process on the graph has inputs and outputs that are unequivocally either information or some other flow (for example, materials).

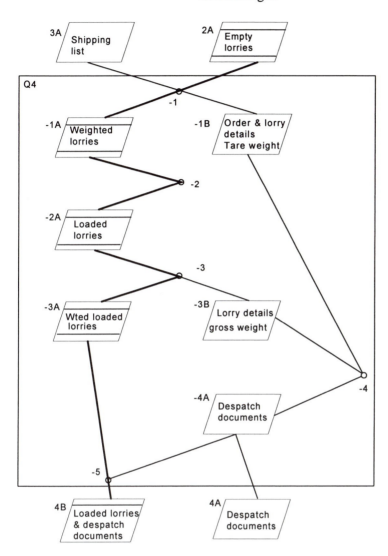

Fig. 6.42: More detailed A-graph

The next step is to identify potential information subsystems relating to groups of users. The subsystems are not supposed to be identified in relation to artificial criteria, such as some common technical aspect, but only in relation to commonality in use. ISAC does not identify any overall information system totality. It is more relevant to identify a number of subsystems. These information subsystems are then classified according to whether they are formalisable or not. An unformalisable information system might be one concerning qualified decisions, informal contacts, know-how and so on. The formalisable subsystems are divided into those that seem sensible to automate, in terms of cost, social desirability and so on, and those that are not. The automated ones are further classified according to whether they involve calculations or simply involve storing and retrieving information. These classifications are the basis for subsequent steps of the methodology.

Each information subsystem is now studied separately, in terms of its costs and benefits. ISAC attempts to do this cost-benefit analysis without making assumptions about technical implementation. To do this, ISAC refers to ambition levels for an information system, rather than particular technologies for implementation. For example, two different ambition levels for the same subsystem might be a one second response time compared with a three hour one. Each ambition level will have a different cost-benefit associated with it.

The steps in this phase are as follows:

- *Analysis of contributions:* This is a study of the benefits (not quantified) expected to accrue from the change. It is a refinement of the earlier work done in change analysis and the results are documented in a property table. It is emphasised that this analysis must be performed in the context of the environment and the way in which the environment uses the information. This may require a more detailed analysis of the environment than has been done up to this point.

- *Generation of alternative levels of ambition:* A number of alternatives are generated for each subsystem and documented. The alternatives must be realistic. There is no point in generating ambition levels that do not fulfil minimum requirements in terms of, for example, frequencies or volumes.

- *Test of ambition levels:* Here the ambition levels are tested to see if they are practical. ISAC envisages a number of ways that they might be tested, for example, if similar information systems exist elsewhere, then it is likely that such a system can be created. Prototyping is also suggested, but not a prototype of the technology, rather one of the provision of information to the user.

- *Cost-benefit analysis:* This is a conventional cost-benefit study of each identified level of ambition.

- *Choice of ambition level:* The results of the cost-benefit analyses are evaluated in conjunction with the human and social analyses performed at an earlier stage, and a choice of ambition level made. One result may be that the development of an information system is discontinued.

- *Co-ordination of information subsystems:* This concludes the activity studies and is, in effect, the project plan. Special emphasis is given to the inter-relationships between the different subsystems in order that they are sensibly co-ordinated. Priorities, resources and schedules are allocated for the developments, and the plans are documented.

3 Information analysis

The transition from change analysis to activity studies would not be made unless the agreed proposal for change included the development of an information system. Similarly, the transition from activity studies to information analysis is made only if one or more information systems have been identified as formalisable. The techniques used in information analysis assume a formalisable and automatable information system, although it is indicated that a limited degree of information analysis might be appropriate for non-automatable systems.

For each information system, the input and output information sets are extracted from the A-graphs for the system. At this time, an iterative process of function and data analysis is performed.

The ISAC term for functional analysis is precedent analysis, because ISAC recommends that it be performed by reasoning about the precedents for each information set. If the output information set from an information system is clearly derivable from its input set, then precedence analysis stops. If, however, the derivation is not clear, then the information set that immediately precedes the output information set must be deduced. If the derivation of this set from the inputs to the system is not clear, then precedence analysis continues. The precedents from each information set are analysed until the input sets are reached. At each stage of precedence analysis the information system (considered as a set of processes) has been refined to a lower level of detail. Precedence analysis is in this way equivalent to functional decomposition in other methodologies. The reasoning process, however, is different in that instead of enquiring about the logical structure of a process, ISAC concentrates on the transformation that must have been necessary to produce the output information set currently being studied. If this transformation is not clear, then a simpler problem must be solved. The question is asked: 'what was the nature of the information set that immediately preceded the transformation of the current output set?'. In this way the definition of processes is implicit. At any stage a process is always a black box (or using ISAC notation a black dot).

The result of precedence analysis is a set of information precedence graphs (I-graphs). I-graphs describe information sets and precedence relations between information sets. They are more precise than A-graphs in that they not only show input and output sets but also show relationships between sets. Figure 6.43 shows an example of an I-graph derived from the A-graph of figure 6.42. It shows the input and output information sets and the precedence relations. For example, it shows that in order to derive the 'accepted despatch' information set, we first need the 'order details', 'customer details' and 'product details' information sets.

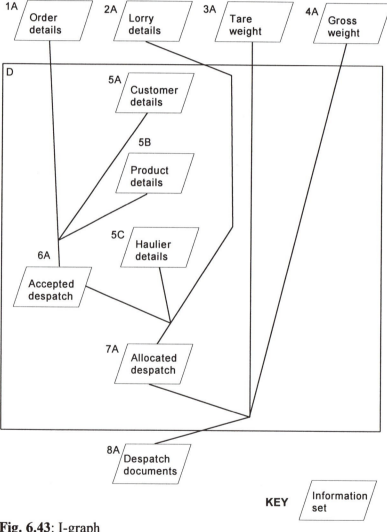

Fig. 6.43: I-graph

Reasoning about the transformations that need to be performed on information sets requires knowledge of the structure of information, and that is why component analysis is performed at the same time as precedence analysis.

In component analysis the structure of the information sets is studied. An information set is either a data flow or a data store. An information set will either have been specified as a basic input to, or output from, the information system being studied; or will have been from a preceding process or a set of permanent information. An information set may be compound, that is, it may itself contain information sets. An elementary information set consists of one or more messages, where each message consists of an identification term and one property term. An 'almost elementary' information set consists of a number of elementary information sets with common identification terms. Thus an almost elementary information set corresponds to a logical record with a key (identification term) and a number of data-item types (property terms).

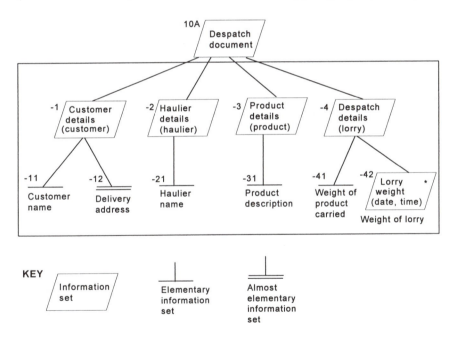

Fig. 6.44: C-graph

ISAC documents information sets by means of a C-graph (component graph). This graph is a hierarchy showing the decomposition of an information set into subsets. Figure 6.44 is an example of a C-graph for the 'despatch document' information set. The lowest level on the graph shows either elementary or almost elementary sets.

A further step in information analysis concerns process logic analysis. This has been ignored by ISAC during precedence analysis in order not to complicate things, but must now be addressed. Process analysis means the detailed description of the information processes in the information system. An information process is the transformation of one or more information sets into some other information set or sets. These transformations are at the logical level, and do not depict representation or implementation aspects.

The processes which have previously appeared simply as nodes or black dots on the I-graphs are now identified and named. The relationships between the processes have already been described in the C-graphs, so that all that needs doing is for the content of the process to be specified individually. This is done using a process table that defines the prerequisites for the process, the conditions, permitted states for conditions and required actions. A process table is effectively a decision table (section 4.9) constrained by prerequisites.

The final stage of information analysis is property analysis. Precedence and component analysis are structural descriptions of an information system. Requirements that are specific to the environment in which the system is used must also be specified. Examples of such properties are volumes, response times, frequency and security requirements.

4 Data system design

Transition to data system design implies a fundamental change in the application of the ISAC methodology. Up to this point the ISAC activities of change analysis, activity studies and information analysis have concentrated on producing a specification of requirements for information systems. Data system design is the first part of data-orientated systems work, the purpose of which is to design a technical solution to meet the requirements specification. A data system is a means of implementing the requirements of one or more information systems. A data system will usually contain both manual and automated parts, both of which must be designed. Data system design is followed in ISAC by an activity known as equipment adaptation. Data system design, therefore, assumes types of equipment but not specific equipment.

The design activity starts with a study of processing philosophy. Processes that are to be performed on a computer are differentiated from those that are not. A decision is made on the mix of batch and on-line processing and centralised versus distributed processing. The results of activity studies are particularly relevant to determination of processing philosophy and the identification of processes performed in information analysis enables preliminary 'process collections' to be made grouped by some common requirement (for example, response time).

The next stage of data system design is data structure design and program delimitation and design. Information analysis has typically decomposed data and functions below the level of files and programs, and appropriate groupings are now made.

The design of a permanent data set is performed by consolidating or aggregating more primitive objects (for example, elementary information sets) into higher-level groupings, on the basis of functional dependency, and second, by considering access requirements and search paths for efficient retrieval and storage.

Program delimitation consists of putting a boundary around a group of processes defined on an I-graph. The number of processes so grouped will clearly partially determine the size and complexity of the program, and these two factors are a constraint on the delimitation. The other important constraint is the nature of the decisions that have been made about file and database design.

Once programs have been delimited and files or a database have been designed, the overall structure of the system can be recorded in a D-graph (a form of program run-chart). The next step is to specify each program, which is completed in some detail in the ISAC approach, and the JSP method is recommended.

The final stage in data system design is the design of manual routines. Some information processing activities are naturally manual, and all computer based systems will have associated manual routines. ISAC recommends that users design their own work routines.

5 Equipment adaptation

The preceding phase of data system design has produced an 'equipment-independent' solution. This is now adapted to fit particular equipment. Equipment adaptation consists of equipment study, adaptation of computer-based routines and the creation of side-routines.

The equipment study involves collecting and evaluating technical specifications and cost and performance characteristics, and formulating an equipment strategy. Possible options might be, for example, the use of existing computer facilities, the purchase or rent of new equipment or the use of bureau facilities. Accurate sizing of the configuration required is performed at this stage. The final choice of equipment is documented in an E-graph, which is a mapping of the D-graph onto a particular hardware configuration.

Adaptation of computer-based routines consists of two tasks. First, physical data structures must be designed, and second, the program specifications must be adapted. Files and databases are mapped to specific storage devices, and retrieval and linkage mechanisms are specified. Outputs are mapped to specific output peripherals, such as monitors or printers. Input formats are mapped to specific methods of data capture. Any of these mappings may alter the data structures of computer programs, and thus the structure of the programs must be adapted. Finally, side-routines are specified. These are work-routines that are a necessary consequence of the choice of a particular set of equipment. For example, side-routines might be specified for data preparation, output handling or computer operation.

The emphasis on the methodology is placed on analysis and design aspects of information systems development where appropriate. The methodology seeks to identify the fundamental causes of users' problems in the present system. These problems may be overcome or the situation improved by analysis of these activities and the initiation of various changes. The authors of this methodology believe that the people best equipped to do this analysis, in terms of their knowledge, interest and motivation, are the users themselves. The methodology attempts to facilitate this by providing a series of work or method steps and a series of rules and techniques which, it is claimed, can be performed by these users. It is accepted that this might lead to a series of self-contained application systems which might be regarded as 'inefficient' from some points of view.

The methodology does not assume that the development of a computer information system is necessarily the solution to the problem. Need is established only if it is seen that an information system benefits people in their work. It is traditional in Scandinavian countries, sometimes backed-up by legislation, that technology is only implemented with the approval of the workers in that workplace. This people-orientated methodology has a wide view of the stakeholders of an information system, including users, managers, workers, and also those usually thought of as outside of the organisation, such as customers and funders. Recent development of ISAC concerns broadening the view towards business process development.

6.10 EFFECTIVE TECHNICAL AND HUMAN IMPLEMENTATION OF COMPUTER-BASED SYSTEMS (ETHICS)

ETHICS is an acronym, but the name of this approach is meant to imply that it is a methodology that embodies an ethical position. ETHICS, devised by Enid Mumford (see Mumford, 1995), is a methodology based on the participative approach to information systems development (discussed in section 3.13). In addition, it encompasses the socio-technical view that for a system to be effective the technology must fit closely with the social and organisational factors. In particular, this means that an improved quality of working life and enhanced job satisfaction of the users must be a major objective of the systems design process. This is not simply to guard the interests of the users in the introduction of computing and technology, although this is obviously of major importance, but it is an essential prerequisite to achieve effective systems as far as the organisation and its management is concerned. To support her case, Mumford points to the failure of many traditionally-performed system implementations, where technical and economic objectives were the only consideration.

The philosophy of ETHICS is different from most information system development methodologies and is also explicitly stated, which is not common in most methodologies. The philosophy is one which has evolved from organisational behaviour and perceives the development of computer systems not as a technical issue but as an organisational issue which is fundamentally concerned with the process of change. It is based on the socio-technical approach of the social sciences as developed by a number of authors, one of the most influential being Davis (1972). Mumford (1983a) defines the socio-technical approach as:

> one which recognizes the interaction of technology and people and produces work systems which are both technically efficient and have social characteristics which lead to high job satisfaction.

Elsewhere, in Mumford and Weir (1979), job satisfaction is defined as:

> the attainment of a good 'fit' between what the employee is seeking from his work - his job needs, expectations and aspirations - and what he is required to do in his job - the organisational job requirements which mould his experience.

In order to ascertain how good this fit is, a theory for measuring job satisfaction has been developed based on the various views of what is important in job satisfaction and these have been integrated into a framework derived from Parsons and Shils (1951). Five areas of measurement are identified as follows:

1. *The knowledge fit:* a good fit exists when the employees believe that their skills are being adequately used and that their knowledge is being developed to make them increasingly competent. It is recognised that different people have widely different expectations in this area, some wanting their skills developed, others wanting to remain static and opt for an 'easy life'.

2. *The psychological fit:* a job must fit the employee's status, advancement and work interest (some of the Herzberg (1966) motivators). These needs are recognised to vary according to age, background, education and class.

3. *The efficiency fit:* this comprises three areas. First, the effort-reward bargain, which is the amount an employer is prepared to pay (as against the view of the employee about how much he is worth). Although, this is probably the prime area of importance to management, it is in practice sometimes way down the list of employee needs. Second, work controls, which may be tight or loose but need to fit the employee's expectations. Third, supervisory controls, such as the necessary back-up facilities, for example, information, materials, specialist knowledge and supervisory help.

4. *The task-structure fit:* this measures the degree to which the employee's tasks are regarded as being demanding and fulfilling. Particularly important are the number of skills required, the number and nature of targets, plus the feedback mechanism, the identity, distinctiveness and importance of tasks, and the degree of autonomy and control over the tasks that the employee has. This measure is seen to be strongly related to technology and its method of employment. Technology can affect the task-structure fit substantially and, it is argued, has reduced the fit by simplification and repetitiveness. However, it is also seen as a variable which can be improved dramatically by designing the technical system to meet the requirements of the task-structure fit.

5. *The ethical fit:* this is also described as the social value fit and measures whether the values of the employee match those of the employer organisation. In some organisations, performance is everything, whilst others value other factors, for example, service. Some firms are paternal or welfare-orientated, others aim to achieve the characteristics of 'success', and so on. The better the match of an

organisation's values with those of the employee, the higher the level of job satisfaction.

A second philosophical strand of the ETHICS methodology is participation. This is the involvement of those affected by a system being part of the decision-making process concerning the design and operation of that system. Those affected by a system include the direct users and also the indirect users, such as management, customers, suppliers and so on. Of course there are limits to this. For example, competitors will be affected, but it is unlikely that they will be asked to participate. Participation is important in many methodologies, but has been described as vital in ETHICS (Hirschheim, 1985). In some other methodologies no more than lip-service is paid to participation, sometimes being regarded simply as 'allowing the users to chose the colour of the terminals that they use'. In ETHICS, users are involved in the decisions concerning the work process and how the use of technology might improve their job satisfaction.

In ETHICS the development of computer-based systems is seen as a change process and therefore it is likely to involve conflicts of interest between all the participants or actors in that process. These conflicts are not simply between management and worker but often between worker and worker and manager and manager. The successful implementation of new systems is therefore a process of negotiation between the affected and interested parties. Obviously major affected and interested parties include the users themselves and if these people are left out of the decision-making process, the process of change is unlikely to be a success. This is not just because of resulting disaffection amongst the user group but, more positively, because they have so much to contribute in making the implementation a success. They are probably the most knowledgeable about the current workplace situation and the future requirements. Mumford (1983a) summarises:

> All change involves some conflicts of interest. To be resolved, these conflicts need to be recognised, brought out into the open, negotiated and a solution arrived at which largely meets the interests of all the parties in the situation....successful change strategies require institutional mechanisms which enable all these interests to be represented, and participation provides these.

It is recognised in practice that participation means different things to different people and that the parties involved may have quite different reasons for wanting participation and quite different expectations concerning the benefits. Management may see it as a way of achieving changes that would otherwise be rejected. This is perhaps not the ideal view for management to take but if the resulting participation is real then so be it, although the end result may not exactly turn out as they expect. The point being that it is not a prerequisite for everybody to hold the view that participation is a moral or ideological necessity, enlightened self-interest will do just as well.

The philosophical commitment to participation outlined above begs the question of exactly how it is to be achieved. There appears to be quite a degree of freedom involved and although there exist 'ideal' types of participation, in practice, a variety of forms are acceptable for it still to be 'ETHICS'. In fact, it can be used by an expert group to design a system for another, non-participating, group. However, this is not recommended. Nevertheless, it shows that ETHICS is a methodology which has quite a level of flexibility. It is better, Mumford argues, to use it in some form, rather than not at all. The implication being, that its use, even stripped of some of its most important participatory trappings, is better than other more traditional methods which concentrate purely on technical and economic objectives.

Mumford distinguishes between structure, content, and process. Structure is the mechanism of participation which, as discussed in section 3.13, can be consultative, representative or consensus. Consultative participation involves the participants giving evidence to the decision makers which, possibly, will influence the decision makers but does not bind them in any way. This is the weakest form of participation and not recommended for detailed design. Representative participation is a structure where selected or, preferably, elected representatives of the various interests are involved in the decision-making process. This is most appropriate for the tactical or middle management type of decision making. In computing terms, this might be at the system definition stage where the system outline and boundaries are discussed and a fairly wide spectrum of interests are involved. The third form of participation is consensus, where all the constituents are involved in the decision making. This is most suitable at the detailed design stage where the decisions probably affect the day-to-day work practices of the people involved. Clearly it is difficult to involve everybody in everything, and what usually happens is that design groups are formed to do the work and present alternatives to the whole constituency, which takes the final decisions.

The content of participation concerns the issues and the boundaries of activities that are within the remit of participation. Generally, prior to any participation, management will want to keep certain things as their own prerogative. One objective of the process of participation is the gaining of relevant knowledge and information by the participants. In general, the users involved in participation will not have previously had the necessary knowledge, information and, perhaps, confidence to discuss issues and make decisions. Without this, participation is only of a very limited kind. The users must have as much information and knowledge as is necessary to make informed decisions, or at least as much as anyone else. Without the acquisition of this information and knowledge they will be at a disadvantage and subject to undue influence from more powerful groups. True participation means equal knowledge and thus, it might be argued, equal power for all groups. Training and education of users is therefore a very important aspect of ETHICS.

Participation usually involves the setting up of a steering committee and a design group or groups. The steering committee sets the guidelines for the design group and consists of senior managers from the affected areas of the organisation, senior managers from management services and personnel, and senior trade union officials (if the organisation is unionised). It is recommended that the steering committee and the design group meet once a month during the course of the project. The design group designs the new system including:

- Choice of hardware and software
- Human-computer interaction
- Workplace re-organisation
- Allocation of responsibilities.

All major interests should be represented, including each section and function, grade, age group and so on. The design group includes systems analysts, although their role is not the normal one of analyst and designer, but one of educator and adviser. This often involves the analysts in a learning process themselves. If the area of the design is large, involving many departments or sections, then a design group may first design in outline, and then hand over to detailed design groups. A participative design requires the appointment of a facilitator to help the design group manage the project and educate the group in the use of ETHICS. The role is multi-faceted and concerns motivation and confidence-building of the design group; it is not one of decision making or persuading. For this reason, the facilitator must be neutral and preferably external, if not to the organisation, then to the department. The role is very important. In one situation that Mumford quotes, the facilitator withdrew, and the confidence of the design group declined and the importance of the group in the eyes of the management also declined.

Depending on which sources are referred to, the actual steps of ETHICS differ somewhat in their names and numbers. However, the content remains much the same. The fifteen step version outlined here corresponds to those described in Mumford (1983a). A six stage, twenty-five step version of ETHICS is described in Hirschheim (1985). The major difference is that there is a greater separation of the technical and social issues. This might mean that the technical issues of design could be addressed by a separate, more technically experienced, design group. ETHICS has also been amended slightly, although the principles are the same, for use by small businesses thinking of purchasing a computer system for the first time. This version is described in Mumford (1986). Unless stated otherwise, the work in the steps that follow are performed by the design group.

1 Why change?
The first meeting of the design group considers this rather fundamental question and addresses the current problems and opportunities. The result should be a convincing statement of the need for change. Presumably, if no convincing statement for change is arrived at, the process stops there, although this is not made explicit.

2 System boundaries
The design group identifies the boundaries of the system it is designing and where it interfaces with other systems. Four areas are considered: business activities affected (for example, sales, finance and personnel); existing technology affected; parts of the organisation affected (for example, departments and sections); and parts of the organisation's environment affected (for example, suppliers and customers).

3 Description of existing system
This is to educate the design group as to how the existing system works. In practice, it is found that people will know the detail of their own jobs and those that they interact with directly, but will probably have little knowledge of the whole system. In this step, two activities are undertaken. First, a horizontal input/output analysis is described with inputs on the left, activities in the middle and outputs on the right. Second, a vertical analysis of the design area activities is made at five different levels. The lowest level is of the operating activities, that is, the necessary activities of a day-to-day nature. These should have appeared in the horizontal analysis. The problem prevention/solution activities are also identified. These are the key problems or variances that occur and how they are corrected. Third,

the co-ordination activities are identified. These are activities that have to be performed together or in a particular sequence or at a particular time. These are both inter-departmental and intra-departmental co-ordinations. Fourth, the development activities are recorded. These are the things or areas that need improving. Fifth, the control activities are identified, indicating how the system is controlled, how it is judged to be meeting targets or objectives and how it is monitored.

4, 5 and 6 Definition of key objectives and tasks
Three questions are asked in order to help define the key objectives. First, why do particular areas exist, what is their role and purpose? Second, given this, what should be their responsibilities and functions? Third, how far do their present activities match what they should be doing? From this, the key objectives can be listed and these form the design objectives of the new system. In addition, the key tasks that need to be performed to achieve the key objectives are defined in outline, along with their key information needs.

7 Diagnosis of efficiency needs
Weak links in the existing system are identified and documented. Mumford talks about them as variances, which is a 'tendency for a system or subsystem to deviate from some desired or expected norm or standard' (Mumford, 1983a). Mumford identifies two types of variance. First, systemic or key types, which are inherent in the system and cannot be completely overcome. They can only be eased. An example is provided by the variances connected with the *financial desire* to keep stocks small and the *service desire* to be able to supply customers with what they want. The second type of variance is operational. These are variances due to poor design or lack of attention to changing circumstances, and can usually be completely eliminated in the new system. Examples could include bottle-necks, insufficient information and inadequate equipment.

8 Diagnosis of job satisfaction needs
This step measures the job satisfaction needs. This is achieved by use of a standard questionnaire provided in the ETHICS methodology. The design group may alter the questionnaire to fit their organisation and requirements. The results are discussed democratically and the underlying reasons established for any areas where there are poor job satisfaction fits. In addition, formulations for improving the situation in the new design are made and everybody is encouraged to play a major part in this design work. Where there have been knowledge or task-structure problems of fit,

these are susceptible to improvement by a redesign of the system. Other areas of poor fit, such as effort-reward or ethical, may be improved somewhat in this way, but will probably require changes in personnel policies, or more radically, organisational ethos.

9 Future analysis

The new system design needs to be both a better version of the existing system and able to cope with future changes that may occur in the environment, technology, organisation or fashion. Thus an attempt is made to try and identify these changes and to build a certain amount of flexibility into the new system. This may involve the design group in interactions with people outside the organisation in order to identify and assess some of the potential changes.

10 Specifying and weighting efficiency and job satisfaction needs and objectives

Mumford identifies this as the key step in the whole methodology. Objectives are set according to the diagnosis activities of the three previous steps. The achievement of an agreed and ranked set of objectives can be a very difficult task and must involve everyone, not just the design group itself. Often objectives conflict and the priorities of the various constituencies may be very different. These differences may not all be resolved, but one of the stated benefits of ETHICS is that at least these differences are aired. Ultimately, a list of priority and secondary objectives is produced. The criterion for the systems design is that all priority objectives must be met along with as many of the secondary ones as possible. At this stage a certain amount of iteration is recommended, to review the key objectives and tasks from steps 4 and 5.

11 The organisational design of the new system

If possible, this should be performed in parallel with the technical design of step 12, because they inevitably intertwine. The organisational changes which are needed to meet the efficiency and job satisfaction objectives are specified. There are likely to be a variety of ways of achieving the objectives, and between three and six organisational options should be elaborated. The design group specifies in more detail the key tasks of step 5 and addresses the following questions, the answers forming the basic data for the organisational design process:

- What are the operating activities that are required?
- What are the problem prevention/solution activities that are required?

- What are the co-ordination activities that are required?
- What are the development activities that are required?
- What are the control activities that are required?
- What special skills are required, if any, of the staff?
- Are there any key roles or relationships that exist that must be addressed in the new design?

Each organisational option is rated for its ability to meet the primary and secondary objectives of step 10, and should identify the sections, sub-sections, work groups and individuals, their responsibilities and tasks. In order to meet the job satisfaction objectives, it is almost inevitable that the design group will have to consider the socio-technical principles of organisational design and be provided with information and experience in relation to design. The socio-technical approach is the antithesis of Taylorism (Taylor, 1947) which is to break each job down into its elemental parts and rearrange it into an efficient combination. The traditional car assembly line which requires its operators to perform small, routine, repetitive jobs, is regarded as the ultimate example of Taylorism in action. The requirements of the machine are given priority over the requirements of the human being. This has, it is argued, inevitably led to a bored, disaffected and ultimately inefficient workforce.

Although ETHICS uses aspects of socio-technical design, the socio-technical school assumes a given technology, whereas, in ETHICS, the technology is part of the design. Further, they assume shop-floor situations, rather than the office and high-level organisational situations which concern ETHICS.

Mumford recommends the consideration of three types of work organisation patterns. The first is task variety, and involves giving an individual more variety in work by providing more than one task to be performed or by rotating people around a number of different tasks. This is the more traditional approach, but is limited especially where the expectations of job satisfaction are more sophisticated. In this case, the principles of job enrichment might be appropriate. This is where the work is organised in such a way that a number of different skills, including judgmental ones, are introduced. In particular, it involves the handling of problems and the organisation of the work by the individual without supervision. This may require an increased skill level on behalf of the individual, but leads to enhanced job satisfaction. A further stage in job enrichment is the incorporation of development aspects into a job. This means that the individual has the freedom to change the way the job is performed. This leads to constant review and the implementation of new ideas and methods. Obviously this cannot be introduced into every job, but there are probably more opportunities than at first imagined.

As important as individual jobs is the concept of what Mumford calls self-managing groups. Here, groups are formed that have responsibility for a relatively wide spectrum of the tasks to be performed. These groups are preferably multi-skilled, so that each member is competent to carry out all the tasks required of the group. They are encouraged to organise themselves, their work, and their own control and monitoring, which may include their own target setting. This can provide a very stimulating and satisfying work environment for the group members. Again, self-managing groups are not always possible and require a good deal of management goodwill at first, but nevertheless can prove very effective.

12 Technical options
The various technical options that might be appropriate, including hardware, software and the design of the human-computer interface, are specified. Each option is evaluated in the same way as the organisational options, that is, against efficiency, job satisfaction and future change objectives. As mentioned in step 11, the organisational and technical options should be considered simultaneously, as often one option implies certain necessary factors in the other. It is advised that one option should exist which specifies no change in technology, so as to be able to see how much could be achieved simply with organisational changes.

The organisational and technical options are now merged to ensure compatibility, and are evaluated against the primary objectives and the one that best meets the objectives is selected. This selection is performed by the design group with input from the steering committee and other interested constituencies.

13 The preparation of a detailed work design
The selected system is now designed in detail. The data flows, tasks, groups, individuals, responsibilities and relationships are defined. There is also a review to ensure that the detail of the design still meets the specified objectives. Obviously, the design detail includes the organisational aspects as well as the technical.

14 Implementation
The design group now applies itself to ensuring the successful implementation of the design. This involves planning the implementation process in detail. This will include the strategy, the education and training,

the co-ordination of parts, and everything needed to ensure a smooth changeover.

15 Evaluation

The implemented system is checked to ensure that it is meeting its objectives, particularly in relation to efficiency and job satisfaction, using the techniques of variance analysis and measures of job satisfaction. If it is not meeting the objectives, then corrective action is taken. Indeed, as time progresses, changes will become necessary and design becomes a cyclical process.

Quite a common reaction to ETHICS is for people to say that it is impractical. First, it is argued, that unskilled users cannot do the design properly, and second, that management would never accept it, or that it removes the right to manage from managers. In answer to the first problem Mumford argues that users can, and do, design properly. They need some training and help along the way, but this can be relatively easily provided. More importantly, they have the skills of knowing about their own work and system, and have a stake in the design. This is much more than many traditional analysts and designers. To answer the second point, managers have often welcomed participation and can be convinced of its benefits. There are many success stories. It is not always the management that needs to be convinced, sometimes it is the users who are sceptical about participation, seeing it as some sort of management trick. The job of a manager is to meet the corporate objectives, not simply oversee people and make every last decision. This is often counter-productive to achieving those objectives, often resulting in very high staff turnover rates, which is not productive.

Mumford does admit that it will not be easy, quite the reverse, but the benefits are, she claims, worth it. She has also produced some publications relating to experiences using ETHICS. They illustrate many of the problems that are encountered and the solutions. They show how users can design their own systems and how they come to terms with their design roles. The first book shows how a group of secretaries at Imperial Chemical Industries (ICI) in the UK designed new work systems for themselves in the wake of the introduction of word processing equipment (Mumford, 1983c). The second example shows how a group of purchase invoice clerks helped design a major on-line computer system (Mumford, 1983b). One of the most interesting aspects, and most telling concerning the power of ETHICS that emerges from this study, is the fact that the clerks designed three different ways of working with the computer system to do essentially the same thing. The one used depends on the clerk. Few professional systems design teams would design a number of alternative ways to achieve the same task.

ETHICS has more recently been used by a number of large companies to assist the building of very large systems. The first major use of ETHICS in the development of a large system was Digital Equipment Corporation's XSEL, an expert system for their sales offices which helped to configure DEC hardware systems for particular customers (Mumford, 1989).

In some situations it has become a method of requirements analysis in particular, and a version has been defined which is referred to as QUICKETHICS (QUality Information from Considered Knowledge). In order to create and maintain managers' interest, it can be organised as a drama having four 'acts':

- Self-reflection
- Self-identification
- Group discussion
- Group decision.

In this approach, each manager describes his or her work role and relationships, along with information needs ranked as 'essential', 'desirable' and 'useful' on an individual basis. This provides an opportunity for self-reflection. Meeting then as a group, each manager gives a short description of his or her mission, key tasks, critical success factors and major problems. This provides an opportunity for self-identification and encourages questions and discussion. Then managers may write the essential information needs on cards placed on a magnetic board explaining the reasons for their importance. This provokes group discussion because it soon becomes apparent that managers have many overlapping needs. These common needs can be agreed as forming a 'core' module of the proposed information system, delivering essential information needs in the group-decision process. Once this is implemented, 'desirable' needs can guide future development. The requirements analysis phase may only involve two days of management's time. ETHICS has changed over time, as for all information systems development methodologies, but the importance of user involvement and participation in systems design has endured the process of change.

6.11 SOFT SYSTEMS METHODOLOGY (SSM)

As discussed in section 3.2, general systems theory attempts to understand the nature of systems. Scientific analysis breaks up a complex situation into its constituent parts for analysis. Although this works in the physical sciences, it is less successful in the social sciences and in management science. One tenet of systems thinking is that the whole is greater than the sum of the parts: properties of the whole are not explicable entirely in terms of the properties of its constituent elements. Human activity systems are complex and the human components, in particular, may react differently when examined singly as when they play a role in the whole system. Something is lost when the whole is broken up in the 'reductionist' approach of scientific analysis.

The systems principle also implies that we must try to develop application systems for the organisation as a whole rather than for functions in isolation. It may take only a few hours by Concorde to cross the Atlantic, but this progress is partly lost if it takes as many hours to get from home to Heathrow Airport and from JFK Airport to the hotel in New York. It is the transport system we should be looking at, not the airline system in isolation. Further, organisations are 'open systems' and therefore the relationship between the organisation and its environment is important. We should always be looking at 'the system' in terms of the wider system of which it is part.

Systems theory would also suggest that a multi-disciplinary team of analysts is much more likely to understand the organisation and suggest better solutions to problems. After all, specialisms are a result of artificial and arbitrary divisions. In the information systems context, a systems approach should prevent an automatic assumption that computer solutions are always appropriate. It will also help in problem situations which have been studied from only one narrow point of view. Such an approach is not appropriate in the study of large and complex problem situations.

Checkland (1981) has attempted to adapt systems theory into a practical methodology. By methodology he means the study of methods to achieve certain purposes. 'For any particular problem situation, the study will lead to a subset of principles which can be applied for that particular situation'. He argues that systems analysts apply their craft to problems which are not well defined. These 'fuzzy', ill-structured or soft problem situations, usually also complex, are common in organisations. The description of one category of systems, human activity systems, also acknowledges the importance of people in organisations. It is relatively easy, it is argued, to model data and processes (the emphasis placed in many of the preceding methodologies discussed), but to understand the real world it is essential to include people in the model, people who may have different and conflicting objectives, perceptions and attitudes. This is difficult because of the unpredictable nature of human activity systems. There is no such thing as a repeatable experiment in this context. The claims for the soft systems approach are that a true understanding of complex problem situations is more likely using this approach than if the more simplistic structured or data-orientated approaches are used, which address mainly the formal or 'hard' aspects of systems.

Wilson (1990) gives an analogy. He considers two examples of problems. The first is a flat tyre. The solution is clear. The second is 'What should the UK government do about Northern Ireland?' (a more recent example might be 'What should the United Nations do about Bosnia?'). The solution to these problems are not clear and it would be difficult to find any solutions that satisfied all the interested parties. Wilson suggests that hard methodologies, that may be suitable for solving 'burst tyre type problems' are inappropriate for organisational problem situations. It is not only a question of techniques and tools, but also concepts and language.

Another difference between hard and soft systems thinking is that in hard systems thinking a goal is assumed. The purpose of the methods used by the analyst (or engineer) is to modify the system in some way so that this goal is achieved in the most efficient manner. Hard systems thinking is concerned with the 'how' of the problem. In soft systems thinking, the objectives of the system are assumed to be more complex than a simple goal that can be achieved and measured. Systems are argued to have purposes or missions rather than goals. Understanding is achieved in soft systems methods through debate with the actors in the system. Emphasis is placed on the 'what' as well as the 'how' of the system. The term 'problem' in this context is also inappropriate. There will be lots of problems, hence the term 'problem situation' - 'a situation in which there are perceived to be problems' (Wilson, 1990).

The methodology of Checkland has been developed at Lancaster University. It was developed through 'action research' whereby the systems ideas are tested out in client organisations. The analysts neither dictate the way the action develops nor step outside as observers: they are participants in the action and results are unpredictable. The practical work provides experience which can be used to draw conclusions and to modify these ideas. This proves to be a successful approach, because, as we have said, it is not possible to develop a good 'laboratory model' of human activity systems and set up repeatable experiments.

Each action research project therefore furthers knowledge which can be used in future soft systems work, as well as to provide some benefits in a particular problem situation. Change is therefore achieved through the learning process as theory and practice meet and affect each other. Checkland's book is aptly titled 'Systems Thinking, Systems Practice'!

Checkland has carried out extensive studies, both theoretical and practical, on the analysis of organisations in his action research projects. The central focus of the methodology is the search for a particular view (or views). This *Weltanschauung* (assumptions or world view) will form the basis for describing the systems requirements and will be carried forward to further stages in the methodology. The world view is extracted from the problem situation through debate on the main purpose of the organisation concerned - its *raison d'être*, its attitudes, its 'personality' and so on. Examples of world view might be : 'This is a business aimed at maximising long-term profits' or 'This is a hospital dedicated to maintaining the highest standards of patient care'.

The original version of soft systems methodology is given in Checkland (1981). We will outline this version (mode 1) below. This will be followed by a brief outline of the changes made to the approach (mode 2) as found in Checkland and Scholes (1990).

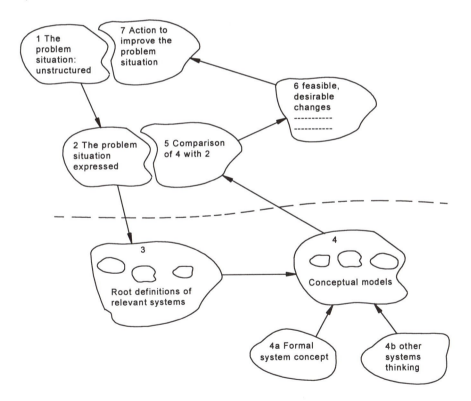

Fig. 6.45: SSM in summary (after Checkland, 1981)

Figure 6.45 shows the seven stages of SSM. Stages above the dotted line are 'real-world' activities involving people in the problem situation, whereas stages below the dotted line are activities concerned with thinking about the problem situation. Checkland argues that this is a logical sequence for description purposes, but that it is possible to start a project at a place other than at point 1. Further, the analyst is likely to be working on a number of stages simultaneously, and backtracking and iteration are essential. Thus the diagram should be taken as a framework rather than as a prescription which must be followed. This is not the case for many of the methodologies described earlier in this chapter which are expected to be followed fairly rigorously.

Stages 1 and 2 are about finding out about the problem situation. This unstructured view gives some basic information from the views of the individuals involved. The application of the CATWOE criteria gives some structure to the expressions of the problem situations and in stage 3 the analyst selects from these those views which he or she considers gives insight to the problem situation. Stage 4 is to do with model building, that is, what the systems analyst might *do* (as against what the system *is* - the root definition). There must be one conceptual model for each root definition. Stage 5 compares the conceptual models from stage 4 and the root definitions formed at stage 2. This comparison process leads to a set of recommendations regarding change, and stage 6 analyses these recommendations in terms of what is feasible and desirable. Stage 7, the final phase, suggests actions to improve the problem situation, following the recommendations of stage 5.

1 The problem situation: unstructured

The first two stages are concerned with finding out about the problem situation from as many people in that situation as possible. There will be many different views as it is unlikely that the views of the problem owner, that is, the person or group on whose behalf the analysis has been commissioned, the other people taking part as 'actors' in the problem situation and other stakeholders, will coincide. There will be different views that the analyst can take regarding the problem situation and at this stage it is important to reveal as wide a range of them as possible. The analyst will also look at the structure of the problem situation in terms of physical layout, reporting structure and formal and informal communication patterns. These activities carried out in the problem situation are also studied along with how these activities are controlled. The 'climate' of the situation, that is, the relationship between structure and process can also be very revealing.

2 The problem situation: expressed

Given the informal picture of the problem situation gathered in stage 1, it is feasible to attempt to express it in some more formal way. Checkland does not prescribe a method of doing this, but he and many users of the approach tend to draw rich picture diagrams of the situation (section 4.2).

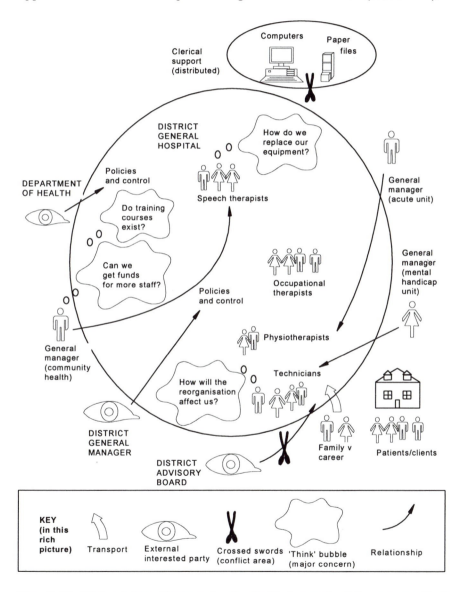

Fig. 6.46: Rich picture diagram (early draft) for part of the paramedical services of health district

Such diagrams show the structures of the processes and their relation to each other. Elements of this rich picture will include the clients of the system, the people taking part in it, the task being performed, the environment and the owner of the system. This picture can be used as an aid to discussion between the problem solver and the problem owner, or may simply help the problem solver better understand the problem situation. This stage is concerned with finding out about the problem situation.

The rich picture can be used as a communication technique between the analysts and the users of the system and therefore uses the terminology of the environment. It will usually show the people involved, problem areas, controlling bodies and sources of conflict. From the rich picture, the problem solver extracts problem themes - things noticed in the picture that are, or may be, causing problems. The picture may show conflicts between departments, absences of communication lines, shortages of supply and so on. Rich pictures are intended to help in problem identification, not in the process of recommending solutions. In general, SSM concentrates on understanding problem situations, rather than developing solutions.

Rich pictures prove to be a very useful way of getting the user to talk about the problem situation. They may stimulate the drawing out of some of the parts of the 'iceberg' which normally lie hidden when using traditional 'methods of investigation. Figure 6.46, an example rich picture chart (from Avison and Catchpole, 1987) highlights areas of conflict and concerns in the problem situation, a branch of the community health service in the UK.

Although mainly intended as a communication technique between the analysts and the users of the system, it will be possible to glean a lot of information about the system, including the main people involved (the 'people' figures in figure 6.46), problem areas (the 'question bubbles'), controlling bodies (the 'big eyes') and sources of conflict ('crossed swords'), by looking at the rich picture. This interpretation of rich pictures is developed from Avison and Wood-Harper (1990).

3 Root definitions of relevant systems

Taking these problem themes, the next stage of the methodology involves the problem solver imagining and naming relevant systems. By relevant is meant a way of looking at the problem which provides a useful insight, for example:

Problem theme	=	conflicts between two departments
Relevant systems	=	system for redefining departmental boundaries.

Several different relevant systems should be explored to see which is the most useful, but, in any case, this may be changed afterwards. It is at this stage that debate is most important. The problem solver and the problem owner decide which view to focus on, that is, how to describe their relevant system. For example, will the conflict resolution in our health example be 'a system to impose rigid rules of behaviour and decision making in order to integrate decisions and minimise conflict' or will it be 'a system to integrate decisions of actors through increased communication and understanding between departments' or even 'an arbitration system to minimise conflict between departments by focusing disagreements through a central body'?

After constructing a rich picture, a root definition (section 4.3) is developed for the relevant system. A root definition is a kind of hypothesis about the relevant system, and improvements to it, that might help the problem situation.

> The root definition is a concise, tightly constructed description
> of a human activity system which states what the system is
> (Checkland 1981).

Using the CATWOE checklist technique (section 4.3), the root definition is created. A root definition for a hospital administration system could be: 'to provide a service which gives the best possible care to the patients and which balances the need to avoid long waiting lists with that to avoid excessive government spending'. But alternative root definitions could have been 'a system for employing medical and administrative staff', 'a system to generate long waiting lists to illustrate the high status of consultants', or 'a system to encourage the use of private health facilities'.

4 Building conceptual models

When the problem owners and the problem solvers are satisfied that the root definition is well formed, a conceptual model (section 4.4) can be developed using this root definition (figure 6.47). In this context, the conceptual model is a diagram of the activities showing what the system, described by the root definition, will do. (The term 'conceptual model' is used in some other methodologies to refer to entity modelling). This stage in the methodology is about model building, but the model is meant to describe something relevant to the problem situation, it is not meant to be a model of the situation itself (otherwise it stifles radical thought about the problem situation).

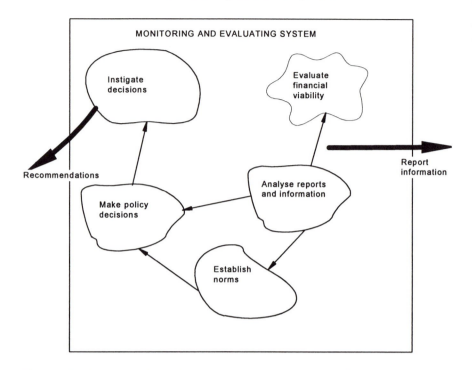

Fig. 6.47: Conceptual model for a monitoring and evaluation system for community health (early draft)

This interpretation is different to that implied in a mathematical model, or an architect's model. The conceptual model, such as that shown in figure 6.47, can again be used as a debating point so that the actors can relate the model to the real world situation. Analysts need also to check the conceptual model against a general model of any human activity system (the 'formal system' model stage 4a in figure 6.45). Checkland includes amongst these pre-requisites a purpose, measure of performance, a decision-taking process, connectivity and an environment. Checkland also suggests evaluating the conceptual model through looking at the work of other systems thinkers, such as Beer, Churchman and Vickers, and thereby improve the model (stage 4b).

Usually there is a conceptual model drawn for each root definition and the drawing up of several root definitions and conceptual models becomes an iterative process of debate and modification towards an agreed root definition and conceptual model. An agreed statement is never easy to achieve as the process described is meant to draw out the different ideologies and conflicts, so that the final version represents an understanding of the problem situation. Similar processes were described in the ETHICS methodology. There is a danger that these final versions represent a conservative compromise. This is not what is intended. In SSM, if the ideological conflict is central to the problem situation, it has to be represented. The approach is meant to represent a holistic view to consider a complex problem, not a simplistic view representing a 'political' compromise.

5 Comparing conceptual models with reality
The next stage concerns the comparison of the problem situation analysed in stage 2 through rich pictures alongside the conceptual models created in stage 4. It is also about a comparison of views, and since these views are those of human activity systems, made by people, we may not be comparing similar things. The debate about possible changes should lead to a set of recommendations regarding change in order to help the problem situation.

6 Assessing feasible and desirable changes
On assimilating these views, stage 6 concerns an analysis of proposed changes from stage 5 so as to draw up proposals for those changes which are considered both feasible and desirable. Checkland's methodology does not limit itself to changing or developing new information systems, though this would be valid in the context of this book.

7 Action to improve the problem situation
This final stage is about recommending action to help the problem situation. Note that the methodology does not describe methods for implementing solutions. The methodology helps in understanding problem situations rather than provide a scheme for solving a particular problem.

Although we have discussed rich pictures, root definitions and conceptual models, the methodology relies much less on techniques and tools than most other methodologies, particularly 'hard' methodologies. These provide tools for use in particular situations at particular times in the development of the information systems project. SSM provides all actors, including the analysts, opportunities to understand and to deal with the problem situation, that is, the human activity system. The analysts are perceived as actors involved in the problem situation, as much as those of the client and problem owner - they are not perceived as outside onlookers providing objectivity.

The process is iterative and the analysts learn about the system and are not expected to follow a laid-down set of procedures. This does present problems: it is difficult to teach and to train others, and it is difficult to know when a stage in the project has been satisfactorily completed. However, these features are also its strengths, because it does not have any pre-conceived notions of a 'solution' and use of the methodology gives a better understanding of the problem situation.

One possible way that SSM can be fitted in the information systems development process is by using it as a 'front end' before proceeding to the 'hard' aspects of systems development. This would seem to be appropriate because SSM concerns analysis whereas the harder methodologies tend to emphasise design, development and implementation. Thus attempts have been made to develop a combination of SSM and SSADM called Compact (CCTA, 1989). The Multiview methodology, discussed in the next section, also draws on SSM in the early parts of the systems definition process.

SSM mode 1, described in Checkland (1981) is still the version most commonly referred to and possibly the most useful in an information systems context. However, there is an alternative version proposed in Checkland and Scholes (1990) based on lessons learnt from further action research. This sees the seven stage methodology given above as just one option in a more general approach.

In mode 2, the problem situation is seen as a product of history and can be looked at in many different ways. People using SSM will follow two strands of enquiry which together should lead to changes being implemented which improve the problem situation. These two strands are described as a 'logic-driven stream of enquiry' and a 'cultural stream'.

The logic-driven stream considers models of human activity systems and a comparison of these models is made with an examination of our views of the real world and the ensuing debate concerns change. The cultural stream examines three aspects of the problem situation: the intervention itself, the situation as a social system and the situation as a political system. The two streams are seen to interact and the process viewed as a continual one.

Mode 2 can be considered as more of a framework of ideas for exploration rather than a methodology, although mode 1 was never proposed to be used prescriptively nor without frequent iterative steps. It is perhaps this which is a key to interpreting SSM: mode 2 is really about suggesting that mode 1 should not be used as a step-by-step prescription and that practitioners will tend to use some sort of compromise between mode 1 and mode 2. Essentially this represents a loose interpretation of mode 1. SSM, more than any other methodology discussed so far, is very dependent on the particular interpretation followed by those who use the approach. This concept of framework rather than methodology and the role of the interpreter is taken further in an information systems context in Multiview.

6.12 MULTIVIEW

Multiview perceives information systems development as a hybrid process involving computer specialists, who will build the system, and users, for whom the system is being built. The methodology therefore looks at both the human and technical aspects of information systems development. In this aspect and others, it has been greatly influenced by the work of Checkland and Mumford, but has fused these ideas with those found in 'hard' methodologies, such as STRADIS and IE.

The approach adopted has been used on a number of projects, and the methodology itself has been refined using 'action research' methods, that is the application and testing of ideas developed in an academic environment into the 'real world', also mentioned in the context of SSM. Further, unlike many of the methodologies discussed in this chapter, it has mainly been used for small scale computing projects, particularly for PC applications using software packages rather than developing 'bespoke' systems, which is the intention of users of most information systems development methodologies. It is a contingency approach in that it will be adapted according to the particular situation in the organisation and the application.

The section is modified from Avison & Wood-Harper (1990). There are a number of projects using Multiview discussed in that book but these are not put forward as 'textbook' examples, showing how the application of the Multiview methodology 'worked perfectly'. The search for a perfect methodology, it is argued, is somewhat illusory. On the contrary, the cases expose some of the difficulties and practical problems of information systems work.

The authors are concerned to show that information systems development theories should be contingent rather than prescriptive, because the skills of different analysts and the situations in which they are constrained to work always has to be taken into account in any project. Avison and Wood-Harper (1986) describe Multiview as an *exploration* in information systems development. It therefore sets out to be flexible: a particular technique or aspect of the methodology will work in certain situations but is not advised for others.

The methodology will be seen by readers of this text as truly 'multi-view', because it includes many of the techniques used by the other methodologies, and also its stages parallel those of other methodologies. The authors of Multiview claim, however, that it is not simply a hotchpotch of available techniques and tools, but a methodology which has been tested and works in practice. It is also 'multi-view' in the sense that it takes account of the fact that as an information systems project develops, it takes on different perspectives or views: organisational, technical, human-orientated, economic and so on.

The five stages of Multiview are as follows:
- Analysis of human activity
- Analysis of information (sometimes called information modelling)
- Analysis and design of socio-technical aspects
- Design of the human-computer interface
- Design of technical aspects.

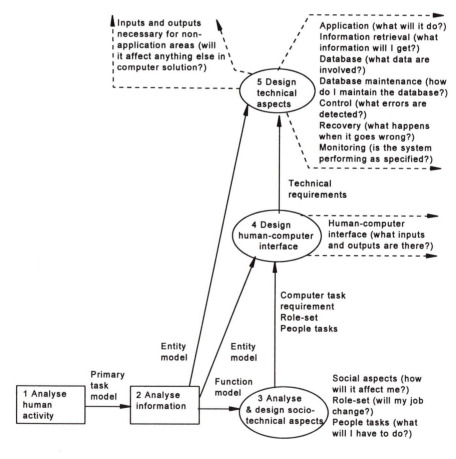

Fig. 6.48: The Multiview framework

They incorporate five different views which are appropriate to the progressive development of an analysis and design project, covering all aspects required to answer the vital questions of users. These five views are necessary to form a system which is complete in both technical and human terms. The five stages move from the general to the specific, from the conceptual to hard fact and from issue to task. Outputs of each stage either become inputs to following stages or are major outputs of the methodology. The Multiview methodology is shown in outline in figure 6.48. The two analysis-orientated stages are shown in boxes and the three design-orientated stages in circles. The arrows represent information passing between stages and the dotted arrows represent outputs of the methodology. The outputs of the methodology, shown as dotted arrows in figure 6.48, are listed in figure 6.49, together with the information that they provide and the questions that they answer.

OUTPUTS	INFORMATION
Social aspects	How will it affect me?
Role-set	Will my job change? In what way?
People tasks	What will I have to do?
Human-computer interface	How will I work with the computer?
	What inputs and outputs are there?
Database	What data are involved?
Database maintenance	How will I maintain the integrity of data?
Recovery	What happens when it goes wrong?
Monitoring	Is the system performing to specification?
Control	How are security and privacy handled?
Information retrieval	What information will I get?
Application	What will the system do?
Inputs and outputs for non-application areas	Will it affect anything else on the computer system?

Fig. 6.49: Multiview methodology outputs

The authors argue that to be complete in human as well as in technical terms, the methodology must provide help in answering the following questions:

1. How is the computer system supposed to further the aims of the organisation installing it?

2. How can it be fitted into the working lives of the people in the organisation that are going to use it?
3. How can the individuals concerned best relate to the machine in terms of operating it and using the output from it?
4. What information system processing function is the system to perform?
5. What is the technical specification of a system that will come close enough to doing the things that have been written down in the answers to the other four questions?

Too often, the authors argue, methodologies and role players have addressed themselves to only a limited subset of these questions: for example, computer scientists to question 5, systems analysts to question 4, users to question 3, trade unions to question 2 and top management to question 1. Multiview attempts to address all these questions and to involve all the role players or stakeholders in answering all these questions. The emphasis in information systems, it is argued, must move away from 'technical systems which have behavioural and social problems' to 'social systems which rely to an increasing extent on information technology'.

Because it *is* a multi-view approach, it covers computer-related questions and also matters relating to people and business functions. It is part issue-related and part task-related. An issue-related question is: 'What do we hope to achieve for the company as a result of installing a computer?'. A task-related question is: 'What jobs is the computer going to have to do?'.

The distinction between issue and task is important because it is too easy to concentrate on tasks when computerising, and to overlook important issues which need to be resolved. Too often, issues are ignored in the rush to 'computerise'. But you cannot solve a problem until you know what the problem is! Issue-related aspects, in particular those occurring at stage 1 of Multiview, are concerned with debate on the definition of system requirements in the broadest sense, that is 'what real world problems is the system to solve?'. On the other hand, task-related aspects, in particular stages 2-5, work towards forming the system that has been defined with appropriate emphasis on complete technical and human views. The system, once created, is not just a computer system, it is also composed of people performing jobs.

Another representation of the methodology, rather more simplistic, but useful in providing an overview for discussion, is shown in figure 6.50. Working from the middle outwards we see a widening of focus and an increase in understanding the problem situation and its related technical and human characteristics and needs. Working from the outside in, we see an increasing concentration of focus, an increase in structure and the progressive development of an information system. This diagram also shows how the five questions outlined above have been incorporated into the five stages of Multiview.

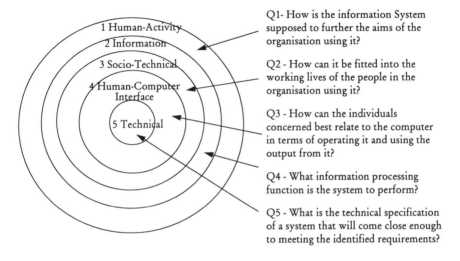

Q1- How is the information System supposed to further the aims of the organisation using it?

Q2 - How can it be fitted into the working lives of the people in the organisation using it?

Q3 - How can the individuals concerned best relate to the computer in terms of operating it and using the output from it?

Q4 - What information processing function is the system to perform?

Q5 - What is the technical specification of a system that will come close enough to meeting the identified requirements?

Fig. 6.50: The Multiview framework

The first stage looks at the organisation - its main purpose, problem themes, and the creation of a statement about what the information system will be and what it will do. The second stage is to analyse the entities and functions of the problem situation described in stage one. This is carried out independently of how the system will be developed.

The philosophy behind the third stage is that people have a basic right to control their own destinies and that if they are allowed to participate in the analysis and design of the systems that they will be using, then implementation, acceptance and operation of the system will be enhanced. This stage emphasises the choice between alternative systems, according to important social and technical considerations. The fourth stage is concerned with the technical requirements of the user interface. The design of specific conversations will depend on the background and experience of the people who are going to use the system, as well as their information needs.

Finally, the design of the technical subsystem concerns the specific technical requirements of the system to be designed, and therefore to such aspects as computers, databases, control and maintenance. Although the methodology is concerned with the computer only in the latter stages, it is assumed that a computer system will form at least part of the information system. However, the authors do not argue that the final system will necessarily run on a large mainframe computer. This is just one solution, many cases show applications being implemented on a microcomputer.

1 Analysis of human activity

This stage is based on SSM (mode 1). The very general term 'human activity' is used to cover any sort of organisation. This could be, for example, an individual, a company, a department within a larger organisation, a club or a voluntary body. They may all consider using a computer for some of their information systems. The central focus of this stage of the analysis is to search for a particular view (or views). This *Weltanschauung* will form the basis for describing the systems requirements and will be carried forward to further stages in the methodology. This world view is extracted from the problem situation through debate on the main purpose of the organisation concerned.

First, the problem solver, perhaps with extensive help from the problem owner, forms a rich picture of the problem situation. The problem solver is normally the analyst or the project team. The problem owner is the person or group on whose behalf the analysis has been commissioned. This picture can be used to help the problem solver better understand the problem situation. The rich picture diagram is also very useful to stimulate debate, and it can be used as an aid to discussion between the problem solver and the problem owner. There are usually a number of iterations made during this process until the 'final' form of the rich picture is decided. The process here consists of gathering, sifting and interpreting data which is sometimes called 'appreciating the situation'. Drawing the rich picture diagram, examples of which are given in figures 4.2 and 6.46, is a subjective process. There is no such thing as a 'correct' rich picture. The main purpose of the diagram is to capture a holistic summary of the situation.

From the rich picture the problem solver extracts problem themes, that is, things noticed in the picture that are, or may be, causing problems and/or it is felt worth looking at in more detail. The picture may show conflicts between two departments, absences of communication lines, shortages of supply and so on. Taking these problem themes, the problem solver imagines and names relevant systems that may help to relieve the problem theme. Several different relevant systems should be explored to see which is the most useful. Once a particular view or root definition (also described, with examples, in section 4.3) has been decided upon, it can be developed and refined. Thus, by using the CATWOE check-list, the root definition can be analysed by checking that all necessary elements have been included. For example, have we identified the owner of the system? all the actors involved? the victims/beneficiaries of the system? and so on.

When the problem owner and the problem solver are satisfied that the root definition is well formed, a conceptual model (or activity model) of the system is constructed by listing the 'minimum list of verbs covering the activities which are necessary in a system defined in the root definition.....'. Examples of conceptual models are seen in figures 4.3, 4.4 and 6.47. At this stage, therefore, we have a description in words of what the system will be (the root definition) and an inference diagram of the activities of what the system will do (the conceptual model).

The completed conceptual model is then compared to the representation of the 'real world' in the rich picture. Differences between the actual world and the model are noted and discussed with the problem owner. Possible changes are debated, agendas are drawn up and changes are implemented to improve the problem situation.

In some cases the output of this stage is an improved human activity system and the problem owner and the problem solver may feel that the further stages in the Multiview methodology are unnecessary. In many cases, however, this is not enough. In order to go on to a more formal systems design exercise, the output of this stage should be a well formulated and refined root definition to map out the universe of discourse, that is, the area of interest or concern. It could be a conceptual model which can be carried on to stage 2, the analysis of entities, functions and events.

2 Analysis of information
The purpose of this stage is to analyse the entities and functions of the application. Its input will be the root definition/conceptual model of the proposed system which was established in stage 1 of the process. Two phases are involved: (a) the development of the functional model and (b) the development of an entity model.

(a) *Development of a functional model* The first step in developing the functional model is to identify the *main* function. This is always clear in a well formed root definition. This main function is then broken down progressively into sub-functions (functional decomposition), until a sufficiently comprehensive level is achieved. This occurs when the analyst feels that the functions cannot be usefully broken down further. This is normally achieved after about four or five sub-function levels, depending on the complexity of the situation. A series of data flow diagrams, each showing the sequence of events, is developed from this hierarchical model. This stage is therefore greatly influenced by the process modelling theme (section 3.7). The hierarchical model and data flow diagrams are the major inputs into stage 3 of the methodology, the next stage, which is the analysis and design of the socio-technical system.

(b) *Development of an entity model* In developing an entity model, the problem solver extracts and names entities from the area of concern. Relationships between entities are also established. Again, this stage is greatly influenced by the data modelling them (section 3.8). The preceding stage in the methodology, the analysis of the human activity systems, should have already given this necessary understanding and have laid a good foundation for this second stage. The entity model can then be constructed. The entity model, following further refinement, becomes a useful input into stages 4 and 5 of the Multiview methodology.

3 Analysis and design of the socio-technical aspects

The philosophy behind this stage (influenced by ETHICS (section 6.10)) is that people have a basic right to control their own destinies and that, if they are allowed to participate in the analysis and design of the systems that they will be using, then the implementation, acceptance and operation of the system will be enhanced. It takes the view therefore that human considerations, such as job satisfaction, task definition, morale and so on, are just as important as technical considerations. The task for the problem solver is to produce a 'good fit' design, taking into account people and their needs and the working environment on the one hand, and the organisational structure, computer systems and the necessary work tasks on the other.

An outline of this stage is shown in figure 6.51. The central concern at this stage is the identification of alternatives: alternative social arrangements to meet social objectives and alternative technical arrangements to meet technical objectives. All the social and technical alternatives are brought together to produce socio-technical alternatives. These are ranked, first, in terms of their fulfilment of the above objectives, and second, in terms of costs, resources and constraints - again both social and technical - associated with each objective. In this way, the 'best' socio-technical solution can be selected and the corresponding computer tasks, role-sets and people tasks can be defined.

The emphasis of this stage is therefore *not* on development, but on a statement of alternative systems and choice between the alternatives, according to important social and technical considerations.

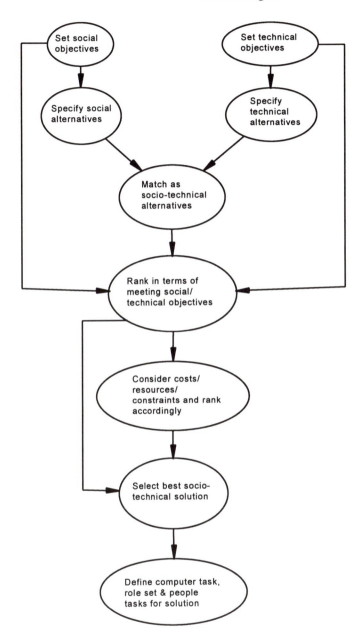

Fig. 6.51: Outline of socio-technical analysis and design

It is also clear that, in order to be successful in defining alternatives, the groundwork in the earlier stages of the methodology is necessary and, in order to develop and implement the chosen system, we must continue to the subsequent stages. An important technique applicable to this stage is future analysis (Land, 1982). This aids the analyst and user to predict the future environment so that they are better able to define and rank their socio-technical alternatives.

The outputs of this stage are the computer task requirements, the role-set, the people tasks and the social aspects. The computer task requirements, the role-set and the people tasks become inputs to the next stage of the methodology, that is, the design of the human-computer interface. The role-set, the people tasks and the social aspects are also major outputs of the methodology.

4 Design of the human-computer interface

Up to now, we have been concerned with what the system will do. Stage 4 relates to how, in general terms, we might achieve an implementation which matches these requirements. The inputs to this stage are the entity model derived in stage 2 of the methodology, and the computer tasks, role-set and people tasks derived in stage 3. This fourth stage is concerned with the technical design of the human-computer interface and makes specific decisions on the technical system alternatives. The ways in which users will interact with the computer will have an important influence on whether the user accepts the system.

A broad decision will relate to whether to adopt batch or on-line facilities. In on-line systems, the user communicates directly with the computer through a terminal or workstation. In a batch system, transactions are collected, input to the computer, and processed together when the output is produced. This is then passed to the appropriate user. Considerable time may elapse between original input and response.

Decisions must then be taken on the specific conversations and interactions that particular types of user will have with the computer system, and on the necessary inputs and outputs and related issues, such as error checking and minimising the number of key strokes. There are different ways to display the information and to generate user responses. The decisions are taken according to the information gained during stages 1 and 2 of Multiview.

Once human-computer interfaces have been defined, the technical requirements to fulfil these can also be designed. These technical requirements become the output of this stage and the input to stage 5, the design of technical subsystems. The human-computer interface definition becomes a major output of the methodology.

5 Design of the technical aspects

The inputs to this stage are the entity model from stage 2 and the technical requirements from stage 4. The former describes the entities and relationships for the whole area of concern, whereas the latter describes the specific technical requirements of the system to be designed.

After working through the first stages of Multiview, the technical requirements have been formulated with both social and technical objectives in mind and also after consideration of an appropriate human-computer interface. Therefore, necessary human considerations are already both integrated and interfaced with the forthcoming technical subsystems.

At this stage, therefore, a largely technical view can be taken so that the analyst can concentrate on efficient design and the production of a full systems specification. Many technical criteria are analysed and technical decisions made which will take into account all the previous analysis and design stages. The final major outputs of the methodology might include:

- *The application subsystem* which is concerned with performing the functions which have been computerised from the function chart
- *The information retrieval subsystem* which is for responding to enquiries about data stored in the system
- *The database* in which all the data are organised
- *The database maintenance subsystem* which permits updates to the data and provides the information necessary to check for data errors
- *The control subsystem* which checks for user, program, operator and machine errors and alerts the system to their presence
- *The recovery subsystem* which allows the system to be repaired after an error has been detected
- *The monitoring subsystem* which keeps track of all system activities for management purposes.

Figure 6.52 shows a schematic of these requirements for the technical specification. These subsystems cover all the things that have to be done by the computer system and the people operating it. These parts, or subsystems, may be implemented in different ways and in different combinations. For example, the information retrieval subsystem may be just another aspect of the database management system and this may also include many of the necessary functions for control and recovery. The Multiview authors have separated them out because it is necessary to be sure that each one of them is catered for in the system.

Following full testing of all aspects of the system, there is a recognition that there will still be changes required and this should be regarded as 'the norm'. Information systems will develop and this requires an on-going relationship between users, analysts, and system creators or owners. The authors recommend that the Multiview framework be applied for these changes (at least with a 'token run') so as to ensure that the system is still meeting its real objectives.

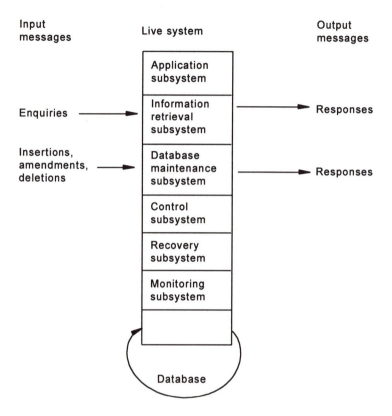

Fig. 6.52: Outline of the requirements for the technical specification

6.13 PROCESS INNOVATION

Business process re-engineering was discussed as a theme in section 3.4. It necessitates a radical rethink about the business processes and therefore about the organisation. Successful process re-engineering leads to major behavioural as well as structural change. It is still in its early days and because of its radical nature it is not surprising that it has not been embraced by all organisations. The work of Tom Davenport, expressed in Davenport and Short (1990) and Davenport (1993), which provides a number of examples where it has been used in Europe as well as the United States, does most to tie business process re-engineering with information technology where each is regarded as the key to thinking about the other. Information technology is seen as the primary enabler of process innovation (as Davenport refers to process re-engineering).

It is argued that only radical change to business processes will lead to significant gains. In the earlier days, computing was used to help automate existing processes. The aim was to use IT to improve efficiency and control. Yet, as we saw in section 2.4, there are basic weaknesses in the way IT has been used conventionally, and it has not always been successful even in these aims However, IT can be used to change processes completely and Davenport's view is that it should be used to support this more radical change implied by process innovation, perhaps along with other major change factors, such as those in human resource management.

The earlier paper proposes a five stage approach to re-engineering processes with IT:

- Develop the business vision and process objectives
- Identify the processes to be redesigned
- Understand and measure the existing process
- Identify the IT levers
- Design and build a prototype of the new process.

The content and sequence of the five stages are different in the Davenport book and the paper by Davenport and Short, although the two approaches are not fundamentally different.

1 Develop the business vision and process objectives

Davenport and Short argue that it is not enough to attempt to eliminate obvious bottlenecks and inefficiencies from processes; it is necessary to redesign the entire process according to a business vision. In this phase, the organisational strengths and weaknesses need to be identified, along with an analysis of the market and the opportunities it provides. A knowledge of the innovative activities of competitors will also be useful. But a business vision will only come as a result of the creative thinking of executives and others.

There needs to be an effective linkage between business strategy and business processes. Where strategy implies radical business change, this suggests radical changes to business processes and process innovation rather than the more usual incremental change. Examples given of such a vision include developing systems with a customer perspective, improving product quality and taking best practice from the industry. Process innovation may lead to more complex processes. Davenport argues that process simplification or rationalisation only leads to marginal change and therefore implies a lack of vision. A more radical vision, it is argued, will

imply objectives which might include cost reduction, time reduction, increasing the quality of products and empowering staff. Key activities in developing process visions include:
- Assess existing strategy with respect to processes
- Consult with process customers for performance objectives
- Set up performance objectives and functionality targets.

2 Identify the processes to be re-engineered

At this stage the major processes are identified, along with their boundaries. The critical processes of the organisation are considered for IT-enabled re-engineering. Processes which are of high impact, of great strategic relevance or presently conflict with the business vision in some way are selected for consideration and a priority attached to them. It is unlikely that they can all be re-engineered in parallel. There may be somewhere between ten and twenty processes identified for innovation. For businesses, these might include:
- Customer contact
- Inventory management
- Product design
- Personnel support
- Product marketing
- Production
- Supplier management
- Customer feedback
- Human resource management
- Financial management.

The processes are classified according to beginning and end points, interfaces, owners, functions, users and departments involved. It is important that the process owner, usually a senior manager, is motivated towards making the change. If there are difficulties in identifying these processes, senior manager workshops may help, as will interviewing senior managers. An alternative approach is to consider re-engineering all processes, but this may neither be feasible nor efficient. For example, there may be constraints preventing re-engineering of some processes due to the necessity of supporting some legacy systems. Some of these may need to be kept because of the degree to which they are embedded in the organisation. It is rare that a 'clean slate' may be assumed.

3 Understand and measure the existing process

Processes cannot be redesigned before they are understood. The present processes need to be documented. This will help communications within the group studying the process. It will also help to understand the magnitude of the change and the associated tasks. Understanding existing problems should help to ensure that they are not repeated. It also provides measures which can be used as a base for future improvements. For example, measuring the time and cost consumed by process areas that are to be redesigned can suggest initial areas for redesign in a process. However, although designers should be informed by past process problems and errors, they should work as if in virgin territory, otherwise, processes will be tampered with, rather than redesigned. Key activities in this stage are:

- Document the current process flow
- Measure the process in terms of new process objectives
- Measure the process in terms of new process attributes
- Identify problems and weaknesses of the existing process
- Identify short term improvements in the process.

4 Identify the IT levers

The accepted view in information systems is that the business requirements should be determined first before considering IT solutions. However, the benefits of simply automating existing processes are likely to be minimal. IT can be used to change processes completely. Davenport and Short argue that an awareness of IT capabilities can influence process redesign and should be considered in the early stages.

IT capabilities can enable better information access and co-ordination of processes. IT can make new process design options feasible, rather than simply to support them. One distinguishing aspect of this approach compared to most well-used information systems development methodologies, is that the latter concentrate on the development and implementation of information systems, whereas business innovation sees IT as the most powerful design tool providing opportunities for process re-engineering which is fundamental and broader.

A list of eight IT capabilities, along with organisational impacts, are suggested:

- *Transactional:* IT can transform unstructured processes into routine transactions
- *Geographical:* IT can transfer information rapidly and across long distances
- *Automating:* IT can reduce the need for human intervention in processes
- *Analytical:* IT can enable complex analytical methods to be incorporated into processes

- *Integrating:* IT can support the co-ordination of tasks and processes
- *Informational:* IT can bring in vast amounts of information to be included into a process
- *Sequential:* IT can reorder the operation of tasks and allow them to be processed in parallel where appropriate
- *Knowledge orientating:* IT can be used to capture and disseminate knowledge to improve the process
- *Tracking:* IT enables monitoring the status, inputs and outputs of tasks
- *Simplifying:* IT can be used to simplify communication so that, for example, intermediary stages are not required.

5 Design and build a prototype of the new process

In this final stage, the process is designed and the prototype built through successive iterations. Design comes from a review of the information collected in the first four stages. It is suggested that the design team consist of key process stakeholders as well as those from the IT side who debate possible design alternatives.

Key activities at this stage are:
- 'Brainstorm' design alternatives
- Assess feasibility, risk and benefit of design alternatives and select the preferred process design
- Prototype the new process design
- Develop a strategy for changing to the new process
- Implement the new organisational structures and systems.

Davenport suggests process design at three levels: a process level, sub-process level and activity level. At the process level, the inputs, outputs, interfaces, flow and measures are specified. At the sub-process level, the objective, performance metrics, the performers, IT enablers, information needs and activities in the process are defined. At the activity level, the information needed, decision points, the performers and value added are defined.

Through the use of CASE tools, IT is seen as a design tool to draw process models rapidly and generate any computer code necessary. Such a design needs to satisfy general design criteria, such as satisfying the objectives set, simplicity, control mechanisms and the generalisation of tasks which can be executed by more than one person. It is suggested that prototypes are more likely to lead to systems which are accepted by the users as well as being produced faster than the conventional approach.

Davenport discusses the potential role of IT in process innovation. As well as in the design and build stage where the impact of IT is most obvious both in the design and the prototyping stages (using tools suggested in Chapter 5), all the phases can be supported by information systems and information technology. Executive information systems should provide managers with information about current business performance and market factors. Computer supported conferencing may help in the brainstorming activities. When identifying the processes to be redesigned, information about the performance of present processes can be provided and simulation packages may help to identify alternative approaches which are potentially more successful.

Although the above would seem to concentrate on the role of IT in process innovation, Davenport in his 1993 book does not over-stress its importance and he also discusses the major role of organisational and people factors, for example, references (albeit brief) are made to the work of Checkland and Mumford. Greater empowerment and participation in decision making about process operations are suggested. He also stresses the importance of a team approach to process re-engineering and of human resource management generally. It can help in this process, for example, if project control techniques are used (section 5.2). On the other hand, there is an awareness of differences in companies and much anecdotal evidence is given showing how different factors played different roles in different situations. Indeed, the approach as a whole is more steeped in business experiences than theoretical depth.

6.14 RAPID APPLICATION DEVELOPMENT

The goal of rapid development of applications has been around for some time and with good reason, as the objective of speeding up the development process is something that has been on the agenda of both general management and information systems management for a long time. The need to develop information systems more quickly has been driven by rapidly changing business needs. The general environment of business is seen as increasingly competitive, more customer-focused and operating in a more international context. Such a business environment is characterised by continuous change, and the information systems in an organisation need to be created and amended speedily to support this change. Unfortunately, information systems development in most organisations is unable to react quickly enough and the business and systems development cycles are substantially out of step. In such a situation, the notion of rapid application development (RAD) is obviously attractive.

Once again, the exact definition or nature of the term is not clear, and authors and vendors use the term in a variety of different ways with different meanings and emphases. RAD is probably best known from the works of James Martin who has defined a RAD methodology which we discuss in this section. Martin is of course also known for the development of the Information Engineering (IE) methodology (section 6.4), and it comes as no surprise to find that his version of RAD is set firmly in the context of IE. However, other writers use the concepts of RAD within the context of other methodologies Yet others use RAD as a separate and stand-alone methodology in its own right. In the UK, a group of RAD users and suppliers have formed a consortium to define standards and a framework for RAD called Dynamic Systems Development Method (DSDM). One interesting aspect of this is the notion of a consortium of companies getting together to define a framework, or a methodology, for a particular type of systems development. This is a new and novel way for methodologies to emerge (see Chapter 7 for further discussion).

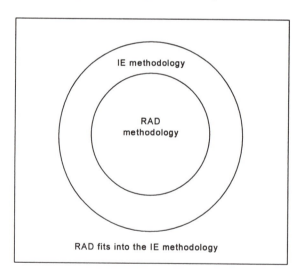

Fig. 6.53: Rapid application development and Information Engineering

It is early days for this consortium and DSDM, and there is little which is both defined and in the public domain as yet. Therefore for the purposes of this book we base our discussion of RAD on that of James Martin, the primary source being Rapid Application Development (Martin, 1991). As mentioned above, Martin's RAD is set firmly in the context of IE, and he illustrates this as in figure 6.53. As IE has already been addressed, we attempt to concentrate in this section, as far as possible, on RAD independently of any other methodology.

RAD is actually a combination of techniques and tools that are, for the most part, already well known and dealt with elsewhere in this book. We identify the following as the most important RAD characteristics:

- It is not based upon the traditional life cycle (section 2.2) but adopts an evolutionary/prototyping approach (section 3.11)
- It focuses upon identifying the important users and involving them via workshops at early stages of development

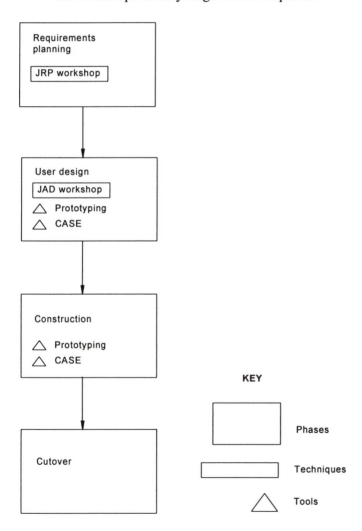

Fig. 6.54: The phases of the RAD approach

- It focuses on obtaining commitment from the business users
- It requires a CASE tool with a sophisticated repository (section 5.5 and 5.7).

RAD has four phases which we shall describe in turn (figure 6.54):

1. Requirements planning
2. User design
3. Construction
4. Cutover.

1 Requirements planning

RAD devotes a lot of effort to the early stages of systems development. This concerns the definition of requirements. There are two techniques used in this phase, both of which are really workshops or structured meetings. The first is joint requirements planning (JRP) and the second is joint application design (JAD). In some other RAD methods, these are not separated, indeed, Martin points out that they may in practice be combined with some of the functions of JRP being subsumed into JAD.

The role of JRP is to identify the high level management requirements of the system at a strategic level. The participants in JRP are senior managers who have a vision and understanding of the overall objectives of the system and how it can contribute to the goals and strategy of the organisation. If this understanding does not already exist, the workshop may be used to help determine such an understanding or vision. The JRP is a creative workshop that helps to identify and create commitment to the goals of the system, to identify priorities, to eliminate unnecessary functions and so on. Martin separates JRP from JAD because there are different people involved. In JRP, the participants need to have a combination of overall business knowledge and specific knowledge about the area that the proposed system is addressing along with its requirements. They also need to have the necessary authority and seniority to be able to make decisions and commitments. Applications often cross traditional functional boundaries, and ensuring the right people are involved is difficult but absolutely critical. Martin suggests that if the right people are not available, the workshop should not take place. The participation of substitute personnel, without the authority to make decisions, who have frequently to refer back to their superiors, could negate one of the main workshop objectives which is to get the requirements identified and agreed in the shortest possible time.

The detail of the workshops will be discussed in the next section as they are the same as JAD workshops.

2 User design

JAD is the main technique of the user design phase, indeed, it appears to contain little else. Readers will find many links with the participative theme (section 3.13) in JAD sessions. In fact, user design is in reality both analysis and design. As mentioned above, the JRP workshop may be combined with JAD in situations where the overall requirements are already well established. Normally, however, JAD would follow on from JRP. JAD adopts a top-down approach to user design and is based on the recognition that user requirements are difficult to understand and define, and that the traditional requirements analysis techniques of observation, interviews and questionnaires are inadequate. In JAD, the user design is achieved via the combination of the right people, the right environment and the use of good prototyping tools. The prototyping tool allows the quick exploration of processes, interfaces, reports, screens, dialogues and so on. Prototyping may be of the overall system or be used to explore particular parts of the system that are contentious or present particular problems. The user design is developed and expressed using the four diagramming techniques of entity modelling (section 4.5), functional decomposition (section 3.7), data flow diagramming (section 4.7) and action diagrams (section 4.11). The participants in the workshop need to be familiar with these techniques, but the emphasis is on getting the requirements as correct as possible and to reflect the business needs. Therefore, the language used in the workshop and expressed in the diagrams is that of the business and the users, rather than the more technical language of information systems. The results of the user design are captured in an Upper or Integrated CASE tool (section 5.7) which checks both internal consistency and that with other applications and corporate models. Where necessary, the terms used should be discussed and defined and entered into the repository of the tool. The use of a CASE tool enables the speedy, accurate and effective transfer of the results into the next phase, the construction phase.

The use of the diagramming and CASE tools, and to a lesser extent the prototyping tool, is specific to Martin's RAD methodology. In some other approaches the term JAD is used to indicate a specific type of meeting. In this sense, JAD is often used outside the context of RAD and is a useful technique for requirements analysis in general.

The underlying concepts of JAD as a facilitated meeting process are not new. According to Wood and Silver (1989) they were first developed by IBM in 1977 to help elicit requirements. However, it was not picked up by many other organisations until around the mid 1980s when it became popular. It has been estimated that since then there have been over 10,000

JAD-type meetings (Carmel and Nunamaker, 1992) to help design and define information systems. The original concept of JAD was not dependent on CASE tools, but was pencil and paper based.

The typical characteristics of a JAD workshop are as follows:

- *An intensive meeting of business users (managers and end users) and information systems people:* There should be specific objectives and a structured agenda, including rules of behaviour and protocols. The information systems people are usually there to provide assistance on technical matters, for example implications, possibilities and constraints, rather than decision making in terms of requirements. Non-participating observers may also be present. The number of people involved in the workshops varies and this will depend on the type of system, its complexity and its reach. Fifteen participants has been suggested as the ideal number. One of the most important people is the executive owner or executive sponsor of the system.

- *A defined length of meeting:* This is typically one or two days, but can be up to five. The location is usually away from the home base of the users, but most importantly, away from interruptions. Telephones and e-mail to the outside world are usually banned. The participants are expected to attend full-time and cannot drop in and out of the meeting.

- *A structured meeting room:* The layout of the room is regarded as important in helping to achieve the meeting objectives. The round table principle is usually employed, and the walls of the room are typically covered in white boards and pin boards. When CASE and other tools are employed, these are usually placed at the side with the ability to display output on large screens and print when necessary.

- *A facilitator:* This is a person who leads and manages the meeting. He or she is independent of the participants, and specialises in facilitation. This person may be internal to the organisation or brought in from outside, and will understand the psychology of group dynamics and the tasks that the participants are undertaking. A facilitator is responsible for the process and outcomes in terms of documentation and deliverables. He or she will control the objectives, agenda, process and discussion, using a variety of techniques to help move the meeting forward and achieve the objectives. Techniques such as brainstorming, reflection exercises and cooling breaks will be used.

- *A scribe:* This is a person (or persons) responsible for documenting the discussions and outcomes of the meeting (including the use of CASE and prototyping tools when available).

From these characteristics it can be seen that there are a number of principles underlying JAD. First, the user design should be moved forward as quickly as possible. There may be a series of JAD meetings (ideally two, although more may be necessary) which either address different parts of the design area or more commonly take the design from overview to more detailed levels. Often further work is carried out between the meetings, such as the preparation of more sophisticated prototypes, but decisions are taken only at the meetings. The proponents of JAD argue that it replaces cycles of interviews and meetings on an individual basis that normally take many months. This can significantly reduce the elapsed time required to achieve the design goals. In the traditional approach, meetings usually consist of a small group or are held on a one-to-one basis. When analysts find a conflict or discrepancy between users as to requirements or interpretations, they have to re-schedule all these meetings again to try to resolve things. It may be necessary to cycle round the groups more than once. Typically, this takes a great deal of time, because setting up meetings is notoriously difficult in most organisations. JAD seeks to overcome these kind of problems with one or two major workshops.

The second key element is getting the right people together for the workshop. The right people are all those with stakes in the proposed system, including end users, and those with the authority to make binding decisions in the area. This avoids all the time-consuming cycles that are encountered with traditional methods.

A third element is the commitment that the JAD meeting engenders. With traditional meetings, commitment is often dissipated over time and decisions may be taken off the cuff in small meetings where all information is not available and implications are not fully understood. With JAD, it is all out in the open and high profile. Decisions tend not to be taken lightly, but when they are made, they are made with conviction and commitment. In particular, because JAD focuses upon the benefits of the system for the business and users, the commitment is more marked and visible.

The fourth element is the presence of an executive sponsor. This is the person who wants the system, is committed to achieving it and is prepared to fund it. This person is obviously a senior executive who must understand and believe in the RAD approach and, due to the senior position can overcome some of the bureaucracy and politics that tend to get in the way of fast development.

Perhaps the most important single aspect of JAD is the facilitator. This person can make or break a workshop and is critical to determining whether the objectives are achieved. Apart from skills in handling JAD workshops, along with an understanding of group dynamics, it is the independence or neutrality of the facilitator, which is crucial. This enables facilitators to achieve more than any other stakeholder who might be regarded with suspicion by others. A facilitator is able to avoid, and smooth, many of the hierarchical and political issues that frequently cause problems, and will be free from the taint of organisational history and past battles.

3 Construction phase

The construction phase in RAD consists of taking the user designs through to detailed design and code generation. This phase is undertaken by information systems professionals using a CASE tool, for example IEF (section 5.7). Construction in Martin's RAD methodology is highly dependent on the presence of an IE-based CASE tool, and is performed by creating a series of prototypes which are then reviewed by the key users. Thus the screens and designs of each transaction are prototyped and the users then approve them. If they do not approve them, they will request changes and the process goes on through a series of iterations. By prototyping and the use of CASE tools, these iterations are achieved quickly and testing is enabled. Some of these key users will already have been involved in the earlier phase of user design. Construction is performed by small teams of three or four experts in the use of CASE tools. These experts are known as SWAT teams. SWAT stands for 'skilled with advanced tools' and the approach requires them to work quickly, making maximum possible use of re-usable designs that already exist. Teams are kept small so as to reduce the number of interfaces and interactions between people in the teams. One of the problems of traditional development is low productivity which, it is argued, results from the large teams of developers involved and the consequent large communications network and the number of communications. Normally there is a SWAT team member allocated to developing each transaction in a system. In practice, there is often only one developer for a particular part of the system and this reduces the number of potential interactions with other developers for the area to zero. Using this approach, it is argued that the core of a system can be built relatively quickly, typically in four to six weeks, and then it is progressively refined and integrated with other aspects developed by other team members.

Once the detailed designs have been agreed, the code can be generated using the CASE tool and the system tested and approved. Because of the way that the construction has occurred, there should not be any surprises to the users when they see the finished version. All associated documentation is then produced and database optimisation is performed.

4 Cutover

The final phase is cutover, and this involves further comprehensive testing using realistic data in operational situations. The users are trained on the system, organisational changes, implied by the system, are implemented, and finally the cutover is effected by running the old and the new systems in parallel, until the new system has proved itself, and the old system is phased out.

RAD adopts an evolutionary, or timebox approach, to development and implementation (section 7.9). Typically, it recommends implementation of systems in a 90 day life cycle. The objective is to have the easiest and most important 75 per cent of system functionality produced in the first 90 day timebox, and the rest in subsequent timeboxes. This forces users and developers to focus on only those aspects of the system that are necessary and probably most well-defined for development in the first timebox. Everything else is left until later. The knock-on benefit of this is that with experience and use of the basic system, developed in the first timebox, users often find that their requirements evolve in different directions to those originally envisaged. In other words, the benefits of an evolutionary approach accrue. The other advantage of the timebox is that it creates a focus on achieving an implementation in the specified period. In order to achieve this, the functionality must be trimmed accordingly. No simply 'nice to have' but not strictly necessary features can be included, and the development must be made achievable. The timebox approach contrasts with the traditional approach where every conceivable requirement is implemented together and the resulting complexity often causes long delays in implementation.

6.15 KADS

As indicated in section 3.14, where expert systems concepts in general are discussed, the development of expert systems has usually been on a somewhat *ad-hoc* basis with no well-established methodology available. Yet there is beginning to be a realisation that there is a need for such a methodology, and that from an organisational and managerial perspective there are many similarities between information systems development and expert systems development. There is, for example, a similar need for good project management, speedy development, a structured set of tasks, a sensible choice of application that fits with the overall strategy of the organisation, an effective analysis of requirements, good design, acceptable user interface and so on. The main difference is that expert systems additionally require a specific focus on the acquisition and representation of knowledge.

In this section we examine one attempt to define a more formalised approach to expert systems development on the basis that they are an important sub-set of information systems, and although expert systems methodologies are at a very early stage of development, they may have something to offer to more general information systems development. The approach we have chosen to describe is known as KADS (Wielinga *et al.*, 1993, De Greef and Breuker, 1992) which is probably the best known approach in Europe and it is claimed that KADS or its variants have been used in forty to fifty projects.

KADS is the outcome of a European Union research project (ESPRIT) of which the objective was to develop a comprehensive, commercially viable methodology for knowledge-based systems construction. The partners in the project were Cap Gemini Innovation (France), Cap Gemini Logic (Sweden), ECN (the Netherlands), ENTEL (Spain), IBM (France), Lloyd's Register (UK), Swedish Institute of Computer Science (Sweden), Siemens (Germany), Touche Ross (UK), University of Amsterdam (the Netherlands) and the Free University of Brussels (Belgium).

KADS adopts the view that developing an expert system is a modelling activity and that it is not the case that the system has to be filled only with knowledge extracted from a human expert. It is rather a computational model of desired behaviour which may reflect aspects of the behaviour of an expert. It is not the functional and behavioural equivalent of an expert, the system may actually do things in different ways and utilise different approaches to human experts.

Figure 6.55 illustrates the various models that are constructed in KADS. It also shows the relationships between the models in the sense that information in a higher-level model is used in the construction of the lower-level model. The model does not necessarily imply that the sequence has to be followed in the linear top-down fashion that the diagram suggests, in fact KADS advocates a spiral approach rather than a waterfall life cycle. The spiral model is illustrated in figure 6.56, and begins in the centre with the process circling around with each pass adding a degree of functionality or progress and only when a number of circles have been made is the process complete. The breakdown or decomposition of the process into the development of these different models is the way that KADS attempts to address the complexity of expert systems. A model is an abstraction that eliminates certain detail but concentrates on certain key features of the area being modelled. Wielinga *et al.* (1993) call this a 'divide and conquer' strategy which is a term used by James Martin in describing the same concept in the context of Information Engineering (section 6.4).

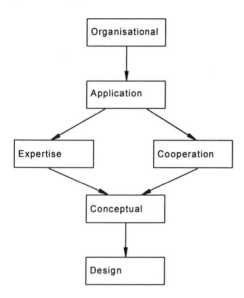

Fig. 6.55: Models and processes of KADS

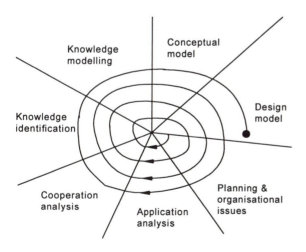

Fig. 6.56: Spiral life cycle model of KADS

1 Organisational model

The first step of KADS is concerned with defining the problem that the expert systems (or knowledge-based system in KADS terms) is addressing in the organisation. This step consists of the construction of the organisational model, which is a socio-organisational analysis of the environment. This is essentially a high-level model of the functions, tasks and problems in the organisation, including an assessment of the effects of the introduction of the expert system on the organisation and the people. Unfortunately, the form that this model takes is not specified in detail but it is a recognition by KADS that expert systems do not exist in an organisational vacuum and that organisational issues are important to the success of expert systems.

2 Application model

Next an application model is constructed which narrows the focus from the organisation to the area of application, and defines what problem the system is going to solve, and the overall function and purpose of the system. Again, we are not told the detail of this stage nor the exact form of the model.

3 Task model

The next stage is to develop the task model which decomposes the overall function of the system described in the application model into a series of tasks and sub-tasks that the system will perform. Each task is defined by the information input to the task and the goal that is achieved by the task. These are known as input/output specifications. The tasks are then assigned to agents which might be the expert system, the user or some other system. This will depend on what or who could best perform each task. Finally, any constraints or specific requirements for the tasks are defined, for example, that the output of the task must be in the terminology of a particular type of user.

4 Model of co-operation

The next step is to define the model of expertise and the model of co-operation. The model of co-operation is the definition of how the system and the user will work together at the task level and interact when using the system in various modes, for example, in solving a problem, seeking advice or requesting an explanation. This is an important stage because in practice the execution of an expert system usually requires a complex set of interfaces and interactions between users or groups of users, and the system. This is typically more complex than that for a traditional information system.

5 Model of expertise

Next the model of expertise is developed. This is the key task in the methodology and the model of expertise is effectively the functional specification of the problem-solving part of the expert system. Unlike many expert systems approaches, this model is a specification of the desired behaviour and the types of knowledge involved, rather than the specification of detailed rules.

In KADS, a process of knowledge identification is undertaken as a preparation phase before construction of the model of expertise. This is a kind of systems investigation of the area of concern prior to the building of the model. Data is collected and tasks identified using a variety of techniques, for example, structured interviews, rational task analysis, work flow analysis and repertory grid analysis. A glossary and lexicon of terms are also developed to help understand and document the area of concern prior to any formal conceptualisation of knowledge. After this, a knowledge modelling phase is undertaken, which involves the identification and definition of four types or layers of knowledge. Some typical techniques and representations that might be used in this process are tree diagrams, decision trees, laddering, data modelling diagrams, think-aloud protocols and scenario simulation. KADS suggests that a prototype of the problem-solving aspect of the system can be implemented as a way of validating the requirements and the knowledge models.

The first type of knowledge identified is the *domain knowledge* which involves the identification of concepts, properties, relationships between concepts and relationships between properties and structures. Concepts are things or objects in the knowledge domain. For example, in the domain of credit rating for loans in a banking application, a concept might be customer, account or application. In practice, these concepts should be as low-level or elementary as possible. A property is an attribute of a concept, for example, a customer might be active or passive, or an application significant or not. Relationships between concepts are identified. For example, an applicant is a customer. Causal relationships between property expressions are also identified, for example, a customer with a transaction in the last year causes customer to be active or a value over £100,000 causes the application to be significant. Structures are also identified, and this is a way of breaking complex objects down into more manageable components. In practice, an account might be broken down into current or deposit, and deposit into short-term and long-term, open or closed and so on. The examples used are illustrative and would in practice be at more elementary levels.

The second type of knowledge defined is *inference knowledge*. Inference knowledge is that which uses the domain knowledge to infer or produce additional information. KADS advocates the separation of domain knowledge from inference knowledge as a matter of philosophy, on the basis that it allows multiple use of the same domain knowledge. Inference knowledge is identified in the form of 'knowledge sources'. These are the processes that generate elementary pieces of information using domain information. For example, within a specific domain, the process by which a particular piece of information is compared to another piece of information in order to see if they are similar, might be a knowledge source, or in terms of our earlier example, the action or reasoning in the rejection of a loan application might be an inference knowledge source. Knowledge sources use domain knowledge, but the process is defined as independently as possible of the information in the knowledge domain and in principle could be applied to a variety of different situations. An extension of this idea is the generation of 'interpretation models' of typical inference knowledge for a particular task which might be reused in a different domain. For example, a credit assessment interpretation model might form a template in the alternative domain of assessment for tenancy agreements. Such interpretation models would guide the knowledge engineer by providing a template for knowledge acquisition in the new domain and perhaps also save significant development time.

The third type of knowledge is *task knowledge* which is information about how the elementary knowledge sources can be combined to achieve a particular higher-level objective. These higher-level processes are called tasks, for example, the combination of various knowledge elements might be the definition of the task of verification of a hypothesis. The tasks are decomposed into a number of sub-tasks which may be inference tasks, problem-solving tasks or transfer tasks. A transfer task is one that requires interaction with an external agent. The structure of the sub-tasks and the control dependencies are described using structured English (section 4.10). Due to the independence of the knowledge sources from the domain knowledge, the task knowledge is also independent and may be thought of as representing fixed strategies for achieving problem-solving goals.

The fourth category of knowledge is *strategic knowledge* which identifies the strategies, goals, high-level rules and tasks that are relevant to the solution of a particular problem. Although the need for strategic knowledge is identified in KADS, little work has been done on the definition of such knowledge, and Wielinga *et al.* (1993) concede that most systems developed with KADS have identified little or no knowledge of this type.

The description of the categories of knowledge relevant to KADS is described in what information systems people might feel is a bottom-up way, and that it might be better if a top-down view were adopted. Figure 6.57 provides this architecture, showing the four levels and their interactions.

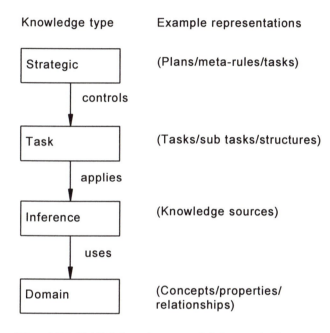

Fig. 6.57: KADS four layer model for modelling expertise

1 Conceptual model
The conceptual model is the combination of the model of co-operation and the model of expertise, and together this provides an application-independent specification of the system to be built.

2 Design model
The design model introduces the computational and representational techniques into the conceptual model. For example, the constraints of the hardware, software and the external world are now defined for the purposes of implementation. This is similar to the separation of the logical and physical models familiar in much information systems development but has not usually been present in approaches to expert systems development. However, the transformation from the conceptual model to

the design model is by no means automatic or indeed well defined. In principle, the designer in KADS is free to proceed in whatever way is thought best providing the conceptual specification is achieved. A completely independent approach can cause problems because it is then difficult to provide explanations of the systems reasoning in domain and conceptual language terms, unless there exists some clear linkages. KADS makes a few recommendations, but the process is essentially still in development and the subject of research. For inference knowledge, a computational technique that can realise the inferences needs to be selected. This requires elements that support a) algorithms, b) input-output data structures, and c) a representation of domain knowledge (usually via production rules, as described in section 3.14). For domain knowledge, a database is required that supports all the domain knowledge elements, for example, concepts, properties and relationships. Most conventional databases are not adequate in this respect. For task knowledge, a control technique for executing tasks is required, which might be a blackboard (section 3.14) or even a simple procedure hierarchy. For strategic knowledge, a production system that handles meta-rules is suggested.

KADS is by no means a mature practitioner-based expert systems development methodology. It is at an early stage of development and is essentially still the output of a major research project. As a result, it tends to focus on principles and models rather than the processes, phases and steps that need to be followed and the tasks that need to be undertaken. Future developments in the KADS-II project include a tool to support the prototyping of knowledge and a CASE tool for the methodology as a whole. There exists a product called Shelley (Anjewierden et al., 1992) which supports some activities of KADS including, for example, a domain text editor, a concept editor and an inference structure editor.

KADS is by no means comprehensive, there are many aspects that are not fully developed and this will no doubt also be addressed in KADS-II. The aim is to make KADS the de facto European standard for expert systems development. However, KADS is a welcome approach in an area that has lacked much emphasis on methodological issues, and it does help expert systems developers to consider and focus upon areas that they might otherwise have neglected.

6.16 EUROMETHOD

With the introduction of the single European market in 1992 and the removal of various barriers to trade, it was predictable that at some point the European Commission would turn its attention to service and procurement standards in the information systems arena. The lack of standards in the area and the fragmentation of the information systems services and tools market place is perceived as a barrier to open competition and the principles of the single or open market. In 1989, the European Commission (EC) established Euromethod as an initiative to facilitate cross-border trading within the European Union (EU) in the acquisition of information systems. Clearly it was also intended to enhance and promote the overall competitiveness of the European information systems industry in a global context and thereby European industry in general. The description below is based on CCTA (1990, 1994a and 1994b), Jenkins (1994) and Stewart (1994).

The objective of Euromethod is to provide a public domain framework for the planning, procurement and management of services for the investigation, development or amendment of information systems. This framework and associated standards will, it is hoped, help overcome the problems posed by the current diversity of approaches, methods and techniques in information systems and help users and service providers to come to common understandings concerning requirements and solutions in information systems projects.

The focus of Euromethod is on the market place, and it seeks to smooth the path for those requiring and procuring information systems services and those potentially providing such services. It seeks to make all suppliers potentially competing on an equal footing, no matter which European country they are from, by providing a common terminology that can bridge the different cultures and methods employed across the EU member states. Euromethod addresses only those arrangements where there is likely to be a contractual relationship between a customer and supplier. It does not address the situation where information systems services are provided to users by an in-house IT facility or department.

The development of Euromethod commenced in 1989 with Phase 1 which was the establishment of the requirement and the proposed programme. Phase 2 in 1991 was a feasibility study to scope the project, and Phase 3 was started in mid-1992 to be the development of Version 1 of Euromethod. Phase 3 was quickly recognised to be a massive undertaking and was divided into two sub-phases. Phase 3a is to develop a prototype Version 1 of Euromethod and if this proves successful then the second sub-phase would be the development of the complete version of Euromethod. Euromethod aims to be compatible with the main European information systems methods and is intended to cover the life cycle at a high level with the requirements specification stage being addressed in more detail. The work is divided into three workpackages and contracted to a consortium of major European IS suppliers as follows:

- Sema Group (France) the main contractor and project leader
- British Telecom (UK)
- Cap-Volmac (the Netherlands)
- CGI (France)
- Datacentralen (Denmark)
- Eritel (Spain)
- Finsiel (Italy)
- INA (Instituto Nacional de Administrao) (Portugal)
- Softlab (Germany)
- EMSC (a Bull, Olivetti, Siemens consortium).

Workpackage 1 consists of the analysis of seven methods, common in Europe, intended to form the basis of Euromethod. These selected methods were:

- SSADM from the UK
- Merise from France
- DAFNE (DAta and Function NEtworking) from Italy
- SDM (System Development Methodology) from the Netherlands
- MEIN (MEtodologica INformatica) from Spain
- Vorgehensmodell from Germany
- IE (Information Engineering) from the UK/US.

SSADM, Merise and IE have been described in this chapter. SDM and Vorgehensmodell are not full methodologies. SDM is more of a framework for project management than a systems development methodology and is typically complemented by other methods, such as, ISAC. It is well known in the Netherlands and is a *de facto* standard with national usage of between 40 and 70 per cent (CCTA, 1990). Vorgehensmodell is also a framework rather than a methodology. DAFNE and MEIN are from countries which are relatively low in their use of methodologies and are not well known even in the host countries.

Workpackage 2 is the design of the initial Euromethod architecture, including the definition of customer and supplier roles and a draft dictionary of concepts. Its emphasis changed and amended to focus on the customer/supplier relationship and the specific support of the tendering process. Workpackage 3 is the production of the prototype or initial version of Euromethod Version 1 which was delivered to the Euromethod Programme Management Board in April 1994. Version 1 is now the subject of trials in various European countries. The results from the trials will be fed back to the Euromethod Project Group and will help shape the final Euromethod Version 1 which it is hoped will be released in about 1996.

The deliverables of the initial Version 1 clarified a number of issues. First, the scope of Euromethod is defined as:

- All stages of procurement through to completion of an information systems adaptation
- The planning and management of the adaptation project.

Second, a number of principles of Euromethod are identified.

An information systems adaptation, to which Euromethod might apply, consists of any development or modification, of an information system, including organisational, human and technical elements, providing that the initial (or current) state and the required final state can be defined. For example, an information system adaptation might be a feasibility study, a system design, an enhancement or a reverse engineering project. Thus Euromethod can be applied at any stage of a project and applied many times in a development.

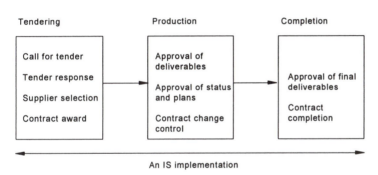

Fig. 6.58: Types of transaction in an IS adaptation

Fig. 6.59: A hierarchy of deliverables

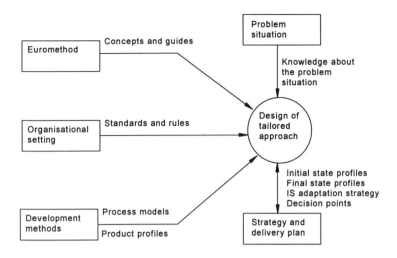

Fig. 6.60: The planning process of Euromethod

Euromethod focuses on the understanding, planning and management of the contractual relationships between customers and suppliers of information systems adaptations. The key transactions required between customers and suppliers are identified. Figure 6.58, for example, summarises the high-level transactions required in the process of tendering, production and completion of an information systems adaptation. This focus results in Euromethod identifying and defining the deliverables required between customers and suppliers. Each transaction implies one or more decision points, each of which needs supporting by a deliverable. Euromethod identifies a hierarchy of deliverables as in figure 6.59. The target domain is the part of the organisation affected by the information systems adaptation and the information systems descriptions are the documentation designs and specifications, whereas the operational items are deliverables that can be installed as part of the information system itself. This could be a screen or a prototype. The project domain is the temporary organisation established to adapt an information system and manage the process. The deliverables on this side are divided into plans for the production process and reports to manage the process. The delivery plan is the key definition of the process of defining the customer/supplier transactions in the overall context of the information systems adaptation, that is, the problem situation. Figure 6.60 represents the planning process to produce the delivery plan.

Euromethod recognises that different types of information systems adaptations require different approaches depending on the situation.

Differences may be due to the complexity or uncertainty of the environment. Euromethod provides guidelines and support for a variety of different situations. Guidelines exist for the identification of important situational factors, the assessment of complexity, uncertainty and risks, and the definition of a strategy for the IS adaptation. The delivery planning process of figure 6.60 illustrates the context of this situation-driven approach.

Euromethod is not another information systems development methodology but is a framework. It seeks to provide guidance as to how to map or bridge a particular method into the basic Euromethod framework. In this sense a method might include a European systems development methodology, for example, SSADM, a project management approach, for example, PRINCE, or a procurement standard, for example, TAP (Total Acquisition Process). (TAP is the UK public sector preferred procurement method for IS/IT services). Guidelines are provided that help bridge between Euromethod concepts and those of the specific method.

These principles result in three Euromethod models. These are the transaction model, the deliverable model and the strategy model. The transaction model helps understand, manage and facilitate customer/supplier interactions across organisational boundaries during an information systems adaptation. It identifies a set of generic customer/supplier transactions in information systems adaptations, including, for each, the goals, the key roles and responsibilities in both the customer and supplier organisations. The deliverable model defines the target domain for an information systems adaptation, what an information system is, the scope of the information systems adaptation and the essential properties. The essential properties are expressed as a set of views as follows:

- *The business information view:* the relevant knowledge about the information resource
- *The business process view:* the relevant knowledge about processes and their use of the information resource
- *The work practice view:* the relevant knowledge about individual actors and their use of the information resource.

These views are further developed and the computer systems required to support these essential properties are identified. The business information view is developed to identify the computer systems' data view, the business process view to the computer systems' function view, and the work practice view to the computer system architecture view.

The strategy model reflects the situation-driven principle above and helps define the approach required which reduces, or contains, the risks of a particular problem situation. The strategy options relate to systems installation, systems construction, systems description (or modelling) and project control.

In practice, the development of a Euromethod delivery plan involves the following tasks:

- Assessment of the problem situation
- Definition of a strategy
- Definition of decision points

The first stage, assessment of the problem situation, helps to clarify the requirements of the information systems adaptation for both customer and supplier. First, the initial state of the adaptation, that is the current state of the existing information system, is defined. A good definition of the initial state is crucial to enable the supplier to understand the starting point of the adaptation. Second, the desired final state of the adaptation is defined. This includes the elements of the initial state that are to be changed, the additional items to be produced, the documentation, the data, the interfaces and so on. The initial and final states are defined in terms of Euromethod essential properties and the views that they represent (see above). As various existing methods and methodologies will have already been defined in these terms, this should enable each final state deliverable to be assessed as to which approach might be most appropriate for its production.

This first stage also involves the determination of the situational factors and an assessment of their complexity and uncertainty. Such factors might include the complexity of the business processes, the heterogeneity of the users, the attitude and ability of the actors involved, the stability of the environment and the requirements and the quality of the specifications.

These factors are now used for the next task, to define an appropriate strategy for the production of the final state deliverables based upon the strategy model. For example, the strategy option chosen for system construction might be evolutionary and participatory in situations where the complexity of requirements is judged to be high and the understanding of the business processes is low. In practice, the analysis is more sophisticated than this, as Euromethod provides heuristics for the suitability of the approach and examples of how to reduce risk.

Finally, once the strategy has been defined, the decision points are identified. These are a by-product of the deliverables exchanged between customers and suppliers and their identification is based on the transaction model. Euromethod identifies three categories of decision: investment, design and systems acceptance. For each decision point, the decision to be made is identified plus the roles of those involved, the transactions they represent and the deliverables that are exchanged. The full documentation of each deliverable is made according to Euromethod product profiles. Both project and target domain deliverables are identified. The profile for target domain deliverables consist of:

- The deliverable type
- The information system view, that is, the essential properties of the transaction model as described above
- The computer system view
- The IS properties within each view
- Non-functional properties
- Scope (of the deliverable)
- State (of the deliverable)
- Degree of formality (of the deliverable).

Some of this information is similar in type to that required for initial and final states, and this reflects Euromethod's approach to standard terms and definitions. These concepts are again used in the method's bridging process and enable relevant external methods for the decision points and deliverables to be identified and used.

The sequence and schedules of the decision points are also identified and the total documentation provides the basis of the contractual agreement for the information systems adaptation between customer and supplier.

Version 1 of Euromethod is not yet released, so it is difficult at this early stage to assess its potential impact. The factors that suggest it could be significant are that it is a Europe-wide initiative that the EC have decided is important and in which they have invested. Its main impact will obviously be in the public sector, and although currently the initial version of Euromethod will not have the status of a formally recognised standard in the EU, if it were eventually to receive this status it might mean that no information systems procurement could take place in the public sector without using Euromethod. This would obviously make it extremely significant and that the private sector vendors would have to embrace it if they wished to win any public sector contracts. In the UK, the government is pushing for the public sector to 'market test' many of its services, including IT, against the offerings of the private sector. Against such a background, Euromethod might prove very effective.

Additionally, the outsourcing of IT activities and services is increasingly being considered as a legitimate option in managing and organising IT in many private sector companies, and although Euromethod has not been designed to address outsourcing as such, it could relatively easily be amended to do so. In such circumstances, it could be envisaged that Euromethod might become a significant method.

On the negative side, Euromethod might be viewed as an expensive aberration by a European bureaucracy that is intent on interfering in issues that are best left to market forces, and that they will shortly get bored or find something new with which to occupy themselves, and Euromethod will wither and die. It may be argued that there is already some evidence of this. The Euromethod Programme Management Board itself commented on the 800 pages of documentation that the initial version of Euromethod required and the complexity that this implied.

There is also concern relating to the trials of Euromethod. Few companies will be able to devote the necessary effort and resources to testing. Further, it is feared that even if organisations can be persuaded to trial the method, they may encounter too many problems and reveal Euromethod to be impractical. Another concern relates to the foundation on what some people regard as outdated methodologies, such as SSADM and Merise. Although IE has been included in the conceptual base, there is little evidence of it having had much impact on Euromethod but even more significant has been the absence of the more recent approaches, such as object-oriented analysis and design and business process re-engineering. The fear is that Euromethod has been saddled with a structure that will mean it is obsolete by the time it is released.

The concept of method bridging is clearly attractive, and it enables Euromethod to be seen not as another information systems development methodology but as a framework which can integrate existing methodologies. However, there is concern as to the practicality of this bridging. The idea is to bridge at the methodology level which may be too high a level, it may make more sense to bridge at the level of individual techniques in a methodology. Currently the Euromethod philosophy is to standardise at the deliverable level without worrying about the techniques used in the creation of the deliverables.

The reality will, as usual, probably be somewhere in the middle of these two views, but only time will tell.

FURTHER READING

STRADIS
Gane, C. and Sarson, T. (1979) *Structured Systems Analysis: Tools and Techniques.* Prentice Hall, Englewood Cliffs, New Jersey.

Yourdon Systems Modelling
Yourdon Inc. (1993) *Yourdon Systems Method: Model-Driven Systems Development.* Yourdon Press, Englewood Cliffs, New Jersey.

Information Engineering
Martin, J. (1989) *Information Engineering.* Prentice Hall, Englewood Cliffs, New Jersey.

SSADM
Eva, M. (1994) *SSADM Version 4: A User's Guide.* 2nd ed., McGraw-Hill, Maidenhead.
Weaver, P. L. (1993*) Practical SSADM Version 4: A Complete Tutorial Guide.* Pitman, London.

Merise
Quang, P. T. and Chartier-Kastler, C. (1991) *Merise in Practice.* Macmillan, Basingstoke (translated by D. E. & M. A. Avison from the French: *Merise Appliquée.* Eyrolles, Paris, 1989).

JSD
Cameron, J. R. (1989) *JSP & JSD: The Jackson Approach to Software Development.* IEEE Computer Society Press, Los Angeles.
Jackson, M. (1983) *Systems Development.* Prentice Hall, Hemel Hempstead.
Sutcliffe, A. (1988) *Jackson Systems Development.* Prentice Hall, Hemel Hempstead.

Object Oriented Analysis
Coad, P. and Yourdon, E. (1991) *Object Oriented Analysis.* 2nd ed., Prentice Hall, Englewood Cliffs, New Jersey.

ISAC
Lundeberg, M., Goldkuhl, G. and Nilsson, A. (1982) *Information Systems Development - A Systematic Approach.* Prentice Hall, Englewood Cliffs, New Jersey.

ETHICS
Mumford, E. (1989) *XSEL's progress.* Wiley, Chichester.
Mumford, E. (1995) *Effective Requirements Analysis and Systems Design: The ETHICS Method.* Macmillan, Basingstoke.

SSM

Checkland, P. (1981) *Systems Thinking, Systems Practice.* Wiley, Chichester.

Checkland, P. and Scholes, J. (1990) *Soft Systems Methodology in Action.* Wiley, Chichester.

Multiview

Avison, D. E. and Wood-Harper, A. T. (1990) *Multiview: An Exploration in Information Systems Development.* McGraw-Hill, Maidenhead.

Process Innovation

Davenport, T. H. and Short, J. E. (1990) The new industrial engineering: information technology and business process redesign. *Sloan Management Review*, 31, 4.

Davenport, T. H. (1993) *Process Innovation: Reengineering Work Through Information Technology.* Harvard Business School, Boston.

Rapid Applications Development

Martin, J. (1991) *Rapid Application Development.* Macmillan, New York.

KADS

Wielinga, B. J., Th. Sterner, A. and Breuker, J. A. (1993) KADS: A modelling approach to knowledge engineering. In: Buchanan, B. G. and Wilkins, D. C. (eds.) *Readings in Knowledge Acquisition and Learning, Automating the Construction and Improvement of Expert Systems.* Morgan Kaufmann, San Mateo, California.

Euromethod

CCTA (1994a) *Euromethod Overview.* CCTA, Norwich.

CCTA (1994b) *Using Euromethod in Practice.* CCTA, Norwich.

Chapter 7
Methodologies: Issues and Frameworks

7.1 THE METHODOLOGY JUNGLE

In the first edition of this book, the area of information systems development methodologies was described as a jungle and we concluded that this was unlikely to change much in the future due to the continuing developments in information technology, information systems and organisations as a whole. Unfortunately, this prediction has been proved correct and the situation is currently even more confused than it was in 1988. At that time we quoted a study that suggested there were over 300 methodologies. In 1994 it has been estimated that there are over 1,000 brand name methodologies world-wide (Jayaratna, 1994). Although we are rather sceptical of such a high figure, there is no doubt that methodologies have continued to proliferate. Many of these are of course similar and are differentiated only for marketing purposes. Others are internal to individual companies and have been developed in-house. Nevertheless, there is a large and confusing variety of methodologies in existence.

Once again this chapter will attempt to review developments and discuss the various issues as a way of providing some help in understanding this 'jungle'. The treatment is theoretical and philosophical as well as practical. The nature of such discussions is that there are few hard and fast facts and much of the chapter involves interpretations and opinions. The authors hope that these are based on sound judgement and experience, but they also recognise that there will be those that make different interpretations. Nevertheless we hope that this chapter contributes to informed debate concerning methodologies.

7.2 WHAT IS A METHODOLOGY?

In Chapter 1, we provided a working definition of the term 'methodology', and although this has been adequate for our purposes, we will now look in more depth at the question 'what is a methodology?'. The term is not well defined either in the literature or by practitioners. There is very little agreement as to what it means other than at a very general level. The term is used very loosely but also very extensively.

This loose use of the term does not of course mean that there are no definitions, simply that there are no universally-agreed definitions. At the general level, a methodology is regarded as a recommended series of steps and procedures to be followed in the course of developing an information system. In a brief *ad-hoc* survey, this proves to be about the maximum that people will agree to, and of course such a definition raises many more questions than it answers. For example:

- What is the difference between a methodology and a method?
- Does a methodology include a specification of the techniques and tools which are to be used?
- Does a collection of techniques and tools constitute a methodology?
- Should the use of a methodology produce the same results each time?

The questions that arise are fundamental as well as numerous. Unfortunately the problem will not be solved here, the most that can be achieved is that the issues will be aired. The information systems community has regularly debated the meaning of the term methodology in an information systems context, and as yet it has not come up with any universal definition.

One of the most useful definitions for the authors is that provided by the British Computer Society (BCS) Information Systems Analysis and Design Working Group as long ago as 1983. They defined an information system methodology as 'a recommended collection of philosophies, phases, procedures, rules, techniques, tools, documentation, management, and training for developers of information systems' (Maddison, 1983). Utilising this definition, we suggest that a methodology has a number of components which specify:

- How a project is to be broken down into stages
- What tasks are to be carried out at each stage
- What outputs are to be produced

- When, and under what circumstances, they are to be carried out
- What constraints are to be applied
- Which people should be involved
- How the project should be managed and controlled
- What support tools may be utilised.

In addition, a methodology should specify the training needs of the users of the methodology.

Apart from the above, we believe that a methodology should also specifically address the critical issue of 'philosophy'. We mean by this the underlying theories and assumptions that the authors of the methodology believe in, and that have shaped the development of the methodology. This identifies those sometimes unwritten aspects and beliefs that make a methodology an effective approach to the development of information systems in the eyes of their authors. We believe that the definition of a methodology should include specific reference to its philosophy as this has a critical bearing on the understanding of a particular methodology. An information systems development methodology is therefore much more than just a series of techniques along with the use of software tools.

In practice, many methodologies, particularly commercial ones, are products which are 'packaged' and might include:

- Manuals
- Education and training (including videos)
- Consultancy support
- CASE tools
- *Pro forma* documents
- Model building templates, and so on.

Some have argued (for example, Flynn, 1992) that the term methodology is not apt in the context of systems development and that the term 'method' is perfectly adequate to cover everything that we mean by a methodology. Indeed Flynn states that 'the term methodology was popular for a time in the 1980s' implying that it is no longer much used. This is contrary to our experience, although it is true that the term method is also used. For us, this seems to substitute one ill-defined word for another. We believe that the term methodology has certain characteristics that are not implied by method, for example, the inclusion of 'philosophy'. Methodology is thus a wider concept than method.

Checkland (1981) has distinguished between the two terms and says that a methodology:

...is a set of principles of method, which in any particular
situation has to be reduced to a method uniquely suited to that
particular situation.

In Checkland, (1985), he argues that information systems development
must be seen as a form of enquiry in the context of the general model of
organised enquiry, which consists of three components: an intellectual
framework, a methodology and an application area. This suggests a
hierarchy of elements to enquiry.

The first element is the intellectual framework, which comprises the
ideas that we use to make sense of the world. This is described as the
philosophy that guides and constrains the enquiry. It consists of
ontological assumptions, that is, beliefs about the fundamental nature of
the physical and social world and the way it operates, and epistemological
assumptions, that is, the theory of the method, or grounds of knowledge.
These terms are discussed in section 7.7. It also consists of ethical values,
which should be articulated, that may serve to guide or constrain the
enquiry.

The second element is the methodology. This is the operationalisation
of the intellectual framework of ideas into a set of prescriptions or
guidelines for investigation which require, or recognise as valid, particular
methods and techniques.

The third element is the application area, that is, some part of the real
world that is deemed to be problematical and worthy of investigation.

This is a useful framework for discussions of research and enquiry, but
we need to know how well it relates to the world of information systems
development methodologies. Working backwards, it would seem that the
application area is that of information systems development in general.
The methodology element is equivalent to the collection of phases,
procedures, rules, methods and techniques that are usually considered to be
a methodology in the information systems world. The intellectual
framework is the element that is often missing from methodologies, or
rather it is not missing, for it exists, but is not explicitly articulated. In our
definition, much of this intellectual framework element is included in what
we term the underlying philosophy, which we include in the definition of
methodology itself. Checkland has it as a separate element, and it
encompasses somewhat wider notions about the underpinnings of
knowledge and beliefs.

7.3 THE RATIONALE FOR A METHODOLOGY

Chapter 2 took a historical view of the demand or needs for methodologies, but it is worth stating more specifically what it is that people are looking for in a methodology. In other words, it is important to discover their rationales for adopting a methodology. Obviously this varies substantially between organisations and individuals, but we can identify three main categories of rationale: a better end product, a better development process and a standardised process.

1 A better end product

People may want a methodology to improve the end product of the development process, that is, they want better information systems. This should not be confused with the quality of the development process, addressed below. It is difficult to assess the quality of information systems produced as a result of using a particular methodology. For example, we cannot know that the use of the methodology produced the particular results. The same results might have occurred if the system had been developed using another methodology or without using a methodology at all. We do not usually have the luxury of developing a control with which we can compare our results.

Even if we had some way of comparing the results of using different methodologies, the elements that are perceived to constitute measures of quality differ considerably from person to person, and there is little agreement within the information systems community on this issue (see, for example, IFIP-NGI, 1986). The following represents our attempt to address some of the components of quality of an information system:

- *Acceptability:* whether the people who are using the system find the system satisfactory and whether it fulfils their information needs. This includes business users and managers and their requirements.
- *Availability:* whether it is accessible; when and where required.
- *Cohesiveness:* whether there is interaction between components (subsystems) so that there is overall integration of both information systems and associated manual and business systems.
- *Compatibility:* whether the system fits with other systems and other parts of the organisation.
- *Documentation:* whether there is good documentation to help communications between operators, users, developers and managers.
- *Ease of learning:* whether the learning curve for new users is short and intuitive.

- *Economy:* whether the system is cost-effective and within resources and constraints.
- *Effectiveness:* whether the system performs and operates in the best possible manner to meet its overall business or organisational objectives.
- *Efficiency:* whether the system utilises resources to their best advantage.
- *Fast development rate:* whether the time needed to develop the project is quick, relative to its size and complexity.
- *Flexibility:* whether the system is easy to modify and whether it is easy to add or delete components.
- *Functionality:* whether the system caters for the requirements.
- *Implementability:* whether the changeover from the old to the new system is feasible, in technical, social, economic and organisational senses.
- *Low coupling:* whether the interaction between subsystems is such that they can be modified without affecting the rest of the system.
- *Maintainability:* whether it needs a lot of effort to keep the system running satisfactorily and continuing to meet changing requirements over its life time.
- *Portability:* whether the information system can run on other equipment or in other sites.
- *Reliability:* whether the error rate is minimised and outputs are consistent and correct.
- *Robustness:* whether the system is fail-safe and fault-tolerant.
- *Security:* whether the information system is robust against misuse.
- *Simplicity:* whether ambiguities and complexities are minimised.
- *Testability:* whether the system can be tested thoroughly to minimise operational failure and user dissatisfaction.
- *Timeliness:* whether the information system operates successfully under normal, peak and every condition, giving information when required.
- *Visibility:* whether it is possible for users to trace why certain actions occurred.

Maximisation of all these criteria is, of course, not attainable, indeed, some actually work against each other. Ideally, an information systems methodology can be 'tuned' so that emphasis can be given to those criteria which are particularly important in the problem situation. Such tuning, or even any consideration of tuning, is rare in methodologies.

2 A better development process

Under this heading comes the benefits that accrue from tightly controlling the development process and identifying the outputs (or deliverables) at each stage. This results in improved management and project control. It is usually argued that productivity is enhanced with the use of a methodology, that is, we can either build systems faster, given specific resources, or use fewer resources to achieve the same results. It is sometimes also argued that the use of a methodology reduces the level of skills required of the analysts, which improves the development process by reducing its cost.

For some, the problems of developing information systems can be improved by adopting the quality standards that have proved popular in manufacturing and industrial processes, for example, BS5750/ISO9001. These standards are designed to ensure the quality of processes rather than the end product, and this sometimes leads to emphasis on conformity to the standard irrespective of whether this helps the quality of the system.

Another problem argued in Avison *et al.* (1994) is that the traditional manufacturing process is quite different to the process of developing software products. For example, the product of software is a 'one-off' rather than a mass replication of a design. There are schemes which recognise these difficulties and attempt to interpret the standards in methods applicable to software development. Although such standards have not yet made a large impact, either in terms of coverage or improved quality, they are helping to raise the issues and level of debate concerning quality in the development of information systems.

3 A standardised process

The needs associated with this category relate to the benefits of having a common approach throughout an organisation. This means that more integrated systems can result, that staff can easily change from project to project, without retraining being necessary, and that a base of common experience and knowledge can be achieved. In short, all the normal benefits of standardisation can be achieved, including the specific benefit of easier maintenance of systems.

All the reasons contained in the above categories have been specified, in some form or other, by the authors or vendors of methodologies as being benefits of adopting their particular approach. In contrast, a survey of benefits that purchasers look for in a methodology shows that the most important factors for them are, according to Gray (1985):

1.   Improved systems specifications.
2.   Easier maintenance and enhancement.

Methodologies rarely directly address improvement of the maintenance task, although vendors often claim it as an indirect benefit, because by using the methodology the information system developed will require less maintenance.

7.4 BACKGROUND TO METHODOLOGIES

In understanding any methodology, an important aspect is its 'background'. Broadly speaking, methodologies can be divided into two categories:

* Those developed from practice
* Those developed from theory.

The methodologies in the first category have typically evolved from usage in an organisation and then been developed into a commercial product. The second category of methodologies have typically been developed in universities or research institutions. These are usually written up in books and journals, although occasionally may have evolved into commercial products.

The commercial methodologies evolving from practice are the most widely known and some claim hundreds, if not thousands, of users. They each have a different history, but it was often typically the case that system developers in an organisation found that particular techniques they were using, or had helped to develop, were more useful than others, and they then concentrated on improving the use and effectiveness of these techniques. Typically, the people concerned would find that the organisation for which they were working was not interested in investing in the development of the technique. Sometimes this resulted in the developers leaving the organisation, and either setting up their own company or joining an existing consultancy company, where the opportunity to develop the techniques and methods was greater, using the clients as guinea pigs. At this stage, it was not the methodology itself, but the consultancy work developing information systems that was sold to clients.

Most of the early methodologies relied on one technique, or on a series of closely related techniques, as the foundation stone of the methodology. Commonly, these techniques were either entity modelling or data flow diagramming, but usually not both. It was only later that methodology authors began to include other techniques and to widen the scope and include prescriptions and phases or stages for development. Slowly these informal and somewhat *ad-hoc* procedures or 'cook-books' evolved into the early methodologies. From time to time the development of the methodologies would go through periods of introspection, where various aspects would be added and others dropped, and then the revised methodology would be put to the test, again by usage.

Sometimes consultants using the same methodology in a consultancy company would discover that they were doing things quite differently to their colleagues. They had their own styles and favourite elements, and yet they were supposed to be applying the same methodology. It was at this stage that some of these consultancy companies realised that it was no longer possible to rely on an ill-defined, inadequately researched, often conflicting, set of procedures and techniques. Rather than selling consultancy, the realisation began to dawn that the methodology itself had to be the product. In one organisation, this happened when it was discovered that no one person had responsibility for the methodology and its content. People in the organisation could add things to it as they thought fit, and they did. The main reason for this state of affairs was the nature of consultancy business. Consultants were charged out on the basis of the amount of time they spent on a client's project and there was no mechanism for accounting for time spent developing a methodology. Thus nobody was responsible for the methodology because at that stage it was not something that was sold

Eventually most organisations with a potential methodology product grasped the nettle and invested resources (people and time) into developing the methodology as a product. This involved ensuring that the methodology was:

- Written up
- Made consistent
- Made comprehensive
- Made marketable
- Updated as needed
- Maintained
- Researched and developed
- Evolved into training packages
- Provided with supporting software.

The consultancy houses had finally invested in their methodologies. As a result of this investment, a number of things have happened.

Filling the gaps: It was realised that most of the methodologies had some gaps in them or, if not complete gaps, they had areas that were treated much less thoroughly than others. This was usually as a direct result of their background of development which had typically involved a concentration on a single, or small set, of development techniques. These gaps needed to be filled because their clients assumed that the methodologies covered the whole spectrum of activities necessary. It was often quite a surprise for users to find that this was not the case. For example, a methodology based on entity modelling techniques might have been very powerful for data analysis and database design, but not so comprehensive when it came to specifying functions and designing the applications, and might not provide any support for dialogue design. Almost all information systems methodologies went through a process of filling the gaps and making the methodology more comprehensive.

Expanding the scope: Another process was that of expanding the scope of methodologies. This occurred because the methodologies did not address the whole of the life cycle of systems development. Frequently, the implementation phase was not included, some did not include design, others did not address analysis. So decisions were taken to expand the scope of the methodologies.

One of the most important aspects of this expansion of scope was for methodologies to expand into the areas of strategy and planning. The traditional life cycle of systems development usually started with a stage termed 'project selection' or 'problem identification', which was characterised by the identification of some problem to be solved, some area of the business where computerisation was a possible option or some application that needed building. The development process was viewed as a one-off, *ad-hoc*, solution to the identified problem, and whilst this may have been a reasonable approach in the early days of systems development, it was now no longer valid. Organisations had had many such 'identified problems' solved in this one-off manner and found that although the individual problem may have been solved to some extent, the existence of a variety of different one-off systems in a business did not lead to harmony nor any general improvement to business processes.

Further, it was realised that these individual problems were not so 'individual', and that almost all areas of the business needed to interact and integrate in some way. In particular, the requirements of tactical and strategic levels of management needed integration across traditional boundaries. A series of systems developed as individual solutions, at different times and without reference to each other, is unlikely to be the ideal starting point for such integration. Yet, for many organisations, this is just what existed, and methodologies were forced to address themselves to this situation if they were to provide more than improved implementations of one-off systems.

Thus, in order to achieve this integration, and because the market was demanding it, some methodologies turned themselves to the topic of information systems strategy. This involved the recognition that:

1. Information systems were becoming a fundamental part of the organisation, and that they could contribute significantly to the success, or otherwise, of the enterprise.
2. Information was increasingly being regarded by organisations as an important, and previously neglected, resource, in the same way as the more traditional resources of land, labour and capital.

It was realised that such a resource was not free, as had been previously assumed, but needed to be controlled, co-ordinated and planned. Further, the controlling and planning had to take place at the highest level in an organisation in order for these resources to contribute effectively to fulfilling the organisation's objectives. Thus, the starting point for effective information system development methodologies was now seen as a strategic plan for the organisation, including a specification of the way in which information systems would contribute to the achievement of that plan.

In practice, it was found that although most organisations, but by no means all, had some kind of strategic or corporate plan, this plan did not usually address the role of the information systems. For this reason, some methodologies incorporated the development of the required strategy plan into their scope. This not only helped to ensure that the information systems met the high level needs of the organisation, but effectively pushed the information system function up the hierarchy of importance in the organisation.

Gaining competitive advantage: Another reason for addressing information systems strategy at a high level in an organisation was the developing management belief that not only could the information system make the running of the business easier and more efficient, but that information systems and information technology could change the position of the organisation in relation to its business competitors. The idea that information systems could enable business change and advantage was prevalent, and this meant that most commercial methodologies introduced strategic and planning phases. This, it was argued, not only ensured that the business and information systems strategy was in alignment, but it also often led to the influencing and determination of business strategy by the identification of new opportunities that information systems could provide.

The process of expanding their scope and 'filling the gaps' has continued for most commercial methodologies. After the introduction of strategic and business planning phases and tasks, the next development was to integrate new and evolving techniques and approaches, such as state transition diagrams (or entity life cycles), the introduction of support tools into the methodology package and most recently the assimilation of object-oriented techniques.

As existing methodology vendors sought to fill the gaps and expand the scopes of their products, a number of other organisations entered the market place with new methodology products. These were only new in the sense that they sought to blend together what were seen as strong features of a variety of existing methodologies, in particular, the combination of entity modelling techniques with data flow diagramming techniques. A typical example is SSADM.

This process of expanding the scope and filling the gaps has resulted in many methodologies becoming extremely large, and often cumbersome. They seek to provide all things to all people, but in doing so they have perhaps lost their original specialist focus and sown the seeds for some of a growing discontent with methodologies that we shall examine later.

The second basic category of methodology has evolved from an academic or theoretical background. These are generally less well known than the commercial methodologies and may have relatively few users, although their influence is often substantially greater than their user base. Academic methodologies were usually developed by individuals who evolved and popularised the methodologies by means of action research and consultancy.

Typically these methodologies started life as research projects in universities or research institutions. Here the researchers took a more theoretical viewpoint and were less concerned by commercial considerations, although they often wanted access to real situations in order to test and experiment. The income from consultancy was certainly also no deterrent. What intrigued the academic researcher, in particular, was that there did not seem to be any standard techniques or approaches based upon sound theoretical concepts. This was clearly a challenge which was taken up by a number of people from a variety of different background disciplines. It was sometimes felt that methods were already available and successfully being used in other disciplines and that these only needed a small amount of adaptation to be useful in the area of information systems development. Mathematicians, psychologists, linguists, social scientists, engineers, sociologists and others turned their attention to the challenge. This did not of course happen all at the same time nor did many people actually get involved in numerical terms, but some of the approaches proved interesting and useful.

In areas other than information systems, the development of techniques and methods by academics has been very influential on practice, particularly in the areas of database design and software engineering. But in the case of information systems development methodologies, at least initially, academics were almost completely ignored.

The reasons for this initial slow take-up are debatable. Some argued that the research-based methods were not good enough, not practical enough or that they were no better than the new methods that the practitioners were already developing themselves. Some of the academic methodologies involved relatively revolutionary changes to current practice which could not be introduced on an evolutionary basis. As Bubenko (1986) observed:

> A very small fraction of these [academic methodologies]... have been applied to practical cases of a realistic size and complexity. The acceptance of 'academic' methods in practice is low and, in general, the rate of transfer of research results and 'know-how' from scientific research to industry is embarrassingly slow.

Some sought to overcome this problem by persuading organisations to try their methods under the control and guiding hand of the academics themselves on a consultancy basis. Others adopted a practice known as action research, which includes the experimentation with, and testing of, the methodology in a practical situation with the academic playing dual roles of participant and researcher.

As this process evolved, the methods became more practical, and in recent times there is evidence that academic methodologies are playing a more important role. The adoption of some elements from academic methodologies into commercial methodologies has sometimes happened, whilst others have attempted to combine the data and process techniques of entity modelling and data flow diagramming from commercial methodologies with academic methodologies, such as SSM, into a more comprehensive approach, for example, Multiview.

7.5 ADOPTING A METHODOLOGY

An organisation thinking of adopting or purchasing a methodology has a number of concerns. These relate to what they get and whether they are guaranteed successful information systems as a result.

What do they get? To address the first question, they get what the vendor or methodology author gives them, and this differs greatly:

- A methodology can range from being a fully fledged product detailing every stage and task to be undertaken to being a vague outline of the basic principles in a short pamphlet.
- A methodology can cover widely differing areas of the development process, from high-level strategic and organisational problem solving to the detail of implementing a small computer system.
- A methodology can cover conceptual issues or physical design procedures or the whole range of intermediate stages.
- A methodology can range from being designed to be applicable to specific types of problems in certain types of environments or industries to an all-encompassing general-purpose methodology.
- A methodology may be potentially usable by anybody or only by highly trained specialists or be designed for users to develop their own applications.
- A methodology may require an army of people to perform all the specified tasks or it may not even have any specified tasks.
- A methodology may or may not include CASE tools.

The variations on this theme are numerous.

It is clear that methodology adopters should be aware of what their needs are and choose their methodology accordingly. This does not always seem to be the case. One organisation that adopted a particular methodology found that they had to write detailed manuals themselves to specify what was required of their development staff. What they purchased was a management overview without any detail.

Some methodologies are purchased as a product, others are available by purchasing a licence, others are obtained through a contract for consultancy work, some come as part of the purchase of the CASE tool and some by a combination of the above. As can be imagined from this, the cost of methodologies varies considerably. Some are effectively free and this is the case for Merise. Often the initial purchase of the methodology product is the least of the investment and it is the training, additional hardware and software and ongoing consultancy costs that mount up and eventually dwarf the initial purchase price. What is also clear is that the potential organisational costs of adopting a methodology, particularly if it is adopted as a company standard across the board, are enormous. These organisational costs are not the purchase price, but the costs of embedding the methodology in the culture of the organisation, the opportunity costs of not doing something else, the costs of training and educating users and managers to participate in the use of the methodology and so on. There may also be large costs associated with changing methodologies, as once a particular methodology becomes embedded in an organisation it is not easy to change the development culture.

These costs are usually seriously underestimated in the evaluation of the cost of a methodology. It also indicates that the IT or systems development department should not necessarily be the ones to make decisions concerning the adoption of a particular methodology. It appears that in the majority of cases, the decision to adopt a standard methodology and the choice of the particular methodology is made by the IT/IS department. This is often because this department assumes it by its role and prerogative. Indeed, it has been suggested that this assumption of technical expertise is the structural basis of IS professionals' power in organisations (Markus and Bjørn-Andersen, 1987). It can legitimately be argued that the users, and business managers in general, should be making such decisions, as they are the ones that have to make the actual investment, not just in terms of money, but in time and effort, and ultimately business success, as a result of methodology decisions. However, the assumption that it is an IT decision is frequently accepted and encouraged by the rest of the organisation. Information systems development is often seen as a technical issue that the technicians should decide upon. After all that is what they are paid for!

However, the authors know of instances where IT departments have attempted to involve users and business managers in decisions concerning methodologies without success. The users and managers preferred to leave it to the IT department despite the potential significant implications on the degree of user involvement in development that the different methodologies under consideration embodied, and ultimately on the overall future success of the business. Equally, there are some cases where users and managers have not been consulted in the choice of methodology at all, and they have refused to co-operate in developments as a result.

Markus and Bjørn-Andersen (1987) argue that the choice of methodology is in practice made by corporate IS rather than by involving individual analysts who have to work with the methodology following the decision. Thus the power structure is argued to be conspiring against the users and the business managers, as discussed above, and against the information systems worker as well.

The answer to the second question posed above, 'are they guaranteed successful information systems as a result of using the methodology?', is clearly 'no'. However, this does raise the question of what a methodology is supposed to produce. If two teams of developers are using the same methodology on a similar project, in the same type of environment, can the same results be expected, and if not, why not?

Clearly the developers may interpret the demands of the methodology differently and thus end up with different results. The methodology may give a lot of leeway to the developers in terms of how they perform the specified tasks and so the results will be different. The developers may have varying skill levels, which will also produce differing results. However, it is sometimes argued that, given these variations, multiple uses of the same methodology in the same circumstances should yield roughly the same results. The tighter, more specific, the methodology, the more reproducible the results, particularly where the methodology specifies the exact techniques and tools to be employed under each circumstance.

This is a highly contentious area, but the implication is that if the results are to some degree reproducible, then we must be sure that the methodology specifies 'a best' way of producing information systems. We cannot say *the* best way, because there may be trade-offs between quality, quantity, the skill levels, reliability, generality and so on. But we want a methodology that will produce good results. This implies that a methodology is not just a helpful set of guidelines that enables the developer to organise the development process more effectively and efficiently, but that it embodies a good way of developing systems. If it is reproducible then it leads to the development of particular solutions, and thus if we adopt a methodology, then this methodology must be, to our minds, a good way of doing things. If it is not, it will lead to problems, because the nature of the methodology will exclude other ways of doing things. For example, the adoption of a data flow diagram-based methodology will, if it is rigidly followed, exclude the kind of analysis that the SSM methodology recommends. There will, for example, be no analysis of conflict between various actors in the existing system.

The debate is not a new one and, for example, we find White (1982) arguing that repeatability is what makes the adoption of a methodology so important:

> The acid test of a methodology is its repeatability. Given the same access to people and facts about the organisation, will

different designers come up with essentially the same solution? If the answer is no, then the business is placing itself in the hands of its systems builders and gambling on their flair and judgement to come up with a satisfactory approach. This is a risk that many organisations are unwilling to take as they progress towards their goal of becoming 'computerised companies'. Instead, they are looking for an approach which can be treated as an engineering discipline and which will provide them with the means of verifying the completeness and correctness of any analysis and of the assumptions made at each stage in the development process. It is this type of approach that will give us a 'repeatable' methodology.

The question of repeatability or reproducibility is obviously not one that can be easily tested in practice. It is impossible to create exactly the same environment in an organisation for two sets of developers to develop systems which we can then compare. Checkland (1987) highlights the problem by challenging developers of systems perceived to be successful to prove that another methodology would not have been better. And of course they cannot. He asks the developers of systems, where there have been problems, to *prove* it was not their incompetence that led to the problems rather than the methodology. Once again they cannot. It has been suggested that two development teams could work independently on developing a system to the same requirements in the same environment enabling a legitimate comparison of the systems at the end. However, the very fact that there are two sets of developers will undoubtedly influence the results. For example, the fact that one set will have talked to a user first will influence the results when the second set of developers talks to that same user. Even if flawed, the results of such experimentation would be interesting, but it seems the practical problems are insurmountable as the authors have no knowledge of any such attempt, let alone systematic study.

This lack of repeatability, or rather the ability to demonstrate repeatability, is often used to suggest that information systems is not really an engineering discipline. However, engineering is also a creative profession. Will two engineers design the same bridge or two electronic engineers design the same circuit for a particular function? Probably not, but the degrees of freedom in designing a bridge are greater than that of designing a particular circuit, that is, the two designs of the circuit will probably be more similar than the two designs of the bridge. With information systems, the degrees of freedom are typically even greater than for the bridge. The adoption of an information systems methodology can thus be argued to be an attempt to reduce the degrees of freedom.

If this is the case, then the implication still holds that the methodology should embody a best way of developing systems. This is not always appreciated when methodologies are being selected and may result in the development of inappropriate systems. The question whether the methodology embodies a best way of developing systems is rarely asked. The more usual questions are:

- Whether it fits in with the organisation's way of working
- Whether it specifies what deliverables (or outputs) are required at the end of each stage
- Whether it enables better control and improved productivity
- Whether it supports a particular set of CASE tools.

As has been observed earlier, there exist probably hundreds of different methodologies. This implies that there are hundreds of 'best' ways of developing information systems. Many of them are probably quite similar when closely examined, but many of them do differ substantially on fundamentals.

7.6 METHODOLOGY COMPARISONS

There are two main reasons for comparing methodologies:

- *An academic reason:* To better understand the nature of methodologies (their features, objectives, philosophies and so on) in order to perform classifications and to improve future information systems development.
- *A practical reason:* To choose a methodology, part of one, or a number of methodologies for a particular application, a group of applications, or for an organisation as a whole.

The two reasons are not totally separate, and it is hoped that the academic studies will help in the practical choices and that the practical reasons will influence the criteria applied in the academic studies. This synergy is a basis for some present-day research. In this section we look at a number of different approaches to the comparison of methodologies which have been attempted and discuss some of the issues that arise from such comparisons.

One of the earliest attempts to compare methodologies was the series of conferences known as CRIS (Comparative Review of Information Systems Design Methodologies). These conferences have been organised by the IFIP (International Federation of Information Processing) WG (Working Group) 8.1. The first conference in 1982 invited authors of methodologies to describe their methodologies and apply them to a case study. The chosen case study was the organisation of an IFIP conference (Olle *et al.*, 1982) which was specified to the authors. The most interesting outcome of this was to illustrate the wide differences between methodologies and what they produced from the same case. Of course a variety of assumptions had to be made by the authors in the interpretation of the case, but even so it was clear that even relatively similar approaches produced quite different results.

The second conference in 1983 consisted of a series of papers which analysed and compared the features of various methodologies (Olle *et al.*, 1983) and the third conference in 1986 addressed the issue of improving the practice of using methodologies (Olle *et al.*, 1986). A fourth conference was held in 1988 which examined the ways in which computers could be used during the construction of information systems (Olle *et al.*, 1988). A book based on the work of these conferences has since been published (Olle *et al.*, 1991).

These conferences have undoubtedly contributed to the knowledge and understanding of systems development methodologies, but as is inevitable in such an endeavour, they have also been criticised. First, they have failed to resolve any of the issues that they set out to achieve and second, that they have not been very influential in the practitioner world.

Catchpole (1987) makes a useful contribution by summarising the views of a number of authors concerning the important areas of concern when comparing methodologies, and suggests the following set of requirements for the design of a methodology:

- *Rules.* Tozer (1985) identifies the need for formal guidelines in a methodology to cover phases, tasks and deliverables, and their ordering, techniques and tools, documentation and development aids, and guidelines for estimating time and resource requirements.
- *Total coverage.* A methodology should cover the entire systems development process from strategy to cutover and maintenance.
- *Understanding the information resource.* Macdonald (1983) suggests that a methodology should ensure effective utilisation of the corporate information resource, in terms of the data available and the processes which need to make use of the data.
- *Documentation standards.* All output, using the methodology, should be easily understandable by both users and analysts.
- *Separation of logical and physical designs.* There should be a separation of logical descriptions and requirements (for example, what an application does, what the interactions are, and what data

are involved), that is, what the system is, from any specific technical or physical design solutions (for example, hardware, software, communication channels and documents), that is, how it is done.

- *Validity of design.* A means is required to check for inconsistencies, inaccuracies and incompleteness.
- *Early change.* Any changes to a system design should be identifiable as early as possible in the development process because costs associated with change tend to increase as the development work progresses.
- *Inter-stage communication.* The full extent of work carried out must be communicable to other stages. Yao (1985) argues that each stage has to be consistent, complete and usable.
- *Effective problem analysis.* Problem analysis should be supported by a suitable means for expressing and documenting the problems and objectives of an organisation.
- *Planning and control.* Careful monitoring of an information system is required and a methodology must support development in a planned and controlled manner to contain costs and time scales. It should be possible to incorporate a project control technique, although this need not necessarily be part of the methodology.
- *Performance evaluation.* The methodology should support a means of evaluating the performance of operational applications developed using it (Wasserman and Freeman, 1976).
- *Increased productivity.* This is a frequent justification for a methodology, sold to management in financial terms (Bantleman 1984).
- *Improved quality.* A methodology should improve the quality of analysis, design and programming products and hence the overall quality of the information system.
- *Visibility of the product.* A methodology should maintain the visibility of the emerging and evolving information system as it develops (Wasserman and Freeman, 1976).
- *Teachable.* Users as well as technologists should appreciate the various techniques in a methodology in order that they can verify analysis and design work.
- *Information systems boundary.* A methodology should allow definition of the areas of the organisation to be covered by the information system. These may not all be areas for computerisation.
- *Designing for change.* The logical and physical designs should be easily modified.

- *Effective communication.* The methodology should provide an effective communication medium between analysts and users.
- *Simplicity.* The methodology should be simple to use, for 'if a methodology is misunderstood, then it will be misapplied' (Tozer, 1985).
- *Ongoing relevance.* A methodology should be capable of being extended so that new techniques and tools can be incorporated as they are developed, but still maintain overall consistency and framework.
- *Automated development aids.* Where possible, automated aids such as data dictionaries and modelling tools should be used since they can enhance productivity.
- *Consideration of user goals and objectives.* The goals and objectives of potential users of a system should be noted so that when an information system is designed, it can be made to satisfy these users and assist them in meeting goals and objectives (Land, 1976).
- *Participation.* The methodology should encourage participation by such attributes as simplicity, the ability to facilitate good communications and so on.
- *Relevance to practitioner.* The methodology has to be appropriate to the practitioner using it, in terms of level of technical knowledge, experience with computers, and social and communications skills. Episkopou and Wood-Harper (1986) describe a framework for selecting an appropriate information systems design strategy based on assessing the characteristics of the problem solver and problem owner.
- *Relevance to application.* The methodology must be appropriate to the type of system being developed, which might be scientific, commercial, real-time or distributed (Macdonald, 1983 and Tozer, 1985).

Land has suggested that to this list should be added:

- *A systematic way of looking into the future.* This should enable possible future changes in requirements to be predicted and the system to be designed in such a way that future changes can be easily accommodated (Land, 1982, Fitzgerald, 1990).
- *The integration of the technical and the non-technical systems.* The methodology should not only address the technical and non-technical aspects of a system but should make provision for their integration.

438   Information Systems Development

- *Scan for opportunity.* The methodology should enable the system to be thought about in new ways. Rather than being viewed as simply a solution to existing problems it should be seen as an opportunity to address new areas and challenges. The scan should identify factors that inhibit this and find ways of overcoming them.

We would also add to Catchpole's list:

- *Separation of analysis and design.* This separation ensures that the analysis of the existing system and the user requirements are not influenced by design considerations.

Other commentators have taken these debates further, and suggested a much broader range of issues that they feel are relevant in the comparison of methodologies. Bjørn-Andersen (1984), for example, identifies a checklist that includes criteria relating to values and society:

1.   What research paradigms/perspective form the foundation for the methodology?
2.   What are the underlying value systems?
3.   What is the context where a methodology is useful?
4.   To what extent is modification enhanced or even possible?
5.   Does communication and documentation operate in the users' dialect, either expert or not?
6.   Does transferability exist?
7.   Is the societal environment dealt with, including the possible conflicts?
8.   Is user participation 'really' encouraged or supported?

Bjørn-Andersen's checklist is useful as it focuses attention on some wider issues that are often ignored. It is of course a subjective list and makes a number of assumptions, for example, that user participation is a desirable feature.

Bubenko (1986) identifies three alternatives to feature analyses which overcome the problems discussed above. These are as follows:

1.   Theoretical investigations of concepts, languages and so on.
2.   Experiences of actually applying the methodology to realistic cases.
3.   Cognitive-psychological investigations.

The first approach is argued to reduce the problems of subjectivity inherent in feature analyses by carrying out studies on well-defined, narrow, subject areas within specific terms of reference. Bubenko quotes Kung (1983) as an example of such a study. This concerns a comparison of the expressive power of various specification languages, and although the narrow focus makes the likelihood of objectivity greater, it seems that the same problems still apply. For example, Kung states that a conceptual model needs to be understandable, and he tests for various elements, one of which is unambiguity. However, we have already seen that the more unambiguous a specification, the more formal it needs to be. Unambiguity is therefore a constraint on informality, and this implies that natural language is not suitable. Thus, despite the study's narrow and carefully controlled domain, it is still highly subjective.

Comparisons of the second type are very difficult and resource-consuming, but can be useful. These are often known as case study approaches or action learning. The objectives of the study need to be fully explicit and adhered to in the performance of the study. It is extremely difficult to account for environmental factors such as an analyst's competence in such studies, and we run again into Checkland's 'proof' dilemma. It is also difficult to make a number of such studies comparable. A further problem is that many such studies are closely linked with consultancy activities, and it is not always the 'experiment' which is of prime importance. However, consultancy work might be the only way to gain real-world access.

Bubenko admits that his third comparison approach is virtually non-explored. He is thinking of areas such as software psychology and human factors in human-computer interaction, although he is aware of the difficulties that these kinds of studies present. He suggests some issues that might be addressed in this way. One is the question of whether graphical techniques give better 'understanding' than formal language approaches, often epitomised by the statement 'a picture is worth a thousand words'. He also adds the question 'what is understanding?', so it is clear that he has not underestimated the difficulties of this approach!

Another comparison framework is NIMSAD (Normative Information Model-based Systems Analysis and Design) (Jayaratna, 1994). This is based on the models and epistemology of systems thinking and to a large degree evaluates and measures a methodology against these criteria. The evaluation has three elements:

- The 'problem situation' (the methodology context)
- The intended problem solver (the methodology user)
- The problem-solving process (the methodology).

The evaluation of the elements is wide ranging and expressed in terms of the kind of questions that require answers. The questions concerning the first element (the problem situation) deal with the way in which the methodology helps the understanding and identification of the following:

- The clients and their understandings, experiences, and commitments
- The problem owners, their concerns and problems

- The situation that the methodology users are facing, its diagnosis as structured or unstructured
- The ways in which the methodology might help the situation
- The culture and politics of the situation, including the risks associated with using the methodology in various circumstances
- The views of the stakeholders concerning 'reality', for example, is there an objective real-world problem out there, and the relationship of this to the methodology's philosophical assumptions about reality
- The dominant perceptions in the problem situation, for example, are they technical, political, social and so on.

The questions concerning the second element (the intended problem solver) ask about:

- The methodology users' beliefs, values and ethical positions
- The relationship of the above to that assumed or demanded by the methodology
- The way in which mismatches in the above two may be handled or reconciled, for example, can the methodology processes be changed and does the methodology help to achieve this
- The methodology users' philosophical views, for example, science or systems based
- The methodology users' experience, skills and motives, in relation to those required by the methodology.

The questions concerning the third element (the methodology itself) ask about the way in which the methodology provides specific assistance for:

- Understanding the situation of concern and the setting of boundaries
- Performing the diagnosis, for example, the models, tools, techniques and the levels at which they operate, how they interact, what information they capture, what is not captured, what happens when people disagree and so on.
- Defining the prognosis outline, for example, the desired states, what constitutes legitimate states, and the handling of conflict
- Defining problems
- Deriving notional systems, for example, are they derived at all, and if so how, and in what ways are they recorded
- Performing conceptual/logical and physical design, including who is involved and what are the implications, for example, does it lead to systems improvements or systems innovation
- Implementing the design, for example, how does it handle alternatives and how does it ensure success.

One feature of this framework is that it recommends that the evaluation be conducted at three stages. First, before intervention, that is, before a methodology is adopted, second, during intervention, that is, during its use, and finally, after intervention, that is, an assessment of the success or otherwise of the methodology. These stages are an important feature of the framework and introduce the important element of organisational learning.

Jayaratna has applied the NIMSAD evaluation framework to three well-known methodologies representing different approaches to systems development. They are Structured Systems Analysis and Systems Specification (DeMarco, 1979), which is not dissimilar to STRADIS, ETHICS (Mumford, 1995), and SSM (Checkland, 1981). These evaluations are useful and thought provoking, and provide a good example and guide to the application of the framework. Obviously the conclusions are in many ways a function of the framework, that is, the issues that the author believes to be important, and in this framework, one of the strengths is its wide-ranging, systems concepts-based, underpinning. This serves to highlight again that all evaluations are subjective and that the choice of methodology evaluation framework is itself a value-laden process.

A number of other commentators have suggested alternative approaches to an overall feature analysis when selecting methodologies. They adopt a more pragmatic approach and suggest that 'there is no benefit to be gained from attempting to find, in isolation, a 'best' methodology, because the approaches are not necessarily mutually exclusive. One or more approaches may be suitable, depending on the circumstances. Thus there should be a search for an appropriate methodology in the context of the problems being addressed, the applications, and the organisation and its culture.

Davis (1982) has advocated the contingency approach. The contingency approach has the selection of an appropriate approach as part of the framework or methodology itself. Davis offers guidelines for the selection of an appropriate approach to the determination of requirements, rather than to the selection of a methodology itself (although the same principles may well apply). Davis suggests measuring the level of uncertainty in a system. This will help ascertain the appropriate methodology. There are four measures:

1.   System complexity or ill-structuredness
2.   The state of flux of the system
3.   The user component of the system, for example, the number of people affected and their skill level
4.   The level of skill and experience of the analysts.

Once the level of uncertainty has been ascertained, the appropriate approach to determining the requirements can be made. For low uncertainty, the traditional method of interviewing users would be appropriate. For high levels of uncertainty, a prototype or an evolutionary approach would be better. For intermediate levels of uncertainty, a process of synthesising from the characteristics of the existing system might be appropriate. Variations of this contingency approach have been suggested by a number of authors (for example, Benyon and Skidmore, 1987 and Avison and Wood-Harper, 1991).

Episkopou and Wood-Harper (1986) argue: 'that a suitable approach should be determined by examining variables within and around the problem situation'. They identify three areas of concern:

1. *The problem content system.* This is the system that contains the problem and its environment, including the problem owner
2. *The problem solving system.* This is the system which makes use of a methodology or approach, including the problem solver
3. *The approach choosing and matching system.* This involves an examination of the ideology of each approach, the associated tools, the inquiry system and the costs.

The interesting aspect of this approach is the importance it gives to the problem-solving system as part of the problem, whereas the Davis approach only involves the problem-content system.

Avison and Taylor (1996) identify five different classes of situation and appropriate approaches as follows:

1. *Class 1:* well structured problem situations with a well-defined problem and clear requirements. Traditional systems development methodologies are regarded as appropriate in this class of situation.
2. *Class 2:* as above but with unclear requirements. A data or process modelling methodology, or a prototyping approach are suggested as appropriate.
3. *Class 3:* unstructured problem situation with unclear objectives. A soft systems approach would be appropriate in this situation.
4. *Class 4:* high user-interaction systems. A people-focused approach, for example, ETHICS, would be appropriate here.
5. *Class 5:* very unclear situations, where a contingency approach, such as Multiview, is suggested.

In addition to some of the conceptual problems of comparing methodologies discussed above, the authors also discovered a number of practical problems in attempting to compare methodologies themselves. First, methodologies are not stable, but are in fact moving targets that are continuing to develop and evolve. Therefore there exists a version problem and it is often difficult to know which version of a methodology is being applied in a particular situation or which is the latest version. Second, for commercial reasons, the documentation is not always published or readily available to people or organisations not purchasing the methodology. Third, the practice of a methodology is sometimes significantly different to that prescribed by the documentation or the author of the methodology. This is sometimes talked about in terms of the espoused version of the methodology and the way that it is actually used in practice. Fourth, consultants or developers using the methodology often interpret aspects of the methodology in quite different ways.

A further problem in undertaking comparisons concerns terminology, in particular, the use of different terms for the same phenomena and similar terms for different phenomena. Information systems continues to exhibit rather more than its fair share of these. It is unhealthy, as it not only leads to confusion and poor communication, but to a restriction of development, due to the inability to enhance and expand upon earlier research work. Progress in most successful disciplines is usually a process of evolutionary development for, 'out of the blue', quantum leaps are rare. Any restriction in evolution is therefore very serious. Bubenko (1986), for example, states that terminological confusion makes 'it extremely difficult to reject "new methods" which, essentially, do not contribute anything new and which do not advance the state of the art'.

A common approach to methodology comparison attempts to identify a set of idealised 'features', followed by a check to see whether different methodologies possess these features or not. The implication being that those that do possess them or at least score highly on a features rating are 'good' and that those that do not, are 'less good'. The set of features must be chosen by somebody and are thus subjective, although often purported to be objective. The most obvious indication of this is the kind of comparison conducted by particular methodology vendors in which their methodology scores highly and the competition poorly. The comparison has been designed to give this result. We are more familiar with this kind of comparison in relation to cars or soap powders.

The authors have identified their own set of comparison criteria listed below in the full knowledge that they are subjective and can be criticised in exactly the same way as all the other attempts discussed above. Nevertheless we believe that in the context of this book these are a relevant and defensible set of features that have stood the test of time, being an expansion of a set of features constructed for an earlier study reported in Maddison (1983).

1. What aspects of the development process does the methodology cover?

2.   What overall framework or model does it utilise? For example, is it systems development life cycle based, linear, spiral or parallel?

3.   What representation, abstractions and models are employed?

4.   What tools and techniques are used?

5.   Is the content of the methodology well defined and described, such that a developer can understand and follow it? This applies not only to the stages and tasks but also to the philosophy and objectives of the methodology.

6.   What is the focus of the methodology? Is it, for example, people, data, process, and/or problem oriented? Does it address organisational and strategic issues?

7.   How are the results at each stage expressed?

8.   What situations and types of application is it suited to?

9.   Does it aim to be scientific, behavioural, systemic or whatever?

10.   Is a computer solution assumed? What other assumptions are made?

11.   Who plays what roles? Does it assume professional developers, require a methodology facilitator, involve users and managers, and, if so, how and to what degree?

12.   What particular skills are required of the participants?

13.   How are conflicting views and findings handled?

14.   What control features does it provide and how is success evaluated?

15.   What claims does it make as to benefits? How are these claims substantiated?

16.   What are the underlying philosophical assumptions of the methodology? What makes it a legitimate approach?

Perhaps the most important aspects of this list in terms of comparing methodologies has been found to be the final one. The features of a methodology are highly dependent upon the philosophy and without this understanding, the methodology is difficult to explain. Some methodologies, especially the more commercial ones, do not always explicitly state their underlying philosophy: it has to be searched for and interpreted and this makes analysis difficult. Others are more explicit, some of the object-oriented methodologies make great play of explaining the reasoning behind the concept of objects. However, relatively few explain their philosophy in Checkland's terms of an intellectual framework of enquiry (section 7.2).

Despite all the difficulties identified, which may imply failure from the outset, comparisons continue to be made, because it is becoming increasingly important in a world where large numbers of widely differing methodologies claim the same promises of universal applicability and overall usefulness. According to Floyd (1986), we must 'view methods themselves as objects of study. In so doing we must develop methods for the investigation of methods, concepts for the description and comparison of methods, and criteria for their evaluation and assessment.' We therefore present a sample comparison (in section 7.8) based on our framework (in section 7.7), in the spirit that Floyd identifies, but in the sure knowledge of the problems inherent in the task.

7.7 A FRAMEWORK FOR COMPARING METHODOLOGIES

The reader should now be aware that comparing methodologies is a very difficult task and the results of any such work are likely to be criticised on many counts. There are as many views as there are writers on methodologies. The views of analysts do not necessarily coincide with users, and these views are often at variance with those of the methodology authors. Thus the following is simply another set of views and is unlikely to satisfy all the players in the methodologies game.

The framework suggested for comparative analysis reflects the concepts and features discussed above, and builds upon the earlier work of Wood-Harper and Fitzgerald (1982) and Fitzgerald *et al.* (1985). In this work six major approaches to systems analysis were identified: (i) general systems theory approach; (ii) human activity systems approach; (iii) participative (socio-technical) approach; (iv) traditional (NCC) approach; (v) data analysis approach; and (vi) structured systems (functional) approach. With the exception of the general systems theory approach, they were all methodologies used in practice to some extent. General systems theory was included because of its important influence on systems analysis and systems thinking in general.

In this text, general systems theory has been examined in section 3.2, human activity systems in section 6.11, participative approaches in section 3.13, the traditional approach in section 2.2, the data analysis approach in section 3.8, and the structured systems approach in section 3.7.

The Wood-Harper and Fitzgerald analysis seeks to contradict the notion that these approaches are simple alternatives and that it does not really matter which one is selected as an information systems development standard. The analysis tries to show that there exist fundamental differences between the approaches and the analysis contributes to the selection of an appropriate methodology for the particular requirements. Further, the paper argues, different approaches might in fact be complementary and usefully exist side by side, or applied in a contingency approach. The analysis is based on a discussion of the approaches in terms of three criteria: paradigm, model and objective.

This analysis is now extended in the following framework to include a variety of other criteria that it is hoped makes the framework not just an academic approach but also provides practical help for comparing methodologies. There are seven basic elements to the framework as follows, some elements are broken down into a number of sub-elements:

1. Philosophy
    - Paradigm
    - Objectives
    - Domain
    - Target
2. Model
3. Techniques and tools
4. Scope
5. Outputs
6. Practice
    - Background
    - User base
    - Players
7. Product.

The framework is not supposed to be fully comprehensive and one could envisage a number of additional features that might usefully be compared for particular purposes, for example:

- The speed at which systems can be developed
- The quantity of the specifications and documentation produced
- The potential for modification by users to suit their own environment.

Indeed, Rzevski (1985) has argued that it is meaningless to compare methodologies without stating the purpose of the comparison in advance, a view with which we concur. However, in this text it is necessary to provide a generalised framework. This gives a set of features that proves to be a reasonable guide when examining an individual methodology and when used as a basis for comparing methodologies.

The headings are not mutually exclusive and there are obviously inter-relationships between them. For example, aspects of philosophy are reflected in some senses in all the other elements.

Each of the above listed elements in the framework will be briefly described and then a sample application of the framework is made to the methodologies of Chapter 6. The reader is also invited to apply the framework to the methodologies of their own choice.

1 Philosophy

The question of philosophy is an important aspect of a methodology because it underscores all other aspects. As discussed in section 7.2, it distinguishes, more than any other criterion, a 'methodology' from a 'method'. The choice of the areas covered by the methodology, the systems, data or people orientation, the bias or otherwise towards computerisation, and other aspects are made on the basis of the philosophy of the methodology. This philosophy may be explicit but in most methodologies the philosophy is implicit. Indeed, many feature analyses have neglected this aspect completely, partly because methodology authors seldom stress their philosophy.

In this context we regard 'philosophy' as a principle, or set of principles, that underlie the methodology. It is sometimes argued that all methodologies are based on a common philosophy that suggests that they wish to improve the world of information systems development. Whilst this is true to some extent, this is not what is meant here by philosophy.

As a guide to philosophy the four factors of paradigm, objectives, domains and applications are highlighted.

(a) Paradigm:

Wood-Harper and Fitzgerald (1982), in their taxonomy of approaches to systems analysis, identify two paradigms of relevance. The first is the science paradigm, which has characterised most of the hard scientific developments that we see in the latter part of the twentieth century, and the second is the systems paradigm, which is characterised by a holistic approach.

The term paradigm is defined here in the sense that Kuhn (1962) uses it as a specific way of thinking about problems, encompassing a set of achievements which are acknowledged as the foundation of further practice. A paradigm is usually regarded as subject free, in that it may apply to a number of problems regardless of their specific content.

As Checkland (1981) summarises it, the science paradigm consists of reductionism, repeatability and refutation:

> We may reduce the complexity of the variety of the real world
> in experiments whose results are validated by their
> repeatability, and we may build knowledge by the refutation of
> hypotheses.

The science paradigm has a long and successful history and is responsible for much of our current world. The systems paradigm has a much shorter and less successful history, but was evolved as a reaction to the reductionism of science and its perceived inability to cope with living systems and those categorised as human activity systems.

Science copes with complexity through reductionism, breaking things down into smaller and smaller parts for examination and explanation. This implies that the breakdown does not disrupt the system of which it is a part. Checkland argues that human activity systems are systems which do not display such characteristics, they have *emergent properties* (that is, the whole is greater than the sum of the parts) and perform differently as a whole or as part of a system than when broken down to their individual components. This led directly to the development of the systems paradigm which is characterised by its concern for the whole picture, the emergent properties, and the inter-relationships between parts of the whole. The science and systems paradigms are closely related to concepts of hard and soft thinking discussed in section 6.11.

In a series of papers Klein and Hirschheim extend the debate, distinguishing between ontology and epistemology. Ontology is concerned with the essence of things and the nature of the world and two extreme positions of realism and nominalism are identified. Realism, according to Hirschheim (1985), 'postulates that the universe comprises objectively given, immutable objects and structures. These exist as empirical entities, on their own, independent of the observer's appreciation of them'. On the other hand, according to Hirschheim and Klein (1989), nominalism is where:

> reality is not a given immutable 'out there', but is socially constructed. It is the product of the human mind. Social relativism is the paradigm adopted for understanding social phenomena and is primarily involved in explaining the social world from the viewpoint of the organisational agents who directly take part in the social process of reality construction.

Epistemology is the grounds of knowledge. The term relates to the way in which the world may be legitimately investigated and what may be considered as knowledge and progress. It includes elements concerned with sources of knowledge, structure of knowledge and the limits of what can be known. Again, two extreme positions are frequently identified: positivism and interpretivism. Positivism implies the existence of causal relationships which can be investigated using scientific method whereas interpretivism implies that there is no single truth that can be 'proven' by such investigation. Different views and interpretations are potentially legitimate and the way to progress is not to try and discover the one 'correct' view but to accept the differences and seek to gain insight by a deep understanding of such complexity.

Lewis (1994) brings these elements together in a framework, shown in figure 7.1, that identifies objectivist and subjectivist approaches as positions between the ontology of realism and nominalism and the epistemology of positivism and interpretivism. We find this a helpful framework for thinking about and identifying the underlying philosophies of methodologies. For example, if we believe in an ontology of nominalism we should not adopt a methodology based on an assumption of realism. A subjectivist approach might help us think about data collected in the analysis of a system not as 'facts' but more as perceptions from a particular point of view. Sales targets are not necessarily a set of facts, but perhaps part of a political process which is negotiated between sales personnel, management and directors. This negotiation may have far-reaching implications, relating to people's lives, their remuneration, job satisfaction, self esteem and so on. Furthermore, we need ways of handling these different perceptions. This may mean, for example, that we need a methodology that focuses on the highlighting of different opinions and interpretations or one that allows a series of different designs in order to accommodate different perceptions.

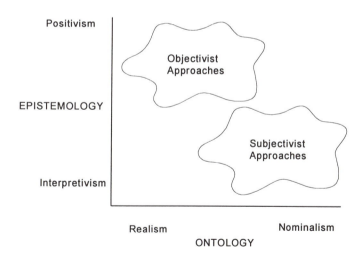

Fig. 7.1: Framework for analysing the underlying philosophies of methodologies (adapted from Lewis, 1994)

It is sometimes argued that not only do methodologies adopt particular philosophical assumptions that we should be aware of but that individual techniques and tools also embody and reflect these concepts. Hirschheim (1985), for example, identifies the technique of entity modelling as objectivist, and Wood-Harper and Fitzgerald (1982) argue that data flow diagrams also embody this assumption. Lewis (1994), however, argues that although data analysis (entity modelling) is generally thought to be objectivist as a technique 'this does not rule out the possibility that it could be employed as part of an alternative (subjectivist) paradigm.' He suggests that some analysts use the technique in this way in practice. They do not think of particular entities and relationships as representations of reality, but as representations of interpretations of reality. They use it as a way of negotiating a shared understanding or interpretation of the problem domain rather than as a way of correcting the user's ambiguities and misunderstandings.

This raises some interesting issues, for if a generally perceived objectivist technique can be used in a subjectivist manner, then this implies that the techniques themselves are not such an important feature in determining the characteristics of a methodology as usually thought. However, we could argue that entity models are not generally used in a subjectivist way but, even if they can be, the technique is still objectivist because it pushes the user of the technique in that general direction. It is a sophisticated developer that does not seek to resolve differing views to one version of reality. It can be argued, however, that it is not so much whether a technique assumes an independent existence of reality or a consensus interpretation of reality but whether it is able to handle differing perceptions of reality that cannot be resolved by consensus.

This is a complex area but the purpose of the discussion has been to indicate the importance of a methodology's underlying philosophies and assumptions and that developers need to be aware that they should match their beliefs to that of the authors of the methodology. As may be appreciated, this discussion of ontology and epistemology is an over simplification of a long history of philosophical debate and the reader is referred to Burrell and Morgan (1979), Everitt and Fisher (1995), Hirschheim (1985), Searle (1995) and Walsham (1995) for more detailed treatments.

(b) Objectives:
One fairly obvious clue to the methodology philosophy, is the stated objective or objectives. Some methodologies state that the objective is to develop a computerised information system. Others, for example, ISAC, have as an objective to discover if there really is a need for an information system. So there exists a difference in that some methodologies are

interested only in aspects that are 'computerisable' whilst others take a wider view and direct their attention to achieving solutions or improvements no matter what this implies. Such improvements might involve manual, procedural, managerial, organisational, educational or political change. This is an important difference, because it determines the boundaries of the area of concern. The problem with concentrating only on aspects to be computerised is that this is an artificial boundary in terms of the logic of the business. There is no reason why the solution to a particular problem should reside only in the area that can be automated. More likely the problem needs to be examined in a wider context.

It has often been found that 'computerisation' *per se* is not the answer. What is required is a thorough analysis and redesign of the whole problem area. It may be viewed as 'putting the cart before the horse' to decide that computerisation is the solution to a particular problem. Clearly a methodology that concerns itself solely with analysing the need for a computer solution is quite a different methodology from one that does not. In choosing and understanding a methodology, it might be a good idea to ask the question: 'could the use of this methodology lead to the implementation of a purely organisational or manual solution?' If the answer is 'no' then the methodology is not addressing the same problems as one to which the answer is 'yes'. It is interesting to note that many of the most widely-used commercial methodologies would seem to answer 'no' to this question, whereas most of the academic methodologies would probably answer 'yes'. An exception would be the recent BPR methodologies, whose focus is not computerisation specifically but improvements to the business processes which, in practice, often requires non-computerised changes.

(c) Domain:
The third factor relating to philosophy is the domain of situations that methodologies address. Early methodologies, such as the conventional life cycle approach discussed in Chapter 2, saw their task as overcoming a particular and sometimes narrow problem. Obviously the solving of the problem, often through the introduction of a computer system of some kind, might be beneficial to the organisation. However, the solution of a number of these kinds of problems on an *ad-hoc* basis at a variety of different points in time can lead to a mish-mash of different physical systems being in operation at the same time.

Even if the developments of solutions to the different problems have been well co-ordinated, and the later systems have been designed with the earlier systems in mind, it is often found that the systems and problems inter-relate and the solution to a number of inter-related problems is different to the sum of the solutions to the individual problems viewed in isolation.

This has led a number of methodologies adopting a different development philosophy. They take a much wider view of their starting point and are not looking to solve, at least in the first instance, particular problems. The argument is a basis of a number of approaches, for example, the systems and strategic approaches described in sections 3.2 and 3.3. In other words, it is argued that in order to solve individual problems, it is necessary to analyse the organisation as a whole, devise an overall information systems strategy, sort out the data and resources of the organisation, and identify the overlapping areas and the areas that need to be integrated. In essence, it is necessary to perform a top-down analysis of the organisation, sort out the strategic requirements of the business, and in this way ensure that the information systems are designed to support these fundamental requirements.

(d) Target:
The last aspect of philosophy is the applicability of the methodology. Some methodologies are specifically targeted at particular types of problem, environment, or type or size of organisation, whilst others are said to be general purpose.

2 Model
The second element of the framework concerns an analysis of the model that the methodology adheres to. The model is the basis of the methodology's view of the world, it is an abstraction, and a representation of the important factors of the information system or the organisation.

The model works at a number of levels: first, it is a means of communication; second, it is a way of capturing the essence of a problem or a design, such that it can be translated or mapped to another form (for example, implementation) without loss of detail; and third, the model is a representation which provides insight into the problem or area of concern.

Models have been categorised into four distinct types (Shubik, 1979):
1.   Verbal
2.   Analytic or mathematical
3.   Iconic, pictorial or schematic
4.   Simulation.

The models of concern in the field of information systems methodologies are almost exclusively of the third type, although there are methodologies, not examined in this book, for example ACM/PCM (Brodie and Silva, 1982) and those mentioned in the formal methods discussions of section 3.12, that have models of the second type. The reason for the current dominance of iconic, pictorial or schematic models is the perceived importance of using the models as a means of communication, particularly between users and analysts. Further important aspects of the models in information systems development are to ensure that the information necessary is captured, at the appropriate stage, and that this information is that required to be able to develop a working system.

It has been stated that a model is a form of abstraction. An abstraction is usually viewed as the process of stripping an idea or a system of its concrete or physical features. Abstraction can be viewed as any simplification of systems and objects at any level, for example, the physical to the logical.

A benefit of abstraction is the easier development of complex applications. It provides a way of viewing the important aspects of the system at various levels, so that high levels have the 'essence' of the system and low levels introduce detail that does not compromise that 'essence'. The process of abstraction loses information and so a model should only lose that information which is not part of the 'essence' of the system.

Abstraction is closely related to hierarchical decomposition and as Olive (1983) states:

> Each level of the decomposition provides an abstract view of the lower levels purely in the sense that details are subordinated to the lower levels. For this to be effective, each level in the decomposition must be understandable as a unit, without requiring either knowledge of lower levels of detail, or how it participates in the system as viewed from a higher level.

The selection of the levels by which a system is defined depends on the purpose of the investigation and design, but it is suggested that there are some 'natural' or 'inherent' levels, and traditionally these are the conceptual level, the logical level and the physical level (see figure 7.2). The conceptual level is a high level description of the universe of discourse (UoD) or the object system from a particular perspective. In our case this perspective might be that of information systems, or business systems or society. The description may be a static or a dynamic one, or a combination of both. Olive (1983) views a conceptual model as:

'..the definition of the problem structure of an information system, like a map defines the problem structure of a transport system. It can also be seen as the "setting up of the equations" of an application. This set of "equations" constitutes a basis for the various processing solutions' ... 'there is no direct correspondence between the state of the object system and the contents of the database of the future information system, nor between derivation rules and its processes.'

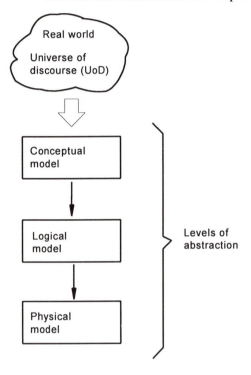

Fig. 7.2: The three-level modelling schema

Thus a conceptual model does not imply any particular database content nor processing activities. Conceptual models cause great confusion, indeed the term has been used in different ways in this text (to reflect different uses in methodologies) and Olive suggests that the role and purpose of the level should be given a great deal more attention and clarification by the authors of methodologies. One confusion is that the language of conceptual modelling is sometimes the same as for logical modelling. An entity model is often described as a conceptual model and it can be so, but it should be remembered that under these definitions it describes the UoD and is not necessarily going to be part of the information system.

The logical level is a description of the information system without any reference to the technology that could be used to implement it. Its scope is the information system itself and it is not concerned with the UoD. Olive describes the main purpose of the logical model as providing the requirements specification of the information system. The physical level is a description of the information system including the technology of the particular implementation.

There are other forms of abstraction than those for defining information systems in general. A well known model for developing database applications is the semantic hierarchy model. This identifies three forms of abstraction: classification, aggregation and generalisation (Smith and Smith, 1977), and Brodie and Silva (1982) identifies a fourth form of abstraction which is called association. These four forms of abstraction allow the integration of both data and process modelling, in that they allow integrity constraints to be defined at the same time as the data.

3 Techniques and Tools:

A key element of the framework is the identification of the techniques and tools used in a methodology, but as these have been extensively discussed in Chapters 4 and 5 they will not be further elaborated here.

4 Scope:

The scope of a methodology is the next element in the framework. Scope is an indication of the stages of the life cycle of systems development which the methodology covers (section 2.2). Further, an analysis of the level of detail at which each stage is addressed is useful. The problem with using the stages of the life cycle as a basis for the examination of scope, is that there is no real agreement on what are, or should be, the stages of the cycle. This is perhaps a problem that can be overcome because most life cycle variations can be mapped onto one another.

A more serious problem is that some notable figures in the information system field have described the life cycle as unsuitable to the needs of systems developers (McCracken and Jackson, 1982). They argue that the development of rapid development tools (prototyping) has changed the traditional picture and that to apply the life cycle restricts development to a rigid pattern and perpetuates the failure of the industry to build an effective bridge between user and analyst.

They are really saying two things:

1.   That the process of development is much speeded up and that the users' perceptions of what is possible and desirable changes during development. Requirements analysis should not terminate early in the development of an information systems project but should permeate through the whole process.

2.   That the end result of prototyping might be the design of the desired system and that it might then be specified for development in a conventional programming language. In this case design comes before specification, quite the reverse to the normal life cycle view.

As a counterbalance to this view of the 'life cycle as harmful', Dearnley and Mayhew (1983) have suggested ways that prototyping could be adopted into the systems life-cycle approach.

Whilst this is an interesting discussion, and prototyping is not the only approach discussed in this text that circumvents the standard life cycle, it need not invalidate the element of scope in the comparison framework, as it would be possible to include the ability to prototype as a feature of scope.

5 Outputs:

The next element in the framework concerns the outputs from the methodology. It is important to know what the methodology is producing in terms of deliverables at each stage and, in particular, the nature of the final deliverable. This can vary from being an analysis specification to a working implementation of the system.

6 Practice:

The next element of the framework is termed the practice and is measured according to: the methodology background (discussed above in terms of commercial or academic); the user base (including numbers and types of users); the participants in the methodology (for example, can it be undertaken by users themselves or must professional analysts be involved); and what skill levels are required. The practice should also include an assessment of difficulties and problems encountered, and perceptions of success and failure. This should be undertaken by investigating the experiences of prior users of the methodology. This will inevitably be subjective, depending on who is consulted, but it can be a revealing exercise. In examining the practice, the degree to which the methodology can be, and is, altered or interpreted by the users according to the requirements of the particular situation should be assessed. It is also important to assess any differences there appear to be between the practice and the theory of the methodology.

7 Product:
The last element of the framework is the product of the methodology, that is, what purchasers actually get for their money. This may consist of software, written documentation, an agreed number of hours training, a telephone help service, consultancy and so on.

7.8 THE COMPARISON

In this section we offer a discussion of some of the methodologies we have described in this text in the context of the comparison framework identified in section 7.7 above. This discussion is selective, for example, we do not include Multiview in the comparison because it combines aspects of many of the other methodologies, RAD because it is so closely associated with IE, KADS because it has a narrow application domain (expert systems development), nor Euromethod, because it is a framework for methodology standards. It is also not intended to be a comprehensive comparison but more a basis for stimulating debate and should not be regarded as a statement of 'facts' about methodologies. It represents only our subjective view.

Element 1 of the framework concerns the identification of the philosophy of the methodology. There are a number of sub-elements to philosophy, the first is that of paradigm, and this first sub-element provides ample illustration of the difficulties in attempting to compare methodologies. In the discussion of the framework, we identified the science paradigm and the systems paradigm to be of critical importance. We suggest that SSM and ETHICS belong to the systems paradigm and that STRADIS, YSM, IE, SSADM, Merise, JSD, OOA and ISAC belong to the science paradigm.

It is clear that SSM adopts the systems paradigm, indeed it is explicitly stated to do so by the methodology author. This is one of the few cases where the issue is addressed as part of the methodology itself. But even if we were not so told, it is clear that the methodology uses many of the systems concepts and does not adopt a reductionist approach.

In ETHICS it is the belief in the interaction of the social and the technical subsystems (the socio-technical approach) that leads to an advocacy of the participative design philosophy. The work system is analysed for variances or weaknesses which prevent the systems objectives being realised. These variances are often discovered at subsystems boundaries, particularly where the social and technical subsystems meet. The ideas of job enrichment and participative design are particular solutions to the more common variances which are encountered. In addition, ETHICS makes no attempt to break down the system into its constituent parts for the purpose of understanding the problems. Thus, the underlying paradigm for ETHICS, it is argued, is also the systems paradigm.

In the analysis of paradigm, ISAC presents the greatest problem and generated the most discussion. ISAC is often regarded as a participative methodology. For example, Maddison (1983), in describing its philosophy, states that:

> ISAC is based on the importance of people in an organisation. This includes all people, for example end users, public users, funders. The activities that people have to perform need to be improved in order to overcome various problems and the best people to do this are the users themselves. The methodology helps people achieve this.

Whilst it is clear that ISAC is firmly in the participative tradition, we believe that this does not mean it automatically incorporates the systems paradigm. The ETHICS methodology is also highly participative, but it is more the socio-technical aspects which lead us to classify that as being of the systems paradigm. ISAC adopts a highly reductionist approach to the understanding of systems. Its authors state that:

> The only way to solve complex problems is to divide them into sub-problems until they become manageable. A requirement for this to work is that the solution to the sub-problems gives the solution for the problem as a whole, that is, that the division in sub-problems is coherent.

This would appear to be a categorical endorsement of the science paradigm and a rejection of the concept of emergent properties. On the other hand, there are a number of areas where ISAC adopts some system thinking notions, such as the hierarchy of systems. It recognises the need to understand wider problems and implications than that specified by the scope of the system. There is other support for the systems paradigm. Iivari and Kerola (1983), for example, suggest that ISAC 'is socio-technical in spirit'. However, for our classification, we believe that the features of the methodology must be the determining factor, not what might be the spirit of the authors. We therefore classify ISAC to be in the science paradigm, although if we accept the notion of a continuum it would not be at the extreme end. We also, on the basis of their clear reductionist approaches and their acceptance of the ontological position of realism, identify STRADIS, YSM, IE, SSADM, Merise, JSD and OOA to belong to the science paradigm.

IE, for example, adopts what they term the 'divide and conquer' approach which is clearly reductionist, and JSD describes its approach to modelling as attempting to reflect the real-world situation, for example, in the entity action step, 'real-world entities are defined' without any discussions of the real world being socially constructed or any problems that might be encountered concerning differing perceptions of the real world.

One of the interesting aspects of applying the comparison framework is not so much whether the classification is right, but the discussion and debate that it generates. The debate proves insightful and causes many significant questions to be asked of the methodologies, some of which are very difficult to answer. It may be suggested that the solution is to turn to the methodology authors themselves, but this would not necessarily result in definitive 'right or wrong' answers to such questions. It may be that the methodology can be, and is, used in ways that the methodology authors had not intended or envisaged. Lundeberg (1983), one of the authors of ISAC, complains that the methodology is often judged on its descriptive techniques which concern documentation, rather than the work methods. He characterises this as 'doing things right' versus 'doing right things' respectively. He maintains that the work methods, that is, 'doing right things', is the more important of the two. Yet it is probable that users of the methodology (using it without his guidance) put their effort into 'doing things right'. This again emphasises the benefit of applying this kind of framework which will help highlight these issues.

The second sub-element of philosophy concerns the objectives of the methodology. There are many objectives that could be discussed, but for the purposes of this framework it is of prime importance whether the objective is to build a computerised information system or whether wider improvements and more general problem-solving are involved. Some of our methodologies indicate their position more clearly than others. ISAC, for example, decides on information systems development as the suitable development measure only if the change analysis indicates that there are problems and needs in the information systems area. In other situations, other development measures are chosen, for example: (i) development of the direct business activities, such as production development, product development or a development of distribution systems; (ii) organisational development; or (iii) development of personal relations (communication training). We therefore see that ISAC is very much more than the development of computerised systems, as are SSM and ETHICS. Process Innovation (PI) also has objectives that are much wider than the development of computerised systems. Its objectives focus on improving and redesigning business processes for an organisation as a whole and although IT is usually regarded as an important enabler of process innovation, many of the specified improvements and redesigned processes are achieved without the construction of computer systems.

On the other hand we classify STRADIS, YSM, IE, SSADM, Merise, JSD and OOA not as general problem-solving methodologies, but as having clear objectives to develop computerised information systems. Some methodologies claim that they are applicable whether the system is going to be automated or not, for example, STRADIS, but we can find no evidence that this is ever put into practice and an examination of the activities of these methodologies illustrates that their main focus clearly embodies an assumption that a computerised system is to be constructed.

Apart from objectives concerning whether a computerised system is the goal or not, there are other objectives of importance in comparisons. For example, in ETHICS there are objectives relating to improving the quality of working life and enhancing job satisfaction of users.

The next sub-element for analysis is that of domain. This is related to the above sub-element of objective but focuses on what aspects or domain the methodology seeks to address. Of particular interest is the distinction between those methodologies that seek to identify business or organisational need for an information system, that is, those which address the general planning, organisation and strategy of information and systems in the organisation, and those concerned with the solving of a specific, pre-identified, problem, for example, the need to provide a wider range of marketing information to the sales force. Often the key to this distinction is the starting point of the methodology. IE, Process Innovation and SSM are identified as being of the planning, organisation and strategy type. In Process Innovation the development of information systems is clearly driven by the identified improvements to the processes required for the benefit of the business and organisation. IE has as its first stage information strategy planning. Here an overview is taken of the organisation in terms of its business objectives and related information needs and an overall information systems plan is devised for the organisation. This clearly implies that it is an approach adopting the philosophy that an organisation needs such a plan in order to function effectively, and that success is related to the identification of information systems that will benefit the organisation and help achieve its strategic objectives.

SSM is quite different to IE and yet we also classify it as a methodology of the planning, organisation and strategy type. Such terminology might not easily be associated with SSM, but it is clearly not a specific problem-solving methodology in the sense that it does not assume that a well-defined, structured, problem already exists. Much of its focus is on the understanding of these wider issues and the context in which the problem situation exists. The term problem situation in SSM is not meant to imply the existence of a well-defined problem, quite the reverse. However, SSM is not usually thought of as a methodology that addresses planning, organisation and strategy, but if we remove the managerial implications from these terms, this is fundamentally what it is about. It is attempting to identify the underlying issues that help in the understanding of the problem situation, including the purpose of the organisation. Later stages of the methodology assess feasible and desirable change and recommend action to improve the situation, the results of which can be the development of information systems.

STRADIS, YSM, SSADM, Merise, JSD, OOA, ISAC and ETHICS are classified as specific problem-solving methodologies, that is, they do not focus on identifying the systems required by the organisation but begin by assuming that a specific problem is to addressed.

The final sub-element of philosophy in the framework is concerned with the target system to be developed, that is, whether the methodology is aimed at particular types of application, types of problem, size of system, environment and so on. This is also a difficult area, because most methodologies appear to claim to be general purpose. Such a claim is clearly made within certain assumptions. In the majority of cases, a large organisation with an in-house data processing department is assumed. Further, it is often assumed that bespoke (tailor-made) systems are going to be developed, rather than, for example, the use of application packages. An exception is IE, where alternative approaches are envisaged. OOA is considered to be general purpose, although it is suggested that it is not very helpful for simple, limited systems or systems with only a few Class-&-Objects. STRADIS is also stated to be general purpose and applicable to any size of system, yet the main technique is data flow diagramming, which is not particularly suitable for all types of application, for example, the development of management information systems. Thus the claims of the methodology authors have to be tempered by an examination of the methodology itself. JSD, for example, has been described as most suitable for real-time processing applications. SSM has been developed to be applicable in human activity situations where very complex (wicked) problem situations exist.

The size of organisations that the methodologies address is also an important aspect of target. Whereas STRADIS, YSM, IE, SSADM, and Merise have all been designed primarily for use in large organisations, Multiview has been designed also to be applicable in relatively small organisations. There is, however, a version of SSADM, called MicroSSADM, which is intended to help develop information systems in smaller environments or where the target system is PC based.

The second element of the framework deals with the model or models that the methodology uses. Section 7.7, in discussing the framework, indicates the various facets of models that can be investigated, the type of model, the levels of abstraction of the model and the orientation or focus of the model.

In this part, we will concentrate on examining the various models of process that methodologies use. The primary process model used is the data flow diagram. In STRADIS it is the primary model of the methodology. It also appears in YSM, SSADM and ISAC as an important model, although not the only one. It is also used in Wilson's (1990) description of SSM (but not in Checkland's version), referred to as 'a "Gane and Sarson" type diagram'. It also features in IE, though in a slightly different form, termed a process dependency diagram, but it plays a less significant role than, for example, in STRADIS. The data flow diagram is predominantly a process model and data is only modelled as a by-product of the processes.

The models in JSD, ETHICS, SSM (Checkland), and Process Innovation are also, in their various ways, process-oriented, but they do not use data flow diagrams. The structure diagram is used in JSD to model aspects of process and we see that this depicts the structure of processes rather than the flow of data between processes, which dramatically changes the focus of what is important in JSD and helps to explain why it is regarded as more suitable for real-time process applications than data flow diagram-based methodologies. In SSM, the rich picture, which is a model of the problem situation, is also, in part, a model of processes, structures and their relationships. ETHICS uses a socio-technical model which involves the interaction of technology and people and the processes performed. It is interesting to note that of the identified model types, this is not a pictorial model type but a 'verbal' or narrative model.

The third element of the framework is that of the techniques and tools that a methodology employs. These have been examined in Chapters 4 and 5 respectively. The techniques to be used in a methodology are usually made explicit by the authors which makes it relatively easy to compare and contrast. Many of the models discussed above are closely reflected in the techniques used but there are sometimes differences in the way the models are used and their importance in the methodology. Many methodologies appear to include the techniques as part of the methodology. STRADIS is an example of a methodology which is largely described in terms of its techniques, but others, such as, YSM, SSADM and JSD also have specific techniques which are regarded as fundamental to the methodology. Other methodologies, for example, ISAC, do not rely on particular techniques quite so much, and it is relatively easy to envisage similar but alternative techniques being used without affecting the essence of the methodology.

Some methodologies, for example, IE, explicitly suggest that the techniques are not a fundamental part of the methodology and that the current recommended techniques can be replaced, or substituted, as better techniques become available, providing of course they address the same fundamentals. This is potentially an important conceptual difference between methodologies, and it is a useful exercise to strip a methodology of its techniques and see what, if anything, is left. For example, are the phases and tasks of the methodology described in terms of when and how to use the techniques? Obviously those methodologies that allow new techniques to be incorporated are somewhat more flexible, but achieving this in practice is not so easy. For example, in a methodology which advocates the clear separation of the modelling of data and processes, such as, SSADM, Merise or IE, it would be quite difficult to accommodate an object-oriented modelling technique, which integrates the two without amending the fundamentals of the methodology. In such cases, the identification of the fundamentals is an important part of any comparison. This also raises interesting issues about how methodologies can legitimately develop and evolve over time without losing the essence of the methodology.

The comparison of the tools of methodologies begins with whether any are specifically advocated. SSM, for example, does not advocate, or even mention, any tools, but most methodologies, YSM, IE, SSADM, Merise, JSD, OOA and Process Innovation, recommend the use of tools to some degree. These range from simple drawing tools through to tools supporting the whole development process, including prototyping, project management, code generation, simulation and so on. The degree of recommendation varies considerably. Some, such as IE, suggest that the process should not be contemplated without the use of tools, the process being too complicated and time consuming. Others, for example, SSADM and OOA, argue that they might be helpful but are not necessarily essential. Coad and Yourdon (1991) argue that CASE tools are overpriced and that 'it's hard to believe that such simplistic software tools are getting so much attention'. An important element of comparison is whether the methodology, like IE, implies the use of a specific brand-name tool, or whether appropriate tools from any vendor can be used. In practice, there is often a trade off to be made between the degree that the tool is specific to the methodology and the degree of lock-in to a particular vendor.

The fourth element of the framework is scope. For the purposes of this text we examine methodologies in terms of the stages of the life cycle they address. We identify nine stages: strategy (planning), feasibility, analysis, logical design, physical design, programming, testing, implementation and maintenance. This is not the only possible approach to the analysis of scope and, depending on the methodologies being compared and the purpose of the comparison, other dimensions may be more appropriate. Any analysis of the scope of a methodology is difficult and subjective. Using the stages of the life cycle may misrepresent those methodologies that are not designed to follow such a structure, for example, prototyping methodologies, in which case a definition of scope based on the spiral model might be more appropriate. It is therefore important that scope is not viewed in isolation from the rest of the framework and that the earlier warnings concerning comparisons are heeded.

Figure 7.3 summarises the results of the analysis of scope. Shaded areas indicate that the methodology covers the stage in some detail which may include the provision of specific techniques and tools of support. An unshaded area means that the methodology addresses the area, but in less detail and depth. In this case there is less guidance from the methodology and more is left for the developers to interpret and perform for themselves. The broken lines indicate areas that are only briefly mentioned in the methodology, but no procedures, techniques or rules are provided although the methodology recognises that the area should be addressed. The identification that a methodology covers a particular stage does not necessarily mean that there exists a defined stage of that name but that within the methodology there are elements that can be construed to be equivalent to that stage.

In our analysis of scope, *strategy* is used to indicate whether the methodology addresses any aspects which relate to an organisation-wide context, and that deals with overall information systems strategy, purpose and planning, rather than just that of the particular system or area of concern. IE and PI are identified as addressing this in some detail, as is SSM which also deals with the wider context, although in a significantly different way to the others. SSADM, Merise and, to a lesser extent YSM, also include aspects of a strategic nature.

The next stage is *feasibility* which is defined as the economic, social, and technical evaluation of the system under consideration. STRADIS, YSM, SSADM and Merise include detailed aspects for the evaluation of feasibility in their methodologies. IE, ISAC and PI also address feasibility, but less comprehensively. However, it should be noted that the way in which they deal with feasibility differs considerably. For example, ISAC does not identify a specific feasibility phase (it is contained in some of the analysis steps) and STRADIS checks and re-checks feasibility at many stages throughout the methodology. ETHICS and SSM are identified as dealing in detail with feasibility, although in quite a different way from the others as it is not financial but social feasibility of change which is addressed. ETHICS focuses on social and technical 'fit' and SSM on feasible and desirable change. In figure 7.3, these methodologies are depicted as open boxes rather than shaded boxes because although their social focus is comprehensive, the other aspects are not.

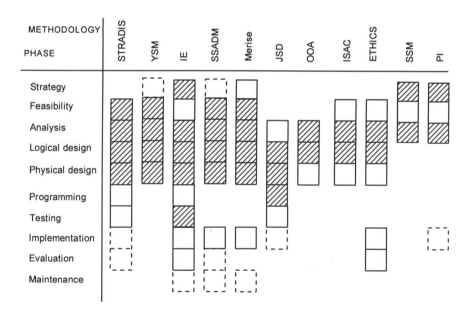

Fig. 7.3: Scope of methodologies

The *analysis* stage, which includes user requirements analysis, is covered in detail, although in a wide variety of different ways by all the methodologies except JSD. JSD does not specifically address user requirements analysis but begins with an assumption that they are given. The *logical design* stages are covered in detail by all the methodologies except SSM and PI which do not cover the stage or indeed any subsequent stages. *Physical Design* is covered in detail by all except OOA, ISAC and ETHICS, which are less explicit. ISAC suggests the use of JSP in the data systems design phase.

The next stage we identify is that of the physical development of the system, characterised as *programming* in figure 7.3. Only STRADIS, IE and JSD cover programming, with JSD judged to address it in most detail. Some methodologies suggest or assume that other approaches are used for the physical development of the system. ISAC, for example, recommends the use of JSP and OOA, an object-oriented language, such as Smalltalk. In such cases we do not include this as being covered in the scope of the methodology. *Testing*, which includes the planning as well as the testing of systems, programs and procedures, is addressed by STRADIS, IE and JSD, but only in detail by IE.

Implementation, in which we include planning and implementation of technical, social and organisational aspects, is covered by IE, SSADM, Merise and ETHICS, although not in as much detail as earlier stages, with STRADIS and PI addressing it, but in even less detail. JSD is a special case, because its interpretation of implementation is simply technical, covering the sharing of processors among processes. Post implementation *evaluation* and review concerns the measurement and evaluation of the implemented system and a comparison with the original objectives. This is only addressed by STRADIS, IE, SSADM and ETHICS, with ETHICS particularly focusing on the evaluation of 'fit'.

The final stage in the analysis of scope is *Maintenance*, which we regard as being covered by a methodology only if it is specifically addressed in terms of tasks. The fact that maintenance may in general be improved by the use of the methodology in earlier stages is not regarded as coverage of maintenance for the purposes of this analysis. With this definition we find that only IE, Merise and ETHICS mention maintenance in any detail.

A glance at figure 7.3 illustrates that the main focus of these methodologies are at the analysis and design stages, but that there are wide scope variations and that to assume all methodologies are the same in this respect, as people often do, is obviously incorrect.

The fifth element of the framework is concerned with outputs. This is an investigation of what is actually produced in terms of deliverables at the end of each stage of the methodology. The outputs of methodologies differ quite substantially, not only in terms of what should be produced but also in the level of detail that the methodology specifies. This is closely related to the degree that the methodology is seen as a blueprint for action, that is, how detailed are the rules about how to proceed and how much is left to the discretion of the analyst.

A related issue concerns how, and to what extent, the analysts know that they are proceeding correctly. As an example, ISAC specifies in some detail the outputs of the change analysis stage (A-graphs), but the process of generating change alternatives are not described in any detail. This is regarded as the creative part of the methodology and not amenable to specification. The identification of such areas in a methodology is regarded as an important contribution to an understanding of that methodology.

The penultimate element of the framework is the practice or use of the methodology. It contains the sub-elements of background, user base, applications and players. The background of the methodology broadly identifies its origins in terms of academic or commercial. STRADIS, YSM, IE, SSADM and OOA lie in the commercial sphere, whereas Merise, JSD, ISAC, ETHICS, SSM and PI have academic backgrounds, at least in part, though this does not mean that they are not now commercial methodologies.

The user base, perhaps surprisingly, is often difficult to discover and is possibly less widespread than vendors would have us believe. Vendors have a habit of suggesting that any company who has expressed interest in their methodology, or has bought an evaluation version, is a user. Wasserman *et al.* (1983) produced a survey of 24 methodologies and found that nearly half had been used on 10 or fewer projects. A survey by Informatics (cited in CCTA, 1990) of about 70 users found that 19 used SSADM, 12 JSD, 9 IE and 8 YSM, with 11 using in-house methods. This coincides with a number of other surveys that suggest that SSADM is the most frequently used methodology in the UK, with over 250 organisations represented in the SSADM User Group. Many organisations do not use a brand name methodology, they are either still using traditional approaches or their own in-house variations or approaches to suit themselves. In France, surveys (cited in CCTA, 1990) suggest that Merise is used in developments in between 20-61 per cent of cases and that in Europe as a whole 55 per cent of organisations are using brand name methodologies.

The last sub-element of practice requires an analysis of the players involved. This requires answers to the questions 'who is supposed to use the methodology?' and 'what roles do they perform?'. It also attempts to identify the skill levels required. The traditional view of information systems development is that a specialist team of professional systems analysts and designers perform the analysis and design aspects and professional programmers design the programs and write the code. The system is then implemented by the analysts. This general view is still taken by STRADIS, YSM, IE, SSADM, Merise, JSD and OOA. The methodologies SSM, ISAC, PI and, in particular, ETHICS, take a different view, and users have a much more proactive role. In ETHICS, the users perform the analysis and design themselves and the data processing professionals are used as consultants as and when required. In addition, ETHICS incorporates a facilitator role, whose task is to guide the users in the use of the methodology. Facilitators do not actually perform the tasks themselves, but smooth the path and ensure that the methodology is followed correctly. The facilitator should be expert and experienced in the use of the methodology.

The levels of skill required by the players varies considerably. In almost all methodologies considerable training and experience is necessary for at least some of the players. Further, many make significant demands on the users, in which case the methodology would be expected to include training aspects. This may significantly increase the time and costs required to develop a project. ETHICS, which makes heavy demands on the users, recognises and addresses this problem. Even where methodologies adopt the more traditional roles, professional analysts and designers can find them quite difficult to learn at first (and some complain laborious as well), with new vocabularies, techniques and tools to work with.

The final element of the framework is that of product. This describes what is supplied when purchasing a methodology and at what cost. Most methodologies have a range of products and services available, which can be taken or not, although there is likely to be a minimum set. The product is also likely to be changing rapidly, mainly due to the increasing numbers of support tools available. The product can range from large and copious sets of manuals, for example, for SSADM, to a set of academic papers and a book, as in the case of SSM. Some methodologies require consultants, facilitators, and/or training courses to be used as part of the product. Some methodologies offer certification of competency for developers. There is, for example, a certificate of proficiency scheme for SSADM.

This discussion of methodologies using the framework is by no means comprehensive, but is intended to be illustrative of the issues that the framework raises. It is likely to be contentious. It is meant to stimulate debate, to open discussions rather than be taken as a statement of facts.

7.9 METHODOLOGY LIMITATIONS

Methodologies were often seen as a panacea to the problems of traditional development approaches, and as we have also seen, they were often chosen and adopted for the wrong reasons. Some organisations simply wanted a better project control mechanism, others a better way of involving users, still others wanted to inject some rigour or discipline into the process. For many of these organisations, the adoption of a methodology has not always worked or been the total success its advocates expected. Indeed, it was very unlikely that methodologies would ever achieve the more overblown claims made by some vendors and consultants. Some organisations have found their chosen methodology not to be successful or appropriate for them and have adopted a different one. For some this second option has been more useful, but others have found the new one not to be successful either. One manager in an organisation stated that 'they had tried them all now, without success!'. This has led some people to the rejection of methodologies in general. In the authors' experience this is not an isolated reaction, and there is something that might be described as a backlash against formalised information systems development methodologies.

This does not mean that methodologies have not been successful. It means that they have not solved all the problems that they were supposed to. Many organisations are using methodologies effectively and successfully and conclude that, although not perfect, they are an improvement on what they were doing previously, and that they could not handle their current systems development load without them.

In this section the case against the use of methodologies is discussed. The criticisms are deliberately couched in generic form and are not related to specific methodologies. We have attempted to be provocative and readers might attempt to form counter-views.

- *Productivity:* The first general criticism of methodologies is that they fail to deliver the suggested productivity benefits. It is said that they do not reduce the time taken to develop a project, rather their use increases systems development lead-times when compared with not using a methodology. This is usually because the methodology specifies many more activities and tasks that have to be undertaken. It specifies the construction of many more diagrams and models, and in general the production of considerably more documentation at all stages. Much of this may be felt by users to be unnecessary. As well as being slow, they are resource intensive. This is true first in terms of the number of people required, from both the development and user side, and second, from the point of view of the costs of adopting the methodology, for example, the purchase costs, training, CASE tools, organisational costs and so on.

- *Complexity:* Methodologies have been criticised for being over complex. They are designed to be applied to the largest and most comprehensive development project and therefore specify in great detail every possible task that might conceivably be thought to be relevant, all of which is expected to be undertaken for every development project.

- *'Gilding the lily':* Methodologies develop any requirements to the ultimate degree, often over and above what is legitimately required. Every requirement is treated as being of equal weight and importance which results in relatively unimportant aspects being developed to the same degree as those that are essential. It is sometimes said that they encourage the creation of 'wish lists' by users.

- *Skills:* Methodologies require significant skills in their use and processes. These skills are often difficult for methodology users and end users to learn and acquire. It is sometimes also argued that the use of the methodology does not improve system development skills or organisational learning.

- *Tools:* The tools that the methodology advocates are difficult to use and do not generate enough benefits. They increase the focus on the production of documentation rather than leading to better analysis and design.

- *Not contingent:* The methodology is not contingent upon the type of project or its size. Therefore the standard becomes the application of the whole methodology, irrespective of its relevance.

- *One-dimensional approach:* The methodology usually adopts only one approach to the development of projects and whilst this may be a strength it does not always address the underlying issues or problems. In some cases the approach needs a more political or organisational dimension.

- *Inflexible:* The methodology might be inflexible and does not allow changes to requirements during development. This is problematic as requirements, particularly business requirements, frequently change during the long development process.

- *Invalid assumptions:* Most methodologies make a number of simplifying yet invalid assumptions, such as a stable external and competitive environment. Many methodologies that address the alignment of business and information systems strategy assume the existence of a coherent and well documented business strategy as a starting point for the methodology. This may not exist in practice.

- *Goal displacement:* It has frequently been found that the existence of a methodology standard in an organisation leads to its unthinking implementation and to a focus on following the procedures of the methodology to the exclusion of the real needs of the project being developed. In other words, the methodology obscures the important issues. De Grace and Stahl (1993) have termed this 'goal displacement' and talk about the severe problem of 'slavish adherence to the methodology'. Wastell (1996) talks about the 'fetish of technique' which inhibits creative thinking. He takes this further and suggests that the application of a methodology in this way is the functioning of methodology as a social defence which he describes 'as a highly sophisticated social device for containing the acute and potentially overwhelming pressures of systems development'. He is suggesting that systems development is such a difficult and stressful process, that developers often take refuge in the intense application of the methodology in all its detail as a way of dealing with these difficulties. Developers can be seen to be working hard and diligently, but this is in reality goal displacement activity because they are avoiding the real problems of effectively developing the required system.

- *Difficulties in adopting a methodology:* Some organisations have found it hard to adopt methodologies in practice. They have found resistance from developers who are experienced and familiar with more informal approaches to systems development and see the introduction of a methodology as restricting their freedom and a slight on their skills. One organisation experienced these problems to the extent that they had to introduce the methodology by setting up a new development team from scratch, recruited from new graduates not tainted with the old ways of doing things. In other cases it has been the users that have objected to a methodology, because it did not embody the way they wished to work and included techniques for specifying requirements that they were not familiar with and did not see a good reason to adopt.

- *No improvements:* Finally in this list, and perhaps the acid test, is the conclusion of some that the methodology has not resulted in better systems, for whatever reasons. This is obviously difficult to prove, but nevertheless the perception of some is that 'we have tried it and it didn't help and may have actively hindered'.

We thus find that the great hopes of some in the 1980s that methodologies would solve most of the problems of information systems development have not come to pass.

Strictly speaking, a distinction should be made in the criticisms of methodologies made above between poor application and use of a methodology and an inadequate methodology itself. A defence made by a number of methodology vendors implies that the methodology is not being correctly implemented by some organisations. Whilst this may be true in some cases, it is not an argument that seems to hold much sway with methodology users. They argue that the two issues are much the same and for whatever reason they have experienced disappointments.

Fitzgerald (1992), in a study of Executive Information Systems (EIS), found that many of the most successful EIS were developed outside the IT or systems development department as a way of freeing the development from the overbearing and stifling effects of the organisation's information systems development methodology. The most senior executives in these organisations were acutely aware of what they saw as the debilitating effects of formal approaches, including methodologies. They regarded development in the IT department as unlikely to lead to development in the timescale required. They were unable to cope with the executives frequently changing their minds as they struggled to establish, through trial and error, what they really needed to run the business. They required a fast, flexible, prototyping/evolutionary approach. They found this could be achieved by setting up a totally separate development team, away from the IT department, staffed and led by business-oriented people with a few highly skilled IT people with the 'right' attitudes.

Whether the backlash against methodologies is justified or not, the reaction of some organisations has been to move away from formalised methodologies to a generally less formalised or non-methodological approach. Some of the approaches that organisations have adopted are as follows:

- *Ad-hoc:* This might be described as a return to the approach of the pre-methodology days in which no formalised methodology is followed. The approach that is adopted is whatever the developers understand and feel will work. It is driven by, and relies heavily on, the skills and experiences of the developers. This is perhaps the most extreme reaction to the backlash against methodology and in general terms it runs the risk of repeating the problems encountered prior to the advent of methodologies as discussed in section 2.1.

- *Contingent:* This is an approach which does not advocate any particular methodology but suggests that the developers and users pick an appropriate approach depending on the characteristics of the project or domain. These characteristics include the type of project, whether it is an operations-level system or a management information system, the size of the project, the importance of the project, the projected life of the project, the characteristics of the problem domain, the available skills and so on. This is a reaction to the 'one methodology for all developments' approach that some companies adopted, and is a recognition that different characteristics require different approaches. There are problems inherent in the contingent approach. First, there is a wide range of different skills that are required to handle many approaches. Second, the benefits of standardisation are lost. Third, the selection of approach requires experience and skills to make the best judgements. Multiview aims to provide a framework which helps people make such contingent decisions.

- *Prototyping:* This has been described in section 3.11 and is seen as an effective alternative to rigid methodologies for situations that require a better focus on the analysis of requirements. This may not be a rejection of methodology as such, because there are now emerging a number of prototyping methodologies.
- *Incremental development/timeboxing:* Evolutionary development has been mentioned earlier in section 3.11 where it was discussed in terms of an ongoing prototype. Incremental development is a closely related concept, but has the characteristic of building upon, and enhancing, the previous versions rather than evolving the whole system each time. As a reaction to the problems of using methodologies, incremental development has been adopted as a way of overcoming a number of specific problems. First, it aims to reduce the length of time that it takes to develop a system. Second, it addresses the problem of changing requirements as a result of learning during the process of development. This particular form of incremental development has also been termed 'timebox' development (section 6.14). The system to be developed is divided up into a number of components that can be developed separately. The most important requirements and those with the largest potential benefit are developed first. Some argue that no single component should take more than 90 days to develop, whilst others suggest 180 days. Whichever timebox is chosen, the point is that it is relatively quick.

The idea of this approach is to compartmentalise the development and deliver early and often (hence the term timeboxing). This provides the business and the users with a quick, but it is hoped, useful part of the system in a refreshingly short time scale. The system at this stage is probably quite limited in relation to the total requirements, but at least something has been delivered. This is radically different from the conventional delivery mode of most methodologies which is a long development period of typically two to three years followed by the implementation of the complete system. The benefits of evolutionary development is that users trade-off unnecessary (or at least initially unnecessary) requirements and wish lists (that is, features that it would be 'nice to have' in an ideal world) for speed of development. This also has the benefit that if requirements change over time, the total system has not been completed and each timebox can accommodate the changes that become necessary as requirements change and evolve. It also has the advantage that the users become experienced with using and working with the system and learn what they really require from the early features that are implemented.

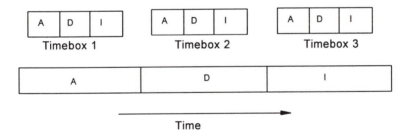

A = Analysis
D = Design
I = Implementation

Fig. 7.4: Comparison of evolutionary/timebox development and traditional development

Figure 7.4 illustrates three chunks of development and although the overall time to achieve the full implementation could be argued to be the same as with a conventional development, the likelihood is that the system actually developed at the end of the three timeboxes will be radically different from that developed at the end of one large chunk as a result of the learning process which leads to change being made to each specification at each stage.

Some proponents argue that if the system cannot be divided into 90 day chunks, then it should not be undertaken at all. Obviously such an approach requires a radically different development culture from that required for formalised methodologies. The focus is on speed of delivery, the identification of the absolutely essential requirements, implementation as a learning vehicle and the expectation that the requirements will change in the next timebox. Clearly such radical changes are unlikely to be achieved using conventional techniques.

* *RAD/JAD:* This is closely related to the ideas of timeboxing and introduces a series of techniques to help achieve rapid delivery of systems. RAD and JAD have been discussed in section 6.14.

* *Package-based development:* Some organisations have decided not to embark on any more major in-house system development activities but to buy-in all their requirements in the form of packages or 'turn-key' systems. This is regarded as a quick and relatively cheap way of implementing systems for organisations that have fairly standard requirements. A degree of package modification and integration may be required which may still be undertaken in-house.

* *Outsourcing:* For some, the continuing problems of systems development and the perceived failure of methodologies to deliver has resulted in the outsourcing of systems development. This is different to buying-in packages or solutions, because normally the management and responsibility for the provision and development of appropriate systems is given to a vendor (Willcocks and Fitzgerald, 1993). The client organisation no longer has any great concern about how the systems are developed. They are now more interested in the end results and the effectiveness of the systems that are delivered. It is argued by some that this can actually be a cheaper approach because you are only paying for what you get and that the payment of real money (often coming directly from the business users) serves to 'focus the mind' and results in more accurate and realistic specifications.

We have described some approaches that organisations are taking as a result of a degree of disappointment with methodologies. However, they are not exclusively adopted as a result of the perceived failure of methodologies, Some have used them to complement conventional methodologies, for example, IE and RAD. They are also not mutually exclusive, as many are used together providing a greater variety of approaches in a more contingent manner.

7.10 THE FUTURE

So where does all this leave the concept of information systems development methodologies?

We expect to see the continuing refinement and improvement of existing commercial information systems methodologies, in terms of the products and techniques, but particularly in the use of CASE tools. We expect the tools to support more and more phases of the methodology and to become more sophisticated, with the development of improved repositories, the use of multi-media and support for distributed development and co-operative working. Code generation systems based on information held in CASE repositories will also continue to develop. This does not mean that programming and programmers will become obsolete, but that they are likely to be of declining importance in the development of standard information systems. We also believe that the development of systems based on bought-in packages (or chunks of code) and then amended and integrated into existing systems will continue to advance and we expect to see the development of methods that support this type of development.

The advent of object-oriented systems development is likely to have an important effect on current commercial methodologies. The market push is certainly in this direction, and we expect to see a number of methods, particularly those that currently rigidly separate the analysis of data and processes, evolve by incorporating object-oriented concepts.

In our view the future will continue to be characterised by a large number of different types of methodologies and we do not expect to see the development of one 'common' methodology, despite the moves in the European Community (section 6.16) for greater harmonisation. There are a number of reasons. First, that no one methodology has proved itself successful enough for it to be considered as a model for all development. Second, if the philosophies and objectives of current methodologies are examined, it becomes obvious that many are attempting to do fundamentally different things (compare, for example, SSADM and SSM). Third, we believe that systems development will become more contingent, and there will continue to exist different methodologies and approaches for different situations and purposes. The development of management information systems, safety critical systems, business critical systems, expert systems and so on, will require different approaches. Finally, if the section on methodology limitations is examined, it is clear that some people, at least, are looking to move away from the constraints of bureaucratic methodologies and are seeking new and alternative directions which should stimulate the growth of new, and currently unconceived, methodologies.

In conclusion, we expect that methodologies will continue to be of great importance in the development of information systems but with perhaps a greater focus on a contingent approach to their use. We do not expect to see the emergence of one standard methodology or approach, nor do we expect to see their disappearance, despite what has been termed a 'backlash' against methodologies. Rather we think that this will create the climate for an increasing range and diversity of methodologies in the future. However, predicting the future is always problematical and has a habit of making fools of those who attempt it!

FURTHER READING

Many topics have been covered in this chapter and it is difficult to suggest readings that cover all the issues, but the following are a good starting point in connection with specific topics:

Comparison of methodologies
Bubenko, J. A. Jr. (1986) *Information system methodologies - a research view*. In: Olle *et al.* (1986).
Olle, T. W., Sol, H. G. and Verrijn-Stuart, A. A. (1982) *Information Systems Design Methodologies: A Comparative Review*. North Holland, Amsterdam.

Philosophical issues
Checkland, P. (1981) *Systems Thinking, Systems Practice*. Wiley, Chichester.
Hirschheim, R. and Klein, H. K. (1989) Four paradigms of information systems development. *Communications of the ACM*, 32, 10.

Contingency
Avison, D. E. and Wood-Harper, A. T. (1990) *Multiview: An Exploration in Information Systems Development*. McGraw-Hill, Maidenhead.

Limitations of methodologies

Fitzgerald, B. (1994) The systems development dilemma: whether to adopt formalised systems development methodologies or not? In: Baets, W. R. J. (ed.) *Proceedings of the Second European Conference on Information Systems,* Nijenrode University Press, Breukelen, the Netherlands.

Bibliography

Ackoff, R. L. (1971) Towards a system of system concepts. *Management Science*, 17.

Ahituv, N., Neumann, S. and Riley, H. N. (1994) *Principles of Information Systems for Management.* 4th ed., Brown, Dubuque, Ia.

Alavi, M. (1993) Making CASE an organizational reality. *Information Systems Management*, 10, 2.

Alter, S. (1992) *Information Systems: A Management Perspective.* Addison-Wesley, Reading, Mass.

Andrews, D. and Ince, D. (1991) *Practical Formal Methods with VDM.* McGraw-Hill, Maidenhead.

Andrews, D. C. (1991) JAD: A crucial dimension for rapid application development. *Journal of Systems Management*, 42, 3.

Andrews, T., Harris, C. and Sinkel, K. (1991) *The Ontos Object Database.* Ontologic, Burlington, Ma.

Angell, I. O. and Smithson S. (1991) *Information Systems Management: Opportunities and Risks.* Macmillan, Basingstoke.

Anjewierden, A., Wielemaker, J. and Toussaint, C. (1992) Shelley: computer aided knowledge engineering. *Knowledge Acquisition*, 4.

ANSI/X3/SPARC (1975) Study group on data base management systems, interim report. *ACM SIGMOD Bulletin*, 7, 2.

Avison, D. E. (1990) *Mastering Business Microcomputing.* 2nd ed., Macmillan, Basingstoke.

Avison, D. E. (1992) *Information Systems Development: A Database Approach.* 2nd ed., McGraw-Hill, Maidenhead.

Avison, D. E. and Catchpole, C. P. (1987) Information systems for the community health services. *Medical Informatics*, 13, 2.

Avison, D. E., Fitzgerald, G. and Wood-Harper, A. T. (1988) Information systems development: a tool-kit is not enough. *Computer Journal*, 31, 4.

Avison, D., Kendall, J. and DeGross, J. (eds.) (1993*) Human, Organisational, and Social Dimensions of Information Systems Development.* North Holland, Amsterdam.

Avison, D. E., Shah, H. U. and Wilson, D. N. (1994) Software quality standards in practice: the limitations of using ISO-9001 to support software development. *Software Quality Journal*, 3.

Avison, D. E. and Taylor, A. V. (1996) Information systems development methodologies: a classification according to problem situations. *Journal of Information Technology,* 11.

Avison, D. E. and Wilson, D. (1991) Controls for effective prototyping. *Journal of Management Systems*, SA-58.

Avison, D. E. and Wood-Harper, A. T. (1986) Multiview - an exploration in information systems development. *Australian Computer Journal*, 18, 4.

Avison, D. E. and Wood-Harper, A. T. (1990) *Multiview: An Exploration in Information Systems Development*. McGraw-Hill, Maidenhead.

Avison, D. E. and Wood-Harper, A. T. (1991) Information systems development research: a pragmatic view. *Computer Journal*, 34, 2.

Bantleman, J. P. (1984) A feature analysis of the LBMS system development method. In: *Structured Methods, State of the Art Report*, 12, 1, Pergamon Infotech, Maidenhead.

Beer, S. (1985) *Diagnosing the System for Organizations*. Wiley, Chichester.

Bemelmans, T. M. A. (ed.) (1984) *Beyond Productivity: Information Systems Development for Organizational Effectiveness*. North Holland, Amsterdam.

Benyon, D. (1990) *Information and Data Modelling*. McGraw-Hill, Maidenhead.

Benyon, D. and Skidmore, S. (1987) Towards a tool-kit for the systems analyst. *Computer Journal*. 30, 1.

Bertalanffy, L. von (1968) *General Systems Theory*. Braziller, New York.

Bjørn-Andersen, N. (1984) *Challenge to certainty*. In: Bemelmans (1984).

Booch, G. (1991) *Object Oriented Design with Applications*. Benjamin/Cummings, Redwood City, California.

Bracchi, G. and Lockermann, P. C. (eds.) (1978) *Information Systems Methodology: Proceedings of the Second Conference*. Springer-Verlag, Berlin.

Bradley, K. (1993) *PRINCE: A Practical Handbook*. Butterworth-Heinemann, Oxford.

Braysher, R. (1994*) Summary of the Euromethod Trialling Meeting*. CCTA, Norwich.

Brinkkemper, S., Engmann, R., Harmsen, F. and van de Weg, R. (1993) *Complexity reduction and modelling transparency in CASE tools*. In: Avison *et al.* (1993).

British Computer Society (1977) Data dictionary systems working party report. *ACM SIGMOD Record*, 9, 4, December.

Brodie, M. L. and Silva, E. (1982) *Active and passive component modelling*. In: Olle *et al.* (1982).

Brooks, F. P. (1987) No silver bullet: essence and accidents of software engineering. *Computer*, 20, 4.

Bubenko, J. A. Jr. (1986) *Information system methodologies - a research view*. In: Olle *et al.* (1986).

Buchanan, B. G. and Smith, R. G. (1993) *Fundamentals of expert systems*. In: Buchanan and Wilkins (1993).

Buchanan, B. G. and Wilkins, D. C. (eds.) (1993) *Readings in Knowledge Acquisition and Learning, Automating the Construction and Improvement of Expert Systems*. Morgan Kaufmann, San Mateo, Ca.

Buckingham, R. A., Hirschheim, R. A., Land, F. F. and Tully, C. J. (eds.) (1987a) *Information Systems Education: Recommendations and Implementation.* CUP, Cambridge.

Buckingham, R. A., Hirschheim, R. A., Land, F. F. and Tully, C. J. (1987b) *Information systems curriculum: a basis for course design.* In: Buckingham *et al.* (1987a).

Bullen, C. V. and Rockart, J. F. (1984) *A Primer on Critical Success Factors.* CISR Working Paper 69, Sloan Management School, MIT, Boston, Mass.

Burns, R. N. and Dennis, A. R. (1985) Selecting the appropriate application development methodology. *Data Base*, 17, 1.

Burnstine, D. C. (1986) *BIAIT: An Emerging Management Engineering Discipline.* BIAIT International, Petersburg, New York.

Burrell, G. and Morgan, G. (1979) *Sociological Paradigms and Organisational Analysis.* Heinemann, London.

Cameron, J. R. (1989) *JSP and JSD: The Jackson Approach to Software Development.* IEEE Computer Society Press, Los Angeles.

Cardenas, A. F. (1985) *Database Management Systems.* 2nd ed., Allyn and Bacon, Boston.

Carlson, W. M. (1979) Business information analysis and integration technique (BIAIT), the new horizon. *Data Base*, 10, 4, Spring.

Carmel, E. and Nunamaker, J. (1992) Supporting joint application development (JAD) with electronic meeting systems: a field study. In: DeGross, J. I., Becker, J. D. and Elam, J. J. (eds.) *Proceedings of the thirteenth International Conference on Information Systems (ICIS),* ACM, Baltimore.

Catchpole, C. P. (1987) *Information Systems Design for the Community Health Services.* PhD Thesis, Aston University, Birmingham.

CCTA (1989) *Compact Manual.* CCTA, Norwich.

CCTA (1990) *Euromethod Project, Phase 2, Deliverable 1, State of the Art Report.* Eurogroup, CCTA, Norwich.

CCTA (1994a) *Euromethod Overview.* CCTA, Norwich.

CCTA (1994b) *Using Euromethod in Practice.* CCTA, Norwich.

Checkland, P. (1981) *Systems Thinking, Systems Practice.* Wiley, Chichester.

Checkland, P. (1985) From optimising to learning: a development of systems thinking for the 1990s. *Journal of the Operational Research Society*, 36, 9.

Checkland, P. (1987) *Systems Thinking and Computer Systems Analysis: Time to Unite.* DEC Seminar Series in Information Systems, presentation given at the London School of Economics, 5th November 1987.

Checkland, P. and Scholes. J. (1990) *Soft Systems Methodology in Action.* Wiley, Chichester.

Chen, P. P. S. (1976) The entity-relationship model - toward a unified view of data. *ACM Transactions on Database Systems*, 1, 1.

Churchman, C. W. (1979) The Systems Approach and its Enemies. Basic Books, New York.

Coad, P. and Yourdon, E. (1991) *Object Oriented Analysis*. 2nd ed., Prentice Hall, Englewood Cliffs, New Jersey.

Codd, E. F. (1970) A relational model of data for large shared data banks. *Communications of the ACM*, 13, 6.

Computer Weekly (1994) Special report: rapid application development. *Computer Weekly*, 10th February.

Computerworld (1994) Why JAD goes bad. *Computerworld*, 25th April.

Cook, S. and Stamper, R. (1980) *LEGOL as a tool for the study of bureaucracy*. In: Lucas *et al.* (1980).

Cooper, R. (1990) A survey of current tools. *Blenheim Online Notes: Proceedings of Software Tools 1990*, Wembley Conference Centre.

Cougar J. D., Colter M. A. and Knapp R. W. (1982) *Advanced Systems Development/Feasibility Techniques*. Wiley, New York.

Cougar J. D. and Knapp R. W. (1974) *Systems Analysis Techniques*. Wiley, New York.

Daniels, A. and Yeates, D. A. (1971) *Basic Training in Systems Analysis*. 2nd ed., Pitman, London.

Daniels, J. and Cook, S. (1992) Making objects stick. In: Tagg, R. and Mabon, J. (eds.) *Object Management*, Aldgate Publishing, Aldershot.

Date, C. J. (1986) *An Introduction to Database Systems*. 4th ed., Addison-Wesley, Cambridge, Mass.

Date, C. J. (1995) *An Introduction to Database Systems*. 6th ed., Addison-Wesley, Reading, Mass.

Date, C. J. and White, C. (1988) *A guide to DB2*. 2nd ed., Addison-Wesley, Reading, Mass.

Davenport, T. H. (1993) *Process Innovation: Reengineering Work Through Information Technology*. Harvard Business School, Boston.

Davenport, T. H. and Short, J. E. (1990) The new industrial engineering: information technology and business process redesign. *Sloan Management Review*, 31, 4.

Davidson, E. J. (1993) An exploratory study of joint application design (JAD) in information systems delivery. In: DeGross, J., Bostrom, R. P. and Robey, D. (eds.) *Proceedings of the Fourteenth International Conference on Information Systems (ICIS)*. ACM, New York.

Davies, C. G. and Layzell, P. J. (1993) *The Jackson Approach to Systems Development*. Chartwell Bratt, Bromley.

Davis, G. B. (1982) Strategies for information requirements determination. *IBM Systems Journal*, 21, 2.

Davis, G. B. and Olsen, M. H. (1985) *Management Information Systems: Conceptual Foundations, Structure and Development*. 2nd ed., McGraw-Hill, New York.

Davis, L. E. (1972) *The Design of Jobs*. Penguin, Harmondsworth, Middlesex.

De Grace, P. and Stahl, L. (1993) *The Olduvai Imperative: CASE and the State of Software Engineering Practice*. Prentice Hall, Englewood Cliffs, New Jersey.

De Greef, P. and Breuker, J. A. (1989) Analysing system-user cooperation. *Knowledge Acquisition*, 4.

Deal, T. E. and Kennedy, A. A. (1982) *Corporate Cultures - the Rites and Rituals of Corporate Life*. Addison-Wesley, Reading, Mass.

Dearnley, P. A. and Mayhew, P. J. (1983) In favour of system prototypes and their integration into the systems development cycle. *Computer Journal*, 26, 1.

DeMarco, T. (1979) *Structured Analysis and System Specification*. Prentice Hall, Englewood Cliffs, New Jersey.

Dertouzos, M. L. and Moses, J. (1979) *The Computer Age: A Twenty Year Review*. MIT Press, Boston, Mass..

Downs, E., Clare, P. and Coe, I. (1988) *Structured Systems Analysis and Design Method: Application and Context*. 2nd ed., Prentice Hall, Hemel Hempstead.

DTI (1992) *Knowledge Based Systems: A Survey*. Touche Ross, London.

Durham, T. (1989) Taking object lessons. *Computing*, 28th September.

Earl, M. J. (1989) *Management Strategies for Information Technology*. Prentice Hall, Englewood Cliffs, New Jersey.

Eason, K. D. (1984) Towards the experimental study of usability. *Behaviour and Information Technology*, 2, 2.

EDP Analyzer (1986) Developing high quality systems faster. *EDP Analyzer*, 24, 6.

Elmasri, R. and Navathe, S. B. (1989) *Fundamentals of Database Systems*. Benjamin/Cummings, Redwood City, Ca.

Episkopou, D. M. and Wood-Harper A. T. (1986) Towards a framework to choose appropriate information system approaches. *Computer Journal*, 29, 3.

Ernst and Young (1993) Vaughan Merlyn of Ernst & Young quoted in *I/S Analyser*, 31, 6.

Espejo, R. and Harnden, R. (eds.) (1989) *The Viable System Model: Interpretations and Applications of Stafford Beer's VSM*. Wiley, Chichester.

Eva, M. (1994) *SSADM Version 4: A User's Guide*. 2nd ed., McGraw-Hill, Maidenhead.

Everitt, N. and Fisher, A. (1995) *Modern Epistemology: A New Introduction*. McGraw-Hill, New York.

Fagin, R. (1977) Multivaried dependencies and a new normal form for relational databases. *ACM Transactions on Database Systems*, 2, 3.

Feigenbaum, E. A. (1982) Knowledge engineering in the 1980's. Department of Computer Science, Stanford University, Stanford, Ca., quoted by Giarratano and Riley (1994).

Fischer, S., Doodeman, M., Vinig, T. and Achterberg, J. (1993) *Boiling the frog or seducing the fox: organisational aspects of implementing Case technology*. In: Avison *et al.* (1993).

Fitzgerald, B. (1994) The systems development dilemma: whether to adopt formalised systems development methodologies or not? In: Baets, W. R. J. (ed.) *Proceedings of the Second European Conference on Information Systems,* Nijenrode University Press, Breukelen, the Netherlands.

Fitzgerald, G. (1990) Achieving flexible information systems: the case for improved analysis. *Journal of Information Technology,* 5, 1.

Fitzgerald, G. (1992) Executive information systems and their development in the U.K. *International Information Systems Journal,* 1, 2.

Fitzgerald, G., Stokes, N. and Wood, J. R. G. (1985) Feature analysis of contemporary information systems methodologies. *Computer Journal,* 28, 3.

Floyd, C. (1986) *A comparative evaluation of system development methods.* In: Olle *et al.* (1986).

Flynn, D. J. (1992) *Information Systems Requirements - Determination and Analysis.* McGraw-Hill, Maidenhead, Berkshire.

Galliers, R. D. and Sutherland, A. R. (1991) Information systems management and strategy formulation: the 'stages of growth' model revisited. *Journal of Information Systems,* 1, 2.

Gane, C. P. and Sarson, T. (1979) *Structured Systems Analysis: Tools and Techniques.* Prentice Hall, Englewood Cliffs, New Jersey.

Giarratano, J. and Riley, G. (1994) *Expert Systems, Principles and Programming.* PWS Publishing, Boston.

Gill, T. G. (1995) Early expert systems: where are they now? *MIS Quarterly,* 19, 1.

Gray, E. M. (1985) An empirical study of the evaluation of some information systems development methods. *Proceedings of Conference of the Information Systems Association,* Sunningdale Park, Reading.

Grindley, C. B. B. (1966) *SYSTEMATICS - A nonprogramming language for designing and specifying commercial systems for computers.* In: Cougar and Knapp (1974).

Grindley, C. B. B. (1968) *The use of decision tables within Systematics.* In: Cougar and Knapp (1974).

Grundén, K. (1986) *Some critical observations on the traditional design of administrative information systems and some proposed guidelines for human-oriented system evolution.* In: Nissen and Sandström (1986).

Hammer, M. and Champy, J. (1993) *Reengineering the Corporation: A Manifesto for Business Revolution.* Harper Business, New York.

Herzberg, F. (1966) *Work and the Nature of Man.* Staple Press, New Hope, Mn.

Hirschheim, R. (1985) *Information systems epistemology: an historical perspective.* In: Mumford *et al.* (1985).

Hirschheim, R. and Klein, H. K. (1989) Four paradigms of information systems development. *Communications of the ACM,* 32, 10.

Hirschheim, R. and Newman, M. (1988) Information systems and user resistance: theory and practice. *Computer Journal,* 31, 5.

Howard, G. S. and Rai, A. (1993) Promises and problems: case usage in the US. *Journal of Information Technology*, 8, 2.

I/S Analyser (1993) The costs and benefits of CASE. *I/S Analyser*, 31, 6.

IBM (1971) *The time automated grid system (TAG)*. In: Cougar and Knapp (1974).

IBM (1975) *Business systems planning*. In: Cougar *et al.* (1982).

IFIP-NGI (1986) Information systems assessment. *Proceedings of IFIP 8.2 Conference on Information Systems Evaluation*, Leeuwenhorst, the Netherlands.

Iivari, J. (1992) The organisational fit of information systems. *Journal of Information Systems*, 2, 1.

Iivari, J. (1993) From a macro innovation theory of IS diffusion to a micro innovation theory of IS adoption: an application to CASE adoption. In: Avison *et al.* (1993).

Iivari, J. and Kerola, P. (1983) *A sociocybernetic framework for the feature analysis of information systems design methodologies*. In: Olle *et al.* (1983).

Jackson, M. A. (1975) *Principles of Program Design*. Academic Press, New York.

Jackson, M. A. (1983) *Systems Development*. Prentice Hall, Hemel Hempstead.

Jankowski, D. (1994) *Computer-aided systems engineering support for structured analysis, in managing social and economic change with information technology*. In: Khosrowpour (1994).

Jayaratna, N. (1994) *Understanding and Evaluating Methodologies, NIMSAD: A Systemic Framework*. McGraw-Hill, Maidenhead.

Jenkins, T. (1994) Report back on the DMSG sponsored UK Euromethod forum '94. *Data Management Bulletin*, Summer Issue, 11, 3.

Jones, R. (1990) Object lessons made to work for business. *Computing*, 20th September.

Kendall, K. and Kendall, J. (1993) *Systems Analysis and Design*. 2nd ed., Prentice Hall, Englewood Cliffs, New Jersey.

Kent, W. A. (1978) *Data and Reality*. North Holland, Amsterdam.

Kent, W. A. (1983) Simple guide to five normal forms in relational theory. *Communications of the ACM*, 26, 2.

Khosrowpour, M. (1994) Managing social and economic change with information technology. *Proceedings of the Information Resources Management Association International Conference*, IDEA Group Publishing, Harrisburg, Pen.

Klein, H. K. and Hirschheim, R. A. A. (1987) A comparative framework of data modelling paradigms and approaches. *Computer Journal*, 30, 1.

Kuhn, T. S. (1962) *The Structure of Scientific Revolutions*. 2nd ed., University of Chicago Press, Chicago.

Kung, C. H. (1983) An analysis of three conceptual models with time perspective. In: Olle *et al.* (1983).

Kurbel, K. and Schnieder, T. (1993) *Integration issues of information engineering based I-Case tools*. In: Avison *et al.* (1993).

Kwon, T. H. and Zmud, R. W. (1987) Unifying the fragmented models of information systems implementation. In: Boland, R. J. Jr. and Hirschheim, R. A. (eds.) *Critical Issues in Information Systems Research.* Wiley, Chichester.

Land, F. (1976) Evaluation of systems goals in determining a design strategy for a computer-based information system. *Computer Journal,* 19.

Land, F. (1982) Adapting to changing user requirements. *Information and Management,* 5 59-75.

Land, F. and Hirschheim, R. (1983) Participative systems design: rationale, tools and techniques. *Journal of Applied Systems Analysis,* 10.

Lederer, A. L. and Mendlelow, A. L. (1989) Information systems planning: incentives for effective action. *Data Base,* Fall.

Lee, B. (1979) *Introducing Systems Analysis and Design.* Vols 1 and 2, NCC, Manchester.

Lewis, P. J. (1994) *Information Systems Development: Systems Thinking in the Field of IS.* Pitman, London.

Liebenau, J. and Backhouse, J. (1990) *Understanding Information: an Introduction.* Macmillan, Basingstoke.

Longworth, G. (1985) *Designing Systems for Change.* NCC, Manchester.

Lucas, H. C., Land, F., Lincoln, T. J. and Supper, K. (eds.) (1980) *The Information Systems Environment.* North Holland, Amsterdam.

Lundeberg, M. (1983) *Some Comments on the ISAC Approach in Connection with the CRIS-2 Papers.* unpublished paper, The Institute for Development of Activities in Organisations, Stureplan 6 4 tr, S-114 35, Stockholm, Sweden.

Lundeberg, M., Goldkuhl, G. and Nilsson, A. (1982), *Information Systems Development - A Systematic Approach.* Prentice Hall, Englewood Cliffs, New Jersey.

McClure, C. (1989) *Case is Software Automation.* Prentice Hall, Englewood Cliffs, New Jersey.

McCracken, D. D. and Jackson, M. A. (1982) Life cycle concept considered harmful. *ACM SIGSOFT,* Software Engineering Notes, 17, 2.

McDermid, D. C. (1990) *Software Engineering for Information Systems.* McGraw-Hill, Maidenhead.

MacDonald, I. G. (1983) System development in a shared environment - the information engineering methodology, *Proceedings of the BCS Conference on Data Analysis Methodologies,* Thames Polytechnic, April 1982.

MacDonald, I. G. and Palmer, I. (1982) *System development in a shared data environment, the D2S2 methodology.* In: Olle *et al.* (1982).

McFadden, F. R. and Hoffer, J. A. (1991) *Database Management.* 3rd ed., Benjamin/Cummings, Redwood City, Ca.

McGaughey, R. E. and Gibson, M (1993) The repository/encyclopaedia: essential to Information Engineering and fully integrated CASE. *Journal of Systems Management,* 44, 3.

McLeod, R. (1993) *Management Information Systems: A Study of Computer-Based Information Systems.* 5th ed., Macmillan, New York.

Maddison, R. N. (ed.) (1983) *Information System Methodologies.* Wiley Heyden, Chichester.

Maddison, R. N. (ed.) (1984) *Information System Development: A Flexible Framework.* British Computer Society, London.

Markus, M. L. and Bjørn-Anderson, N. (1987) Power over users: its exercise by system professionals. *Communications of the ACM*, 30, 6.

Martiin, P., Lyytinen, K., Rossi, M., Tahvanainen, V., Smolander, K. and Tolvanen, J. (1992) Modelling requirements for future issues and implementation considerations. In: DeGross, J. I., Becker, J. D. and Elam, J. J. (eds.) *Proceedings of the thirteenth International Conference on Information Systems (ICIS)*, ACM, Baltimore.

Martin, J. (1982) *Application Development without Programmers.* Prentice Hall, Englewood Cliffs, New Jersey.

Martin, J. (1989) *Information Engineering.* Prentice Hall, Englewood Cliffs, New Jersey.

Martin, J. (1991) *Rapid Application Development.* Prentice Hall, Englewood Cliffs, New Jersey.

Martin, J. and Finkelstein, C. (1981) *Information Engineering. Vols 1 and 2.* Prentice Hall, Englewood Cliffs, New Jersey.

Martin, J. and Leben, J. (1989) *Strategic Information Systems Planning Methodologies.* Prentice Hall, Englewood Cliffs, New Jersey.

Martin, J. and McClure, C. (1983) *Structured Techniques for Computing.* Vols 1 and 2. Savant Institute, Carnforth, Lancashire.

Martin, J. and McClure, C. (1985*) Action Diagrams: Clearly Structured and Program Design.* Prentice Hall, Englewood Cliffs, New Jersey.

Martin, J. and Odell, J. (1992) *Object Oriented Analysis and Design.* Prentice Hall, Englewood Cliffs, New Jersey.

Mason, D. and Willcocks, L. (1994) *Systems Analysis, Systems Design.* McGraw-Hill, Maidenhead.

Moad, J. (1993) Does reengineering really work? *Datamation,* 39, 15.

Mockler, R. J. (1992) *Developing Knowledge-Based Systems Using an Expert Systems Shell.* Macmillan, New York.

Moynihan, E. (1993) *Business Management and Systems Analysis.* McGraw-Hill, Maidenhead.

Mumford, E. (1981) Participative systems design: structure and method. *Systems, Objectives and Solutions,* 1, 1.

Mumford, E. (1983a) *Designing Human Systems.* Manchester Business School, Manchester.

Mumford, E. (1983b) *Designing Participatively.* Manchester Business School, Manchester.

Mumford, E. (1983c) *Designing Secretaries.* Manchester Business School, Manchester.

Mumford, E. (1985) Defining systems requirements to meet business needs: a case study example. *Computer Journal*, 28, 2.

Mumford, E. (1986) *Using Computers for Business.* Manchester Business School, Manchester.

Mumford, E. (1989) *XSEL's Progress.* Wiley, Chichester.

Mumford, E. (1995) *Effective Requirements Analysis and Systems Design: The ETHICS Method.* Macmillan, Basingstoke.

Mumford, E., Hirschheim, R. A., Fitzgerald, G. and Wood-Harper, A. T. (eds.) (1985) *Research Methods in Information Systems.* North-Holland, Amsterdam.

Mumford, E., Land. F. F. and Hawgood, J. (1978) A participative approach to computer systems. *Impact of Science on Society*, 28, 3.

Mumford, E. and Weir, M. (1979) *Computer Systems in Work Design - The ETHICS Method.* Associated Business Press.

Myers, G. (1975) *Reliable Software Through Composite Design.* Petrocelli/Charter.

Myers, G. (1978) *Composite/Structured Design.* Van Nostrand Reinhold, New York.

NCC (1995) *SSADM 4+: Version 4.2.* Volumes 1 and 2, NCC Blackwell, Oxford.

Niessen, P. (1990) Re-development engineering: forming an information blueprint for the 1990s, In: *A Survey of Current Tools, Blenheim Online Notes, Proceedings of 'Software Tools 1990'*, Wembley Conference Centre.

Nissen, H-E. and Sandsröm, G. (eds.) (1986) *Quality of Work versus Quality of Information Systems.* Lund University, Sweden.

Nolan, R. (1979) Managing the Crises in Data Processing. *Harvard Business Review*, 57, 2.

Norman, R. J., Corbitt, G. F., Butler, M. C. and McElroy, D. D. (1989) Case technology transfer: a case study of unsuccessful change. *Journal of Systems Management*, 40, 5.

Nunamaker, J., Dennis, A., Valacich, J., Vogel, D. and George, F. (1991) Electronic meeting systems to support group work. *Communications of the ACM,* 34, 7.

Olive, A. (1983) *Analysis of conceptual and logical models in information systems design methodologies.* In: Olle *et al.* (1983).

Olle, T. W., Hagelstein, J., MacDonald, I. G., Rolland, C., Sol, H. G., Van Assche, F. J. M. and Verrijn-Stuart, A. A. (1991) *Information Systems Methodologies - A Framework for Understanding.* 2nd ed., Addison-Wesley, Wokingham.

Olle, T. W., Sol, H. G. and Verrijn-Stuart, A. A. (1982) *Information Systems Design Methodologies: A Comparative Review.* North Holland, Amsterdam.

Olle, T. W., Sol, H. G. and Tully, C. J. (1983) *Information Systems Design Methodologies: A Feature Analysis.* North Holland, Amsterdam.

Olle T. W., Sol, H. G. and Verrijn-Stuart, A. A. (eds.) (1986) *Information Systems Design Methodologies: Improving the Practice.* North Holland, Amsterdam.

Olle T. W., Verrijn-Stuart, A. A. and Bhabuta, L. (eds.) (1988) *Computerized Assistance During the Information Systems Life Cycle.* North Holland, Amsterdam.

Österle, H., Brenner, W. and Hilbers, K. (1993) *Total Information Systems Management: A European Approach.* Wiley, Chichester.

Page-Jones, M. (1980) *The Practical Guide to Structured System Design.* Yourdon, New York.

Palmer, I. and Rock-Evans, R. (1981) *Data Analysis.* IPC Publications, Surrey.

Parkinson, J. (1991) *Making Case Work.* NCC Blackwell, Oxford.

Parsons, T. and Shils, E. (1951) *Towards a General Theory of Action.* Harvard University Press, Mass.

Porter, M. E. (1980) *Competitive Strategy.* Free Press, New York.

Porter, M. E. (1985) *Competitive Advantage.* Collier-Macmillan, New York.

Porter, M. E. and Millar, V. E. (1985) How information gives you competitive advantage. *Harvard Business Review*, July-August.

Pressman, R. S. (1992) *Software Engineering: A Practitioner's Approach.* 3rd ed., McGraw-Hill, New York.

Quang, P. T. and Chartier-Kastler, C. (1991) *Merise in Practice.* Macmillan, Basingstoke (translated by D. E. and M. A. Avison from the French *Merise Appliquée.* Eyrolles, Paris, 1989).

Rockart, J. F. (1979) Chief executives define their own data needs. *Harvard Business Review*, March-April 1979.

Rogers, E. M. (1983) *Diffusion of Innovations.* Free Press, New York.

Rumbaugh, J., Blaha, M., Premerlani, W., Eddy, F. and Lorensen, W. (1991) *Object Oriented Modelling and Design.* Prentice Hall, Englewood Cliffs, New Jersey.

Rzevski, G. (1985) *On the Comparison of Design Methodologies.* working paper, School of Computer Science, Kingston Polytechnic, Surrey.

Searle, J. R. (1995) *The Construction of Social Reality.* Allen Lane, Penguin Press, London.

Shah, H. U. and Avison, D. E. (1995) *The Information Systems Development Cycle: A First Course in Information Systems.* McGraw-Hill, Maidenhead.

Shneiderman, B. (ed.) (1978) *Improving Database Usability and Responsiveness.* Academic Press, New York.

Shubik, M. (1979) *Computers and modelling.* In: Dertouzos and Moses (1979).

Smith, J. and Smith, D. (1977) Database abstractions: aggregation and generalisation. *ACM Transactions on Database Systems*, 2, 2.

Smyth, D. S. and Checkland, P. B. (1976) Using a systems approach: the structure of root definitions. *Journal of Applied Systems Analysis*, 5, 1.

Stamper, R. (1978) *Aspects of data semantics: names, species and complex physical objects.* In: Bracchi and Lockermann (1978).

Stevens, W. P., Myers, G. J. and Constantine, L. L. (1974) Structured design. *IBM System Journal*, 13, 2.

Stewart, J. (1994) *IS Notice: Euromethod.* 71, CCTA, Norwich.

Stone, J. (1993) *Inside ADW and IEF: The Promise and Reality of Case.* McGraw-Hill. New York.

Stowell, F. and West, D. (1994) *Client-led Design.* McGraw-Hill, Maidenhead.

Strassman, P. (1990) *The Business Value of Computers.* The Information Economics Press, New Canaan, Conn.

Sutcliffe, A. (1988) *Jackson Systems Development.* Prentice Hall, Hemel Hempstead.

Taylor, F. W. (1947) *Scientific Management.* Harper and Row, New York.

Teichroew, D. and Hershey, E. A. (1977) *PSL/PSA: a computer-aided technique for structured documentation and analysis of information processing systems.* In: Cougar *et al.* (1982).

Teichroew, D., Hershey, E. A. and Yamamoto, Y. (1979) *The PSL/PSA approach to computer-aided analysis and documentation.* In: Cougar *et al.* (1982).

Texas Instruments (1992) *Rapid Application Development Overview.* Texas Instruments Inc., Plano, Tx.

Thompson, K. (1990) The generic classification of CASE tools. *Blenheim Online Notes, Proceedings of 'Software Tools 1990'*, Wembley Conference Centre.

Thompson, K. (1992) Software development monitor - what's in store. *Vision Software Engineering*, December.

Tozer, E. (ed.) (1984) *Applications Development Tools - A State of the Art Report.* Pergamon-Infotech, Oxford.

Tozer, E. (1985) *A* review of current data analysis techniques. *Proceedings of the BCS Conference on Data Analysis in Practice*, Huddersfield Polytechnic.

Tsichritzis, D. C. and Klug, A. (eds.) (1978) The ANSI/X3/SPARC framework: report of the study group on data base management systems. *Information Systems*, 3.

Turban, E. (1993) *Decision Support and Expert Systems.* 3rd ed., Macmillan, New York.

Utterback, J. M. and Abernathy, W. J. (1975) A dynamic model of process and product innovation. *Omega*, 3, 6.

Verity, J. W. and Schwartz, E. I. (1991) Software made simple. *Business Week*, September 30th.

Vessey, I., Jarvenpaa, S. L. and Tractinsky, N. (1992) Evaluation of vendor products: CASE tools as methodology companions. *Communications of the ACM*, 35, 4.

Vickers, G. (1970) *Freedom in a Rocking Boat.* Allen Lane, London.

Walsham, G. (1995) Interpretive case studies in IS research: nature and method. *European Journal of Information Systems,* 4, 2.

Wasserman, A. I. and Freeman, P. (1976) *IEEE Tutorial on Software Design Techniques.* IEEE Computer Society, New York.

Wasserman, A. I., Freeman, P. and Porchella, M. (1983) Characteristics of software development methodologies. In: Olle *et al.* (1983).

Wastell, D. (1996) The fetish of technique: methodology as a social defence. *Information Systems Journal,* 6, 1.

Weaver, P. L. (1993*) Practical SSADM Version 4: A Complete Tutorial Guide.* Pitman, London.

Weinberg, V. (1978) *Structured Analysis.* Prentice Hall, Englewood Cliffs, New Jersey.

Welke, R. J. (1987) *The New Architecture of PSL/PSA.* MI:MetaSystems, Ann Arbor, Mi.

Wetherbe, J. C. and Davis, G. B. (1983) Developing a long-range information architecture. *Proceedings of the National Computer Conference,* Anaheim, Ca.

White, P. (1982) Towards a repeatable methodology. *Perspective,* 1, 2.

Wielinga, B. J., Th. Sterner, A. and Breuker, J. A. (1993) *KADS: a modelling approach to knowledge engineering.* In: Buchanan and Wilkins (1993).

Willcocks, L. and Fitzgerald, G. (1993) Market as opportunity? Case studies in outsourcing information technology and services. *Journal of Strategic Information Systems,* 2, 3.

Wilson, B. (1990) *Systems: Concepts, Methodologies and Applications.* 2nd ed., Wiley, Chichester.

Wirth, N. (1971) Program development by stepwise refinement. *Communications of the ACM,* 14, 4.

Wood, J. and Silver, D. (1989) *Joint Application Design: How to Design Quality Systems in 40% Less Time.* Wiley, New York.

Wood-Harper, A. T. and Fitzgerald, G. (1982) A taxonomy of current approaches to systems analysis. *Computer Journal,* 25, 1.

Woodman, M. (1988) Yourdon dataflow diagrams: a tool for disciplined requirements analysis. *Information and Software Technology,* 30, 9.

Wynekoop, J. L., Senn, J. A. and Conger, S. A. (1992) The implementation of CASE tools: an innovation diffusion approach. In: Kendall, K. E., Lyytinen, K. and DeGross, J. (eds.) *The Impact of Computer Technologies on Information Systems Development.* North Holland, Amsterdam.

Yang, F., Shao, W. and Li, W. (1994) JadeBird/III: a collaborative multimedia case environment. In: Zupancic, J. and Wrycza, S. (1994) *Proceedings of the Fourth International Conference on Information Systems Development, Methods and Tools, Theory and Practice.* Moderna Organizacija, Kranj, Slovenia.

Yao, S. B. (1985) *Principles of Data Base Design: Logical Organization.* Volume 1, Prentice Hall, Englewood Cliffs, New Jersey.

Yoon, Y., Guimaraes, T. and O'Neal, Q. (1995) Exploring the factors associated with expert systems success. *MIS Quarterly*, 19, 1.

Yourdon, E. (1989) *Modern Structured Analysis*. Prentice Hall, Englewood Cliffs, New Jersey

Yourdon, E. (1992) *Decline and Fall of the American Programmer*. Yourdon Press, Englewood Cliffs, New Jersey

Yourdon, E. (1994) *Object-oriented Systems Design, An Integrated Approach*. Prentice Hall, Englewood Cliffs, New Jersey

Yourdon, E. and Constantine, L. L. (1978) *Structured Design*. 2nd ed., Yourdon Press, New York.

Yourdon Inc. (1993) *Yourdon Systems Method: Model-Driven Systems Development*. Yourdon Press, Englewood Cliffs, New Jersey.

Zloof, M. M. (1978) *Design aspects of the query by example data base management language*. In: Shneiderman (1978).

Index

Items in *italic* refer to names of cited authors (see also Bibliography), companies, and products. Page numbers in **bold** indicate principal references.